Energy, Agriculture and Waste Management

Energy, Agriculture and Waste Management

Proceedings of the 1975 Cornell Agricultural
Waste Management Conference

Edited by

WILLIAM J. JEWELL

Environmental Engineer
Associate Professor of Agricultural Engineering
Cornell University
Ithaca, New York

ANN ARBOR SCIENCE
PUBLISHERS INC
P.O. BOX 1425 • ANN ARBOR, MICH. 48106

Second Printing, 1977

Copyright© 1975 by Ann Arbor Science Publishers, Inc.
P.O. Box 1425, Ann Arbor, Michigan 48106

Library of Congress Catalog Card Number 75-22898
ISBN 0-250-40113-4

Preface

Mankind has passed from a period of affluence for many to global environmental quality degradation, food and energy shortages, and natural resource depletion. As information is developed to describe this situation, we are continuously faced with contradictory statements regarding the major issues that will decide the future of man. This volume discusses three topics:

—Energy consumed in food production and philosophies that relate to various agricultures.

—Technology and energy costs of pollution control.

—Potential for producing energy from agricultural wastes.

American agriculture is charged with being inefficient and with wasting energy. Comprehensive studies of energy consumed in agriculture conducted in New York, Texas, Michigan, and California are summarized here. They indicate that even in diverse types of agriculture, less than 5% of the total energy is consumed in the farm production of food.

While the U.S. Congress has passed new and comprehensive pollution control legislation, others claim that we do not have the energy or the resources to prevent contamination of our environment with our wastes. Cywin, Tchobanoglous and other distinguished authors show that the energy consumed and costs of pollution control would appear to be acceptable. For example, energy required for water pollution control for each person is equivalent to a continuous burning 15-watt lightbulb.

Finally, the energy crisis and the demand for improved environmental quality sometimes result in contradictory solutions. In some cases, however, the two are complementary. Nearly half the chapters in this text deal with the potential of controlling organic wastes with processes that generate energy. Chapters on the bioconversion of animal manures to methane gas via the anaerobic fermentation process provide a comprehensive description of the history, the present and the future of this technology. The speed with which this topic is expanding was clearly emphasized in a response from a farmer to a statement at the conference that no anaerobic digesters were being used in U.S. agriculture. The farmer was quick to note that although all the "experts" had declined to give him detailed advice, he had been operating an anaerobic digester on a 350-head beef feedlot with tremendous success.

Complete answers to the questions surrounding energy, agriculture and waste management will not be found here. But these chapters by experts in the field provide one of the first attempts to answer them.

William J. Jewell
October 1975

v

The 1975 Agricultural Waste Management Conference was sponsored by Cornell University, New York State College of Agriculture and Life Sciences, A Statutory College of the State University, Ithaca, New York, and National Science Foundation, Research Applied to National Needs, Rann Grant GI 43099.

Any opinions, findings, conclusions, or recommendations expressed in this publication are those of the authors and do not necessarily reflect the views of the National Science Foundation.

Contents

Section III   Energy Reclamation from Agricultural Wastes

INTRODUCTION AND OPENING REMARKS TO THE SEVENTH ANNUAL CORNELL UNIVERSITY CONFERENCE ON ENERGY, AGRICULTURE AND WASTE MANAGEMENT

W. K. Kennedy*

Among lay people, the fear of a severe gasoline and fuel oil shortage is much less today than a year ago, but the current availability of adequate supplies of gasoline only leads the people of this country into a false sense of security. The supplies of oil and other fossil fuels are finite and while new reserves of petroleum and natural gas undoubtedly will be located, we know they will be exhausted in too short a time if we continue to use them as lavishly as we have in recent years. Perhaps the only benefit from the unfortunate conflict in the Middle East has been the development of an awareness in the United States and the other developed countries that inexpensive sources of energy are tremendous treasures which must be used carefully and efficiently until we learn how to utilize other sources of energy.

Modern agriculture has been developed through the use of cheap sources of energy for power on our farms and in our processing plants, and for the production of abundant supplies of nitrogen, other fertilizers, agricultural chemicals and other supplies. By substituting capital, mechanization and the liberal use of energy for labor, the farmers and related agricultural industries in the United States have been able to produce, process and market an abundance of food at prices far below those paid by consumers in most other countries. In retrospect, it is easy to criticize the rapid move in this country toward mechanization and the liberal use of fertilizer and other agricultural chemicals, but at the time these decisions were being made, they were correct in terms of economic conditions and the general attitudes of society. In 1975 we are aware that more attention should have been given to the cost and long-term availability of petroleum and natural gas as we developed our labor efficient, but energy intensive, agricultural systems.

Fortunately we do have alternatives to the continued use of several of our high energy practices. In the immediate future, we probably must depend on petroleum products to fuel our tractors, combines and other field machinery. Timeliness in completing tillage, spraying and harvesting operations is such that we will continue to use sizeable tractors where they are needed. In some cases, the size of our

---

*Dean, New York State College of Agriculture and Life Sciences.

1

machinery can be reduced, such as spray equipment for apple orchards planted with size controlled trees instead of the large trees of the past. Nevertheless the saving of energy for our field machines probably will be modest at best and, in some cases, will continue to increase in order to reduce labor costs. In the case of fertilizer and pesticides, significant savings can be realized through timely applications and more careful control of rates of application. Greater use of animal manures and legumes in our farming practices can reduce the amounts of synthetic nitrogen fertilizer used on our farms. We can reduce pesticide and other chemical usage through more careful monitoring of pest populations and through greater use of other control procedures.

In recent years our scientists have developed techniques for growing plants from single cells. At first glance this appears to be an interesting, but useless bit of knowledge, but the perfection of tissue culture techniques opens up an entire new avenue for the improvement of plants. It is highly unlikely that our scientists would ever discover how to cross two dissimilar plants such as alfalfa and corn, but through the use of isolated cells of alfalfa and corn, and perhaps with the aid of selected viruses, our scientists may be able to transfer the appropriate genetic material from a cell of a legume to a cell of a corn or other non-legume plant. Then through tissue culture techniques, the corn plant cell with appropriate genetic material from the legume cell can be nurtured into a mature corn plant with the capability of supporting nitrogen fixing bacteria (rhizobium) on its roots. If this feat can be accomplished, and I am willing to predict it will be within the next decade, its value to mankind will be tremendous.

The opportunities to convert agricultural wastes into usable sources of energy are unlimited, but I do not wish to imply that the task will be easy. In many cases agricultural waste products have limited value at their place of origin. They may have high water content, they may be difficult to handle, they may be difficult to transport or spread on the land and the cost of utilizing them may be greater than present sources of energy or fertilizer materials. The counter to the high cost of utilization is the cost of disposing of these materials. You and your colleagues must continue to explore ways of turning waste products into productive uses. In many cases your efforts may be unsuccessful, but just a few successes will provide ample repayment for the time and dollars spent in these areas of research and development.

Section I
Energy and Food Production

1.

World Food, Energy, Man and Environment

David Pimentel*

As a result of overpopulation and environmental resource limitations the world is fast losing its capacity to supply adequate food. The world population today is 4 billion humans (1). Based upon current growth rates, and even allowing for reasonable reductions in birth rates in several countries, the National Academy of Sciences Committee estimated that the world population will reach at least 7 billion by the year 2000 (Figure 1). The committee concluded there is no feasible means to stop this explosive increase short of some unwanted catastrophe (1).

If we go back only about 2000 years, the records suggest that humans on earth numbered little more than 200 million (2)—about the density of the population of the United States today. World population was about 500 million as recently as 1650. It was shortly after 1700 that the human population explosion began (Figure 1).

Note how the rapid growth in world population coincides with the exponential use of fossil fuels (Figure 1). In addition to improving the quality of life, some fossil energy was used for disease control operations and to improve agricultural production to feed the growing population. Both the effective control of human diseases and increased food production have contributed significantly to the current rapid growth (1).

Of these two factors, the evidence suggests that reducing death rates with effective public health programs is the prime cause (3). The eradication of malaria-carrying mosquitoes by DDT and in-

---

* Department of Entomology and Section of Ecology and Systematics, Cornell University, Ithaca, New York 14853.

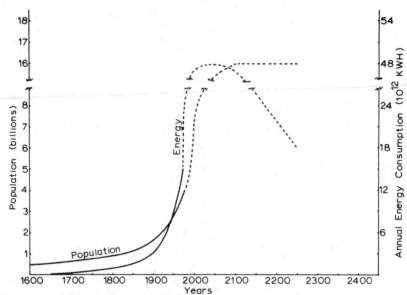

Figure 1. Estimated world population numbers (———) from 1600 to 1975 and projected numbers (— — — —) to the year 2250 (1, 3, 56). Estimated fuel consumption (———) from 1650 to 1975 and projected (— — — —) to the year 2250 (57).

secticides is a good example (note substantial quantities of energy required for production and application). In Ceylon (1946-47) after spraying with DDT, the death rate fell in one year from 20 to 14 per thousand (4). A similar dramatic reduction in death rates occurred after DDT was used in Mauritius where death rates fell from 27 to 15 per thousand in one year and population growth rates increased from 5 to 35 thousand (Figure 2).

Meanwhile in both Ceylon and Mauritius fertility rates did *not* decrease and an explosive increase in population numbers resulted. Recent history documents similar results in other nations where medical technology and medical supplies have significantly reduced death rates (5). It is relatively easy to reduce death rates through public health measures, but birth rates are difficult to change. Birth rates are interwoven with social and religious systems of the people.

With increasing human numbers in the world, many regions could no longer support a hunting-gathering economy. The shift had to be made to a more permanent type agriculture (6). "Slash and burn" or "cut and burn" agriculture was the first technology employed, *i.e.*, cutting trees and brush and burning them on site. This killed weeds and added nutrients to the soil. Crop production was good for a couple of years before soil nutrients were depleted. After

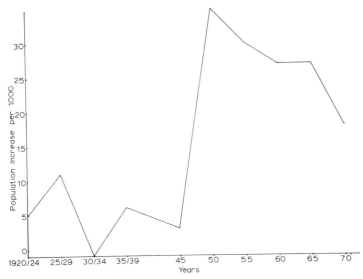

Figure 2. Population growth rate on Mauritius from 1920 to 1970. Note from 1920 to 1945 the growth rate was about 5 per thousand whereas after malaria control in 1945 the growth rate exploded to about 35 per thousand and has since very slowly declined (4, 58). After 25 years the rate of increase is still nearly 4 times the 1920-45 level.

use, it then takes about 20 years for the forest to regrow and for soil nutrients to be renewed.

Cut and burn crop technology required few tools (ax and hoe) and lots of manpower. For example, in a part of Mexico "slash and burn" corn culture was investigated and Lewis (7) reported that a total of 1,144 hours of labor was required to raise a hectare of corn (Table 1). Other than manpower, the only inputs were the ax, hoe, and seeds. Similar data were obtained for corn production in Guatemala (Table 2).

Table 1. Energy inputs in corn production in Mexico using only manpower.

Input	Quantity/ha	kcal/ha
Labor[a]	1,144 hr	622,622
Ax + Hoe[b]	16,500 kcal	16,500
Seeds[c]	10.4 kg	36,508
Total		675,730
Corn yield[a]	1,944 kg	6,842,880
kcal return/kcal input		10.13

[a] Lewis (7). See Table 4.
[b] Ax and hoe assumed to weigh 23 kg. See Table 3.
[c] 10.4 kg $\times$ 3,520 kcal/kg = 36,608 kcal.

Table 2. Energy inputs in corn production in Guatemala using only manpower.

Inputs	Quantity/ha	kcal/ha
Labor[a]	1,415 hr	770,114
Ax + Hoe[b]	16,500 kcal	16,500
Seeds[c]	10.4 kg	36,608
Total		823,222
Corn yield[d]	1,066 kg	3,752,320
kcal return/kcal input		4.56

[a] Corn production in San Pedro Necta, Guatemala infertile Llano soil (8). See Table 4 for labor energy input.
[b] Ax and hoe assumed to weigh about 23 kg. See Table 3.
[c] 10.4 kg × 3,520 kcal/kg = 36,608 kcal.
[d] From reference (8).

The yield of 1,944 kg/ha in Mexico provided about 6,842,880 kcal. Allowing for 3,000 kcal of corn per person per day, this yield was suitable for more than 6 persons. Another way of looking at this is that only one-sixth of a hectare is necessary to feed one person per year with corn. The hours needed then would be about 190 hours per person per year or only about 5 weeks work.

When man started harnessing fossil fuel for crop production, agriculture became revolutionized. Great changes occurred in agricultural production and these are discussed in a later section dealing with energy used in food production.

Arable crop land is in short supply. Of the total of 13 billion hectares of land area in the world (9), only an estimated 7 to 10% is suitable for cultivation (9-13). As Paddock and Paddock (14) point out, "a desert may have fine soil, but it has no rain; the Arctic has moisture but not the right temperature; mountains are too up and down. And so it goes." We are fortunate in the U.S. where about 22% of our land is suitable for cultivation (9). However, South America has only 6% arable land suitable for cultivation (9), for approximately the same number of people. Furthermore, nearly all the arable land of the world is in cultivation (14); bringing the remaining arable hectares in the United States, Canada, and elsewhere in the world into production only an estimated 1% might be added. Even in the United States, which has the greatest amount of arable land of any nation, nearly all the land resources already have been put into use.

To complete the picture on the use of land, mention should be made that about 22% of the land area of the world is used for livestock production and is in pastures, ranges, and meadows (13). Another 30% of the land area is in forests (13).

Although our land resources are vital to us for crop production, these lands are rapidly deteriorating. For example, each year in the U.S., about 3.6 billion metric tons of soil are washed into our streams and ponds, and into the oceans (15). This valuable top soil is lost from our cropland, home building sites, and other areas where soil is left with insufficient plant cover. On bare soil such as construction sites, about 1,120 metric tons of soil per hectare may be lost (16). The average loss of top soil per hectare of corn production is 44.1 metric tons (17). In the corn state of Iowa, the loss averages 36 metric tons annually and the aim is to reduce this loss to 11 metric tons annually (18). We in the U.S. are literally mining our soils for crop cultivation. How long can we continue to abuse our valuable soils?

Water is another vital resource in crop production. The 1974 drought in the Midwest emphasized the importance of water to us. Tremendous quantities of water are necessary to raise a crop. About 122 cc of water per cm^2 are needed to raise corn in the subtropics. This is about 12.2 million liters of water per hectare of corn. One hectare is about 2½ acres.

Only about 13% of the world's cultivated land is now irrigated (13). The use of irrigation could significantly increase the arable crop land in the world (19), but this type of alteration of the ecosystem requires energy. A liter of water weighs 1.0 kg. To pump from a depth of a little over 90 m in order ot supply 122 cc of water/cm^2 to a crop hectare would require about 2,060 liters of fuel (ca. 19.7 million kcal) (20). Because of the high energy-demand of irrigation, it is doubtful that irrigation will be used extensively to increase the arable land of the world (21, 22).

Earlier I mentioned that man has utilized fossil energy resources to increase his population numbers. In fact, the use of energy has been increasing faster than population numbers. For example, while it took about 60 years for the U.S. population to double, the U.S. doubled its energy consumption during the past 20 years. More alarming is that fact that while the world population doubled in the last 30 years, world energy consumption doubled within the past decade.

Energy use in food production has been increasing faster than in many other sectors of the world economy. For example, using corn as an average crop, Pimentel et al. (23) documented that energy inputs in corn production more than *tripled* (Tables 3 and 4) during the last 25 years. Note that the quantity of energy used to produce nitrogen fertilizer during 1970 nearly equalled all the energy inputs for 1945. The other large inputs of energy come from machinery (1,037,400 kcal); fuel (1,971,420 kcal); drying (296,400 kcal); and electricity (765,700 kcal).

Drying corn was one of the factors that increased significantly

Table 3. Average energy inputs in U.S. corn production during different years (all figures per hectare) (23).

Inputs	1945	1950	1954	1959	1964	1970
Labor (hours)[1]	57	44	42	35	27	22
Machinery (kcal)[2]	444,600	617,500	741,000	864,500	1,037,400	1,037,400
Fuel (liters)[3]	140	159	178	187	197	206
Nitrogen (kg)[4]	8	17	30	46	65	125
Phosphorus (kg)[4]	8	11	13	18	20	35
Potassium (kg)[4]	6	11	20	34	46	67
Seeds for planting (kg)[5]	11	13	16	19	21	21
Irrigation (kcal)[6]	103,740	128,440	148,200	170,430	187,720	187,720
Insecticides (kg)[7]	0	.11	.34	.78	1.12	1.12
Herbicides (kg)[8]	0	.06	.11	.28	.43	1.12
Drying (kcal)[9]	9,880	34,580	74,100	163,020	247,000	296,400
Electricity (kcal)[10]	79,040	133,380	247,000	345,800	501,410	765,700
Transportation (kcal)[11]	49,400	74,100	111,150	148,200	172,900	172,900
Corn yields (kg/ha)[12]	2,132	2,383	2,572	3,387	4,265	5,080

1. Mean hours of labor per crop hectare in United States (26, 27).
2. An estimate of the energy inputs for the construction and repair of tractors, trucks, and other farm machinery was obtained from the data of Berry and Fels (28), who calculated that about 31,968,000 kcal of energy was necessary to construct an average automobile weighing about 1,530 kg. In our calculations we assumed that 244,555,000 kcal (an equivalent of 11,700 kg of machinery) were used for the production of all machinery (tractors, trucks, and miscellaneous) to farm 25 hectares of corn. This machinery was assumed to function for 10 years. Repairs were assumed to be 6% of total machinery production or about 15,000,000 kcal. Hence, a conservative estimate for the production and repair of farm machinery per corn hectare per year for 1970 was 1,037,400 kcal. A high for the number of tractors and other farm machinery on farms was reached in 1964 and continues (29, 30). The number of tractors and other types of machinery in 1945 were about half what they are now.
3. DeGraff and Washbon (31) reported that corn production required about 140 liters of fuel per hectare for tractor use—intermediate between fruit and small grain production. Because corn appeared to be intermediate, the estimated mean fuel (liters) burned in farm machinery per harvested hectare was based on U.S. Department of Agriculture (29, 32) and U.S. Bureau of the Census (30) data.
4. Fertilizers (N, P, K) applied to corn are based on USDA (26, 33, 34, 35) estimates.
5. During 1970, relatively dense corn planting required about 21 kg of corn (61,750 kernels or 83,980 kcal) per hectare; the less dense plantings in 1945 were estimated to use about 10.5 kg of seed. Because hybrid seed has to be produced with special care, the input for 1970 was estimated to be 147,840 kcal.
6. Only about 3.8% of the corn grain hectares in the United States were irrigated in 1964 (36), and this is not expected to change much in the near future (37). Although a small percentage, irrigation is costly in terms of energy demand. On the basis of the data of Smerdon (20), an estimated 4,921,166 kcal is required to irrigate a hectare of corn with 30.48 cm of water for one season. Higher energy costs for irrigation water are given by the Report on the World Food Problem (38). Since only 3.8% of the corn hectares are irrigated (1964-

1970), it was estimated that only 187,720 kcal were used per hectare for corn irrigation. The percentage of hectares irrigated in 1945 was based on trends in irrigated hectares in agriculture (36, 39).

7. Estimates of insecticides applied per hectare of corn are based on the fact that little or no insecticide was used on corn in 1945, and this reached a high in 1964 (40, 41).

8. Estimates of herbicides applied per hectare of corn are based on the fact that little or no herbicides were used on corn in 1945 and that this use continues to increase (40, 41).

9. When it is dried for storage to reduce the moisture from about 26.5% to 13%, about 1,008,264 kcal are needed to dry 5,080 kg (42). About 30% of the corn was estimated to have been dried in 1970 as compared to an estimated 10% in 1945.

10. Agriculture consumed about 2.5% of all electricity produced in 1970 (43) and an estimated 424.2 trillion British thermal units of fossil fuel were used to produce this power (44); on croplands this divides to 765,700 kcal per hectare for 1970 (27, 40). The fuel used to produce the electrical energy for earlier periods was estimated from data reported in Statistical Abstracts (45).

11. Estimates of the number of calories burned to transport machinery and supplies to corn hectares and to transport corn to the site of use is based on data from U.S. Department of Commerce (46), U.S. Bureau of Census (30, 36, 44), Interstate Commerce Commission (47, 48, 49), and U.S. Department of Transportation (50). For 1964 and 1970 this was estimated to be about 172,900 kcal per hectare; it was about 49,400 kcal per hectare in 1945.

12. Corn yield is expressed as a mean of 3 years, 1 year previous and 1 year past (25, 39, 51).

from 1945 to 1970. This in part is related to one of the factors involved in increasing corn yields. Having corn with a longer growing season allows the corn to collect more light energy and convert this into corn grain. With a longer growing season the corn stands in the field later in the fall and does not have an opportunity to dry. Hence, the corn has to be dried before it is put into storage.

The 7.1 million kcal of fossil fuel used to raise a hectare of corn (Table 4) represents a small portion of energy when compared with solar energy input. During the growing season, about 5,046 million kcal reaches a hectare of corn, about 1.26% of this is converted into corn and about 0.4% into corn grain (at 6,272 kg/ha) (24). The 1.26% represents about 63.6 million kcal. Hence, when solar energy input is included, fossil fuel input of 7.1 million kcal represents about 11% of the total energy input in corn production. The very good return for corn of 2.52 kcal in corn grain per input kcal of fossil fuel is due in part to the high efficiency of corn to convert light energy into plant material (Table 4). Most crop plants are far less efficient in capturing light energy.

I should mention the tremendous amount of feed energy that goes into animal protein production. For example, to produce 1 kcal of beef protein, a total of 123 kcal of feed (grains, etc.) have to be fed to beef cattle! Milk protein production is much more

Table 4. Energy inputs in U.S. corn production (all figures in kcal) (23).

Inputs	1945	1950	1954	1959	1964	1970
Labor[1]	31,022	23,947	22,859	19,049	14,695	11,974
Machinery[2]	444,600	617,500	741,000	864,500	1,037,400	1,037,400
Fuel[3]	1,339,800	1,521,630	1,703,460	1,789,590	1,885,290	1,971,420
Nitrogen[4]	140,800	299,200	528,000	809,600	1,144,000	2,200,000
Phosphorus[5]	25,520	35,090	41,470	57,420	63,800	111,650
Potassium[6]	13,200	24,200	44,000	74,800	101,200	147,400
Seeds for planting[7]	77,440	91,520	112,640	133,760	147,840	147,840
Irrigation[2]	103,740	128,440	148,200	170,430	187,720	187,720
Insecticides[8]	0	2,662	8,228	18,876	27,104	27,104
Herbicides[9]	0	1,452	2,662	6,776	10,406	27,104
Drying[2]	9,880	34,580	74,100	163,020	247,000	296,400
Electricity[2]	79,040	133,380	247,000	345,800	501,410	765,700
Transportation[2]	49,400	74,100	111,150	148,200	172,900	172,900
Total inputs	2,314,442	2,987,701	3,784,769	4,601,821	5,540,765	7,104,612
Corn yield (output)[10]	7,504,640	8,388,160	9,053,440	11,922,240	15,012,800	17,881,600
kcal return/ kcal input	3.24	2.81	2.39	2.59	2.71	2.52

1. It is assumed that a farm laborer consumes 21,770 kcal per week and works a 40-hour week. For 1970: (22 hours/40 hours) × 21,770 kcal = 11,974 kcal.
2. See Table 1.
3. Fuel, 1 liter = 9,570 kcal (52).
4. Nitrogen, 1 kg = 17,600 kcal, including production and processing (52).
5. Phosphorus, 1 kg = 3,190 kcal, including mining and processing (52).
6. Potassium, 1 kg = 2,200 kcal, including mining and processing (52).
7. Corn seed, 1 kg = 3,520 kcal (53). This energy input was doubled because of the effort employed in producing hybrid seed corn.
8. Insecticides, 1 kg = 24,200 kcal including production and processing (similar to herbicide; see 9).
9. Herbicides, 1 kg = 24,200 kcal including production and processing (54).
10. Each kg of corn was assumed to contain 3,520 kcal (53).

efficient, requiring only 22 kcal of feed for each kcal of milk protein. Incidentally, milk protein nutritionally has higher quality than beef protein.

The total amount of fossil energy used to grow a hectare of corn in 1970 averaged about 742 liters of fuel (7.1 million kcal) (23). An estimated 134 million hectares were planted in crops in 1970 (excluding cotton and tobacco) (25). With an estimated population of 200 million in the United States in 1970, this averages about 0.7 hectares per capita; but since about 20% of our crops are exported, the estimated number of hectares per capita used for domestic food production is about 0.56. In terms of fuel per person for food, employing modern agricultural technology, this is the equivalent of 416 liters of fuel per person (742 liters per hectare × .56 hectares per person = 416 liters). If we include processing, distribution, and home cooking, the total inputs per person for the food system are estimated to be 1,273 liters of fuel equivalents per person per year. If U.S. agricultural technology were used to feed a world population of 4 billion an average U.S. diet for one year, it would require the energy equivalents of 5,092 billion liters of fuel.

To gain some idea about what the energy needs would be if U.S. agricultural technology were employed, an estimate is made of how long it would take to deplete the known and potential world reserves of petroleum. The known reserves have been estimated to be 86,912 billion liters (55). If we assume that 76% of raw petroleum can be converted into fuel (55), this would equal a reserve of 66,053 billion liters. *If petroleum* were the only source of energy and if we used *all* petroleum reserves solely to feed the world population, the 66,053-billion-liter reserves would last a mere *13 years* [(66,053 billion liters)/(5,092 billion liters) = 13 years].

The world population is fast approaching the carrying capacity of the earth. We already mentioned the limitation of energy resources if we attempted to feed a world population of 4 billion the same diet as we consume in the U.S. employing U.S. agricultural technology. However, there is an additional limitation, and this is cropland. About 0.56 hectares of cropland are required to feed an individual in the U.S. with a diet high in animal protein. With 4 billion humans, the amount of cropland available per person amounts to only 0.36 hectares per person. Therefore, already the world population is above a density that we could feed a U.S. diet with the energy and land resources available.

* * *

Paper presented at the Energy Forum entitled, "The Role of Electric Power in Total Energy Needs" sponsored by the College Task Force on Energy and Agriculture. College of Agriculture and Life Sciences, Cornell University, January 14, 1975.

REFERENCES

1. NAS. Rapid population growth. Vol. I, Publ. for NAS by Johns Hopkins Press, Baltimore, Md., 105 pp. (1971).
2. Coale, A. J. The history of human population. Scien. Amer. 231(3): 40 (1974).
3. Freedman, R. and B. Berelson. The human population. Scien. Amer. 231(3): 30 (1974).
4. PEP. World population and resources. A report by PEP. Political & Economic Planning, London, 339 pp. (1955).
5. Corsa, L. and D. Oakley. Consequences of population growth for health services in less developed countries—an initial appraisal. pp. 369-402. *In* Rapid population growth. Vol. II. Research Papers. Natl. Academy of Sciences. Johns Hopkins Press, Baltimore, 690 pp. (1971).
6. Boserup, E. Conditions of agricultural growth. Aldine-Atherton, Chicago, 124 pp. (1965).
7. Lewis, O. Life in a Mexican village: Tepoxtlan revisited. Univ. Ill. Press, Urbana, 512 pp. (1951).

8. Stadelman, R. Maize cultivation in Northwestern Guatemala. Compiled by the Carnegie Institution of Washington. Contributions to American Anthropology and History, No. 33. Carnegie Inst. of Washington Publ. 523: 83 (1940).

9. FAO. Production yearbook, 1960. Vol. 14, FAO of UN., 507 pp. (1961).

10. Hainsworth, R. G. How many people can the earth feed? Foreign Agriculture 17: 23 (1953).

11. Clawson, M., R. B. Held, and C. H. Stoddard. Land for the future. Johns Hopkins Press, Baltimore. (1960).

12. Pawley, W. H. Possibilities of expanding world food production. FFHC Basic Study No. 10. FAO of UN, Rome, 231 pp. (1963).

13. FAO. The provisional indicative world plan for agricultural development. FAO. Rome. Vol. I and II. 672 pp. (1970).

14. Paddock, W. and P. Paddock. Hungry nation. Little, Brown & Co., Boston, 344 pp. (1964).

15. Wadleigh, C. H. Agricultural pollution. Trans. 35th North Amer. Wildlife Conf. pp. 18-25 (1970).

16. Lauer, G. J., D. Pimentel, A. MacBeth, B. Salwen, and J. Seddon. Hudson Basin Project. Report of the Biological Communities Task Group. Unpublished ms., 49 p. & appendices (1974).

17. Miller, M. F. Cropping systems in relation to erosion control. Bull. Mo. Expt. Sta. #366 (1936).

18. Shrader, W. D., H. P. Johnson, and J. F. Timmons. Applying erosion control principles. J. of Soil and Water Cons. 18: 195 (1963).

19. Kellogg, C. E. World food prospects and potentials: a long-run look. pp. 98-111. *In* Alternatives for balancing world food production needs. E. O. Head, ed. Iowa State Univ. Press, Ames. 273 pp. (1967).

20. Smerdon, E. T. Energy conservation practices in irrigated agriculture. pp. 11-15. Sprinkler Irrigation Assoc. Ann. Tech. Conf., Denver, Colorado, Feb. 24 (1974).

21. Addison, H. Land, water and food. Chapman and Hall Ltd., London, 284 pp. (1961).

22. Clark, C. The economics of irrigation. Pergamon Press, London, 116 pp. (1967).

23. Pimentel, D., L. E. Hurd, A. C. Bellotti, M. J. Forster, I. N. Oka, O. D. Sholes, and R. J. Whitman. Food production and the energy crisis. Science 182: 443 (1973).

24. Transeau, E. N. The accumulation of energy by plants. Ohio J. Sci. 26: 1 (1926).

25. USDA. Crop production. 1971 annual summary. USDA, Crop Rep. Bd. State Rept. Ser., 82 pp. (1972).

26. USDA. Changes in farm production and efficiency. USDA, Agric. Research Service. Prod. Econ. Res. Br., 40 pp. (1954).

27. USDA. Changes in farm production and efficiency. USDA, Econ. Res. Ser. Stat. Bull. 233, 31 pp. (1972).

28. Berry, R. S. and M. F. Fels. The production and consumption of automobiles. An energy analysis of the manufacture, discard, and reuse of the automobile and its component materials. Department of Chemistry, Univ. of Chicago, Chicago, Ill., 56 pp. (1973).

29. USDA. Farm power and farm machines. USDA (U.S. Department of Agriculture). Bur. Agr. Econ. Bull. F.M. 101, 35 pp. (1953).

30. USBC. Statistical abstract of the United States. USBC, Washington, D.C., 93rd Ed., 1017 pp. (1972).

31. Degraff, H. F. and W. E. Washbon. Farm-tractor fuel requirements. Cornell Univ., Agr. Econ. 449, 4 pp. (1943).
32. USDA. Liquid petroleum fuel used by farmers in 1959 and related data. USDA, Econ. Res. Ser. Farm Prod. Econ. Div. Stat. Bull. 344, 20 pp. (1964).
33. USDA. Fertilizer used on crops and pasture in the United States. 1954 estimates. USDA, Agric. Res. Ser. Stat. Bull. 216, 55 pp. (1957).
34. USDA. Fertilizer use in the United States. 1964 estimates. USDA, Econ. Res. Ser. Stat. Rep. Serv., Stat. Bull., 408, 38 pp. (1967).
35. USDA. Fertilizer situation. USDA, Econ. Res. Ser. FS-1., 42 pp. (1971).
36. USBC. Census of agriculture, 1964. USBC, Washington, D.C. Vol. II (1968).
37. Heady, E. O., H. C. Madsen, K. J. Nicol, and S. H. Hargrove. Future water and land use: effects of selected public agricultural and irrigation policies on water demand and land use. Report of the Center for Agricultural and Rural Development, Iowa State Univ. of Science and Tech. Prepared for the National Water Comm., PB-206-790 (NWC-EES-71-003) NTIS. Springfield, Va. (1972).
38. PSAC. The world food problem. Rept. of Panel on the World Food Supply. Pres. Sci. Adv. Comm., The White House. U.S. Govt. Printing Off. Vols. I, II, III. 127 pp., 772 pp., 332 pp. (1967).
39. USDA. Agricultural statistics 1970. USDA., Govt. Printing Office, 627 pp. (1970).
40. USDA. Extent of farm pesticide use on crops in 1966. USDA, Agric. Econ. Rep. 147, Econ Res. Ser., 23 pp. (1968).
41. USDA. Quantities of pesticides used by farmers in 1966. USDA, Econ. Res. Ser. Agric. Econ. Rep. 179, 61 pp. (1970).
42. CGG. Corn grower's guide. P-A-G Div., W. R. Grace & Co., Aurora, Ill., 142 pp. (1968).
43. CAHR. Food costs—farm prices. Comm. on Agr., House of Rep., 92nd Congress, 118 pp. (1971).
44. USBC. Statistical abstract for the United States. USBC, Washington, D.C. 92nd Ed., 1008 pp. (1971).
45. USBC. Statistical abstract of the United States. USBC (U.S. Bureau of the Census). Washington, D.C. 86th Ed., 1047 pp. (1965).
46. USDC. Census of transportation. Vol. III, Part 3. Govt. Printing Office, 633 pp., (1967).
47. ICC. Freight commodity statistics, Class I motor carriers of property in intercity service. ICC (Interstate Commerce Comm.). Govt. Printing Office, Washington, D.C., 97 pp., (1968).
48. ICC. Freight commodity statistics, Class I railroads in the U.S. ICC. Govt. Printing Office, Washington, D.C., 40 pp. (1968).
49. ICC. Transportation statistics, Part 1 (402 pp.), Part 5 (64 pp.), Part 7 (165 pp.), ICC. Govt. Printing Office, Washington, D.C. (1968).
50. USDT. Highway statistics. USDT (U.S. Department of Transportation). Govt. Printing Office, 199 pp. (1970).
51. USDA. Agricultural statistics 1967. USDA, U.S. Govt. Printing Office, 758 pp. (1967).
52. Leach, G. and M. Slesser. "Energy equivalents of network inputs to food producing processes," (Strathclyde University, Glasgow, 1973), 38 pp. (1973).
53. Morrison, F. B. Feeds and feeding. Morrison Publ. Co., Ithaca, New York, 1050 pp. (1946).
54. Pimentel, D., H. Mooney, and L. Stickel. Panel Report for Environmental Protection Agency, in preparation. (1975).
55. Jiler, H. Commodity Yearbook. Commodity Res. Bur., Inc., New York (1972).

56. UN. World population prospects as assessed in 1968. Department of Economic and Social Affairs. Population Studies, No. 53, 167 pp. (1973).
57. Hubbert, M. K. Man's conquest of energy: its ecological and human consequences. pp. 1-50. *In* The environmental and ecological forum 1970-1971. U.S. Atomic Energy Commission Office of Information Services. Oak Ridge, Tennessee, 186 pp. (1972).
58. UN. Statistical Yearbooks. Statistical Office of the United Nations, Department of Economics & Social Affairs, New York, N.Y. (1957-1971).

2.

The Emerging Economic Base for Low-Energy Agriculture

Jerome Goldstein*

The recognition of energy as a key issue in our society is having major impacts in what had once been the comfortable domain of economists. As the objectives of this conference set forth, energy has become an over-riding factor in food production, distribution and consumption, as well as in waste management.

Because of the dynamics of the energy issue, the same questions asked a few years ago will most likely not get the same answers now:

How much is a farmer's profit affected by his waste management methods?

How much attention should be given to preserving crop nutrients in wastes?

How much is manure worth to a feedlot owner, or a farmer?

How important are cropping methods and rotations relative to waste management?

Even more significant, the questions being asked at this year's Agricultural Waste Management Conference are most likely far different from those asked years ago. While most of you have been trained and concerned about agricultural wastes, you are about to be or already have been asked about nonagricultural wastes as well:

How effective is sewage sludge in a crop fertilization or soil-improvement program?

---

* Executive Editor, Organic Gardening and Farming Magazine, Emmaus, Pennsylvania.

What kind of equipment and materials-handling procedures are most efficient for farm use of urban wastes?

What incentives have increasing energy prices created for changing the inputs used in crop production and food distribution?

As land disposal becomes common practice, can we continue to maintain the distinctions between agricultural waste management and urban waste management?

Fortunately or unfortunately, depending upon how the problems are viewed, energy—whether its high price or its scarcity—is bringing many of our problems into a more unified perspective. The disadvantage is that no one knows for certain just how to analyze the economics of a problem. The waters of problem-solving, based upon traditional thinking, are more muddied than ever!

The advantages are that alternatives for low-energy agriculture—and new economic incentives—are looking more favorable than they ever have. The need for such alternatives is not expected to disappear. According to the Director of Agricultural Economics for the U.S. Department of Agriculture, "the energy problem is real and long-term. . . . We shall have to learn how to use energy more sparingly and more efficiently in our agricultural operations." (1)

As waste management is seen to be part of the larger issue of *energy management*, the economics of waste management has become interwoven within the entire fabric of our society. Thus when we speak of economics and economic incentives, it is absolutely necessary to include factors which reflect the intricate network of wastes touching such seemingly disparate factors as water quality, land use, oil prices—from creating jobs as wastes are used as resources to the political realities associated with using wastes-on-the-land. This message can be postulated in the following anecdotal equation:

The Economics of Energy + The Economics of Environment +
The Economics of Food = The New Alternatives in Wastes Recovery.

The economics of recycling organic wastes have rarely been attractive, except for some isolated examples. Until the passage of environmental legislation affecting air and water quality, waste management was considered synonymous with waste disposal. While there was some talk among economists of "externalities . . . social costs . . . quantifying public benefits . . . public investment" there was little change toward the direction of resource recovery based upon economic analyses. (2). However, in the last year, high energy costs and high food prices have joined forces with the environment in the push toward effective alternatives. The common factor in these alternatives is a low-energy agricultural and food distribution system.

According to research studies by Pimentel (3), Steinhart (4), Heichel (5), Hirst (6), Stanford Research Institute (7), Black (8), Commoner (9) and others, energy use in food production has been increasing faster than energy use in many other sectors of the world economy. Energy use for transportation has more than tripled in the past 30 years, and the food-processing industry is the fourth largest consumer of energy among American industries. The fact is that foods do vary in energy consumption depending upon where and how they were grown, how much processing, and how far they were transported.

The characteristics of a low-energy agricultural system, based upon the findings of the researchers cited above, include the following:

1. Greater use of labor-intensive, soil-conserving methods. Prime technique is to maintain crop yields by using organic wastes (animal manures, treated sludge, effluent, compost, cannery wastes, methane, etc.) and legumes in rotation to reduce applications of nitrogen and other commercial fertilizers. Diversified farming operations—with both livestock and grain—relate well to these practices, as opposed to methods available on farms which practice monoculture.

Humus-rich soils have also been found to require less tractor fuel consumption during cultivation. Baker and Cook report that soils high in organic matter help the farmer avoid use of fungicides, because the more diverse the soil, the more it is biologically-buffered (10). "Nitrification is one of the most pesticide-sensitive soil microbiological transformations," says J. F. Parr.

2. More direct marketing of food between farmer and consumer. Main feature is a shorter transportation route between producer and consumer, characterized by roadside stands, farm markets, local butchers, custom canneries, food cooperatives and buying clubs. Special emphasis is on unprocessed foods. The shortest "food route" for some 40 million Americans this year is from their own vegetable garden to kitchen. Anyone who grows part of their own food supply becomes well aware of the advantages of shortening the gap between producer and consumer.

3. The need for public acceptance of land application of agricultural and urban organic wastes. Prior to energy and environmental considerations, a feedlot owner or a city public works director performed his responsibilities best by selecting a disposal method that was cheapest to his company or his voting taxpayers. However, with the rejection of incineration, the absence of landfill sites and passage of such legislation as the Water Pollution Act,

the same authorities responsible for waste management quite often must receive public support for methods involving land application. When such land application methods are seen to be part of a low-energy, environmentally-sound system, they have received enthusiastic support.

4. More people able to earn sufficient income while living in rural areas. Consistent with the development of an energy-intensive agricultural system has been the exodus of rural populations to urban centers. Recently, there has been an increase, evident from latest census data, in the number of persons living on farms but who earn salaries in nonagricultural occupations. Most of these farms produce sufficient food for the families engaged in part-time agriculture.

5. A recognition of the interrelationship between farms and cities. More Americans today than ever before are beginning to understand the *philosophy* of using urban organic wastes in a well-run permanent food-producing system. Indeed, an acceptance of that philosophy way well be more important than any technology that has been or will be developed in waste management. Using wastes in agriculture may well never be an attainable goal until it is viewed by us all as both an urban and rural mission. A distinguished professor, author and organic farmer in Kentucky, Wendell Berry, phrased the challenge eloquently when he said:

> "To recognize the extent and the destructiveness of our 'urban waste' is to recognize the shallowness of the notion that agriculture is another form of technology to be turned over to a few specialists. The sewage and garbage problem of our cities suggests, rather that a healthy agriculture is a cultural organism, not merely a universal obligation as well. It suggests that, just as the cities exist within the ecology, they also exist within agriculture. It suggests, that, like farmers city-dwellers have agricultural responsibilities; to use no more than necessary, to waste nothing to return organic residues to the soil." (12)

By describing the following examples of elements of a low-energy agricultural system, I hope to convey the "emerging economic base." I maintain that these examples, however scattered, are part of a growing trend which will relate to the research findings you will hear about from others at this conference. I do not mean to oversimplify in any way the complexities involved in waste management. What I am trying to stress again and again is that waste management is—and must be treated as—part of a larger objective that encompasses our national energy policies and food production/delivery system.

FARMING AND MUNICIPAL SLUDGES

According to University of Minnesota soil scientists, sewage sludge sufficient to produce 160-bushel per acre corn could replace scarce fertilizer under certain conditions. "Growers can apply 10 tons of dry solids per acre each year for at least 2 years on land going into corn or other cereal crops with only a slight metal increase in the grain. . . . Nutrients of dry sludge range from 3.5 to 6.5% nitrogen, 1.8 to 8.7% phosphoric acid, and up to 0.84% potash— equivalent to a 4–7–0 fertilizer." (13)

The Burlington, Wisconsin, Wastewater Treatment Plant actually advertises in the local newspaper that it has for free "well digested odorless aerobic activated sludge in a solid state for easy handling. . . . Beat the high cost of commercial fertilizer," the advertisement tells farmers.

The Metropolitan Sanitary District of Greater Chicago tells local townships about its free fertilizer, Nu-Earth, taken from the sludge dry beds, dewatered and composted, "an excellent soil builder/ fertilizer."

An announcement earlier this year by scientists of the Food and Agriculture Organization began: "Developing countries suffering from shortages of food and unable to pay for expensive chemical fertilizers, even when available, could ease both problems by making much greater use of organic materials as nutrients in their agriculture." The experts considered a wide range of organic materials, including animal manure, human waste and garbage.

At the United States Department of Agriculture's Beltsville, Maryland, research station, sludge from the Blue Plains treatment plant is being composted and shown to be of value in agronomic tests.

Unfortunately, there are very few successful examples of farmers who are using sludge to cut down fertilizer use. One reason is that many farmers who begin using sludge are harassed by local officials who accuse them of violating health ordinances. When farmers must use lawyers to get permission to save energy in food production, not much energy will be saved (14).

Recently the staff of the Northwest District Office of the Ohio Environmental Protection Agency conducted a survey of sludge disposal practices. Following are some excerpts from their findings from a paper, "Farming and Municipal Sludges: They're Compatible," by Manson and Merritt (15):

"Findlay, Ohio, is an industrialized city of 33,000 people located about 50 miles south of Toledo. The City has been disposing of digested sludge by land application for 13 years. Treatment plant personnel utilize a 5000-gallon tank truck and for some years have spread on a 10-acre parcel of land west of the plant. In recent years 6 to 8 farms in outlying areas as well as a nursery have been used. The average haul distance is 3½ miles in one way with a maximum of 8-9 miles one-way on occasion. The sludge normally averages 3½% solids.

"The City has considerable flexibility in sludge handling since the plant includes six drying beds, several acres of lagoons and a vacuum filter in addition to the tank truck. However, direct land application is preferred as it involves handling sludge only one time. Plant personnel prefer fall and winter application as opposed to spring or summer. However, during a reasonably dry year they can readily dispose of all the wastes by land application.

"The basic method of operation calls for one uniform application of digested sludge on any given field per year. The sludge is not plowed under since care is taken to apply the material in a thin even layer. Truck spraying is stopped several hundred feet short of the end of the field in order to avoid having to drive through the applied material on subsequent passes. This is also important to prevent run-off from the field. Plant operators report two possible problems with land application: (1) weather and (2) continued cooperation by the farmers.

"In interviewing farmers and the nurseryman, we noted all were quite anxious to continue receiving the material. In fact, one of the farmers was complaining that a nearby farmer, further from the treatment plant, was receiving a considerable amount of sludge and that it hardly seemed fair to haul it right by his place and not apply any to his land. This particular farm is also of interest in that the main 75-acre field, which receives the waste, is adjacent to and across the street from a small trailer park. There have been no complaints of odors or nuisance from this municipal sludge operation. . .

"Based on our experiences with land application and our conversations with users, we would offer the following conclusions:

A. The biggest limitation to land application is the ability to get on the land either because of wet conditions or a growing crop. All-weather roads, irrigation equipment or grasslands must be available to make the program work.

B. Farmers feel the next major limitation is soil compaction. Balloon tires, irrigation equipment or other means must be used to distribute the load to minimize soil compaction.

C. Most farmers don't give adequate credit to the nutrients available in the material. However, as fertilizers become scarce and prices rise a new prospective will evolve.

D. The biggest advantage most authorities and farmers seem to agree on is the addition of humus material to the soil. Both heavy soils and sandy soils will be improved over the long run by the addition of sufficient sewage sludge."

Twenty farmers with a total of 4900 acres have volunteered to have sewage sludge applied on their land from the Lima, Ohio, treatment plant. According to Roland Nevergall, waste disposal engineer in Lima: "We actually have more applicants than we can supply sludge to. . . . The turning point came when they saw the results of some test strips we made last spring. We spread sludge in a few strips across a corn field before planting. Those strips were up and growing before the other corn. A month or so later, the corn in the strips was a good 8 inches taller than the rest of the field."

GAINING PUBLIC ACCEPTANCE FOR THE
AGRICULTURAL USE OF ORGANIC WASTES

A sure way to arouse the wrath of rural dwellers has always been to propose spreading city sewage sludge or garbage on their farmland, especially if the sludge came from a distant large city. Simply put, the expected response has mostly been: "Don't dump it on us. If the stuff is so good, let 'em use it in Times Square, Broad & Vine, etc." If little or no attempt is made to prove the agricultural worth of sludge, then a negative reaction is entirely predictable.

But, as the experience of Ohio indicates, when farmers are shown how sludge or any other waste can improve profits, the sludge becomes positively desirable. The survey previously cited by Manson and Merritt also indicated that direct land application of sludge was much more acceptable to rural residents who were used to handling and disposing of animal wastes on the land.

Recent experience also indicates that the *composting* method exerts a most significant influence in gaining citizen approval for recycling wastes. Specifically, since composting is a method used by millions of Americans in their own gardens, to convert such things as orange peels into something far more pleasant and useful than garbage, the principle of composting connotes a positive image. And, by definition, composting is the biological decomposition of the organic constituents of wastes under *controlled* conditions (16). The word "controlled" is vital if citizens are to accept the construction of a waste recycling plant in their neighborhood, and the application of treated wastes on land near where they live. According to Golueke, "it is the application of control that distinguishes composting from the natural rotting, putrefaction, or other decomposition, that takes place in an open dump, a sanitary landfill, in a manure heap, in an open field, etc."

A case in point is the newly-built 36-acre Recycling and Composting Center built and operated by the Lehigh County (Pennsylvania) Authority. The facility serves Allentown as well as neighboring communities. Previously the Allentown City Council had voted to build a new incinerator to replace the old smoky one which had been ordered closed by the State Department of Environmental Resources. The vote changed when the proposed incinerator bid came in at three times the original estimate. Landfill was the next choice of the City Council, but the problem was no landfill site within the city limits.

A landfill, no matter how well conceived, would have been unacceptable to the supervisors of the township in which the proposed treatment area was located. But, because composting was the treatment method chosen, the township's environmental activists and ardent home gardeners rallied support of neighbors to get the elected supervisors to vote approval for the county to create a regional recycling installation. The result is that in April 1975, Lehigh County has a $2.1 million composting plant that shreds garbage, spreads it onto windrows 400 feet long on a paved pad, and mechanically aerates the material. The plant's capacity is about 350 tons a day, appears capable of handling close to 600 tons, and may eventually compost sewage sludge as well as garbage.

When they were first told in 1973 that Lehigh County had decided to buy an existing landfill and turn it into a landfill for all the solid waste of the county, citizens of Upper Saucon Township were understandably indignant and ready to pressure their supervisors to fight. "This pleasant green area of well-kept homes, some beautiful old stone houses, forested hills, the Saucon Creek and a few old-fashioned working farms still planting wheat and corn among the housing developments and shopping centers make Upper Saucon Township a very desirable place to live." (17)

But after the county's spokesman presented the finer aspects of composting at several township meetings, many potential opponents became staunch advocates. Instead of the usual cries: "Don't dump it on us!", the neighbors became proud to have an environmentally-rational, economically-sound recycling plant that already is attracting visitors from all over the nation. This is not to imply that everyone in the township was completely enthusiastic; rather there was a significant number of active partisans, and a larger group who accepted the fact that a sensible waste management decision for the county meant that the compost plant had to be located *somewhere*, even if the somewhere was in their immediate neighborhood.

Composting has the important dual feature of being acceptable *semantically* as a process to urban people and *agriculturally* as a product to farmers and gardeners. Thus composting will become far more significant than it has been as a waste treatment method in this nation, as we move more aggressively toward low-energy agriculture.

INCREASING INCOME FOR SMALL AND PART-TIME FARMERS

Another characteristic of a low-energy agriculture system is a greater number of jobs—and an economic base to provide adequate income for those jobs. It is paradoxical to say the least, after learning the findings of Dr. Pimentel's research into "Food Production and the Energy Crisis," and reading the latest unemployment forecasts, to find that American agricultural efficiency is still measured in how few farm workers are needed to supply us with food, fiber and other farm products. We are told that two farm workers supply the needs of 105 persons. Economists at North Carolina State University with great pride proclaim that "man-hours for farm work have declined from 20.5 billion in 1940, to 15 billion a decade later and only slightly over 6 billion in the early 1970's." (18)

To a large extent, agree the economists, capital has been substituted for labor. Labor needed in animal agriculture, according to 1968-72 averages, were 0.7 hour for 100 pounds (46 quarts) of milk, 1.8 hours for 100 pounds of beef, and 1.2 hours for the same quantity of pork. While farm chickens raised for meat still required 3.5 hours per 100 pounds, turkeys were down to 1 hour and commercial broilers to 0.4 hour per 100 pounds of meat produced. Laying hens needed 0.4 hour per 100 eggs.

Most of today's economists appear to take the position that we would all be better off if our agricultural system "improved in efficiency" with the result that only one farm worker could feed 105 persons. But not all economists agree, and one of the earliest critics of our efficiency criteria has been Perelman who cautioned about "Farming with Petroleum." (19)

Thus a case can be made that our society as well as our food production system would be improved if it were possible to *lower* the ratio of farm workers to number of persons supplied, while at the same time *not* increasing costs or lowering yields. And the fact is that such developments are taking place increasingly in the U.S. Some state departments of agriculture have created special programs to aid the marketing of crops grown on smaller, labor-intensive farms.

The West Virginia Department of Agriculture operates four regional markets to aid the small farmer solve his marketing problems. The program is two-phase—selling volume on consignment to supermarkets in and out of the state (at wholesale price) and also retail to local consumers. Total sales in 1974 were $1,684,258, up more than $500,000 from the previous year. Each of the farmers markets are, in effect, state-sanctioned food co-ops run by the farmers themselves with an 8- to 10-man board of directors, elected yearly from a pool of local farmers. Markets are run from a 5% commission charged each farmer from his sales. The market manager is the sole state-paid employee. Farmers receive the same price regardless of volume.

Other programs being developed to help the West Virginia small and part-time farmer include the addition of meat cutting to the courses available at the West Virginia Vocational-Technical Schools.

Carcasses would be bought from West Virginia cattlemen, slaughtered in schools where local persons would be learning a trade and then passed on—at cost—to state hospitals and schools at savings to the state. Another project calls for a one-half million dollar cold storage warehouse—just approved by the legislature—to help West Virginia apple producers from taking a loss because of the perishable nature of fruit. The new warehouse will have a 100,000-bushel capacity.

The West Virginia Director of Agriculture, Guy Douglass, stresses these programs since more than 80% of his state's farmers fit the "small" category. With the number of small farms actually increasing (up 500 over 1973), West Virginia is very conscious of the economic role that they play in the state.

To aid direct marketing in Pennsylvania, the Department of Agriculture has compiled a directory of farmers who operate retail outlets. The directory of some 800 farmers will be sent to all of the food cooperatives and buying clubs in Pennsylvania as part of the state's program to encourage producer-consumer cooperatives. A farm market has been set up by the department in the state fairgrounds parking lot, where 21 farmers in the Harrisburg area can sell directly to consumers.

IMPROVING DIRECT MARKETING
OPPORTUNITIES IN URBAN CENTERS

On a grand scale, national political leaders have called for a United States energy policy that leads to independence. On a more modest scale, there is a grass-roots movement toward regional self-sufficiency—not total necessarily—but a self-sufficiency that would measurably reduce energy inputs.

Whether you travel to Burlington, Vermont, or San Jose, California, Syracuse, New York, or Ann Arbor, Michigan, you will find thousands of people in each area talking about alternatives in food production and distribution. And the talk is resulting in a rapid proliferation of food cooperatives. These people are questioning why a chicken in a supermarket must travel an average of 1200 miles (20), when they know a local poultry farmer who just went out of business. They question why local growers are about to plow under fields of potatoes, while the stores bring in potatoes from several thousand miles away by truck.

In their modest ways, city people are setting up food-buying clubs, food cooperatives, shopping at roadside farms or right-on-the-farm. Each one, and there are thousands of them, help to localize the food sources, give increased economic base to regional farmers, and in general aid the development of low-energy agriculture.

In Chicago, Dorothy Shavers helped to organize the Self Help Action Center, a not-for-profit corporation 3 years ago. In 1974, almost $700,000 worth of food was purchased directly from farmers in Illinois and through southern cooperatives and sold to thousands of poor people in center-city Chicago. The center now has its own warehouse, refrigerated truck, and coordinates 115 food-buying locations.

Homeworkers Organized for More Employment in Orland, Maine, sell food in four retail outlets to summer tourists, and crafts produced at home by members other times of the year. H.O.M.E. has organized a Sheep and Goat Co-op, an association of independent farmers wanting to provide income through the sale of lambs, wool and milk. More than 1000 members belong to the H.O.M.E. cooperative.

Networks of state and regional food cooperatives have developed programs to improve trucking efficiency, farmer contacts and quality control. The Vermont network of food co-ops last year sold more than $1 million worth of food.

CONCLUSION

Waste management is now interwoven within the entire fabric of our society. Today, waste policy now reflects domestic policy, just as food policy is foreign policy. To make the right decision demands more than your technical expertise; the right decision demands your understanding of the interrelationships between our agriculture, waste management, energy and social systems.

We must recognize that to implement a really good decision, one has to be able to tie all the loose ends together—and to accomplish that in waste management requires a tremendous amount of knot-tying.

The main point is that if we want to manage our wastes—whether those wastes are created on the farms, in the cities, or in the factories—we cannot do it with only the approaches of the past. In the final analysis, waste management demands the same all-encompassing considerations as people management.

We see today that where positive far-sighted programs are in practice, where a city, or county or its authority has turned to composting, or wastes-to-fuel, or methane production, that installation is praised for its resource recovery aspects, for its energy-saving qualities, for its job-creating abilities. It seems fair to predict that as you meet again in the future, these Agricultural Waste Management Conferences will offer more and more models of waste management that are synonymous with low-energy agriculture.

REFERENCES

1. Paarlberg, D. "After the Storm," National Farm Institute, Des Moines (March 24, 1975).
2. McFarland, J. "The Economics of Composting Municipal Wastes," Compost Science (July-August 1972).
3. Pimentel, D., L. E. Hurd, A. C. Bellotti, M. J. Forster, L. N. Oka, O. D. Sholes, and R. J. Whitman. "Food Production and the Energy Crisis," Science 182, 443-449 (1973), Cornell University; Pimentel, D., "Food Nitrogen and Energy, NY State College of Agriculture and Life Sciences, Cornell University, Ithaca, NY. Presented at the International Conference on N_2 in Washington State, June 6, 1974.
4. Steinhart, J. S. and Carol E. "Energy Use in the U.S. Food System," Science, April, 1974, p. 307-316, Prof. of Geology and Environmental Studies, University of Wisconsin-Madison; Steinhart, John S. and Carol E., "The Fires of Culture: Energy, Yesterday and Tomorrow," Duxbury Press, North Scituate, MA, 1974.
5. Heichel, G. H. "Energy Use and Crop Production," The Connecticut Agricultural Experiment Station, PO Box 1106, New Haven, CT 06504; Heichel, G. H. "Comparative Efficiency of Energy Use in Crop Production," The Connecticut Agricultural Experiment Station, Bulletin 739, Nov. 1973.
6. Hirst, E. "Energy for Food from Seed to Stomach," Environmental Program, Oak Ridge National Laboratory, Tennessee. Organic Gardening and Farming, May 1974.
7. Brown, S. L. and U. F. Pilz. "U.S. Agriculture: Potential Vulnerabilities," prepared for Office of Civil Defense, Office of the Secretary of the Army. Final Report, January 1969. Stanford Research Institute, Menlo Park, CA; Brown, Stephen L. and Pamela G. Kruzic. "U.S. Agricultural Vulnerability in the National Entity Survival Context," prepared for Office of Civil Defense, Office of the Secretary of the Army. Final Report July, 1970. Stanford Research Institute, Menlo Park, CA.
8. Black, J. N. "Energy Relations in Crop Productions," Annals of Applied Biology, 1971 (British) Vol. 67, pp. 262-285; Perelman, Michael, "Farming With Petroleum," Chico State University, Calif., Environment, Oct. 1972.
9. Commoner, B., M. Gertler, R. Klepper, and W. Lockeretz. "The Vulnerability of Crop Production to Energy Problems," Center for the Biology of Natural Systems, Washington University, St. Louis, MO, Jan. 1975.
10. Baker, K. F. and J. R. Cook. "Biological Control of Plant Pathogens," W. H. Freeman & Co., San Francisco, 1975.
11. Guenzi, W. D., ed. "Pesticides in Soil and Water," Soil Science Society of America, 1974.
12. Goldstein, J. "The New Food Chain, Rodale Press, Inc. 1974.
13. Anon. "160-Bushels Corn with Sewage Sludge," Successful Farming, Nov-Dec, 1974.
14. Lear, J. "Organic Farmer Goes on Trial," Organic Gardening and Farming, August 1974.
15. Manson, R. and C. Merritt. "Farming and Municipal Sludge," Compost Science (to be published in 1975).
16. Golueke, C. G. "Composting," Rodale Press, 1972.
17. Adams, R. C. "Use it Again, Sam" (unpublished).
18. Anon. North Carolina State University news release, 12/31/74.
19. Perelman, M. "Farming with Petroleum," October, 1972.
20. Hightower, J. "Eat Your Heart Out," Quadrangle Books, 1975.

3.

Energy Utilization on Beef Feedlots and Dairy Farms

D. W. Williams, T. R. McCarty, W. W. Gunkel,
D. R. Price and W. J. Jewell*

The multiple concerns of the energy crisis and environmental problems caused by agricultural wastes have focused attention on agriculture as a possible source of energy. It has been demonstrated that the photosynthetic energy contained in animal manures and other organic matter can be converted to a useful energy form, methane gas, via anaerobic digestion. What has not been done is to evaluate the process as it fits into whole farm systems in the U.S.

The Cornell-NSF Bioconversion Project was designed to meet this objective: to evaluate the overall feasibility of anaerobic digestion, including the aspects of energy production and utilization, nutrient conservation, and pollution control. The subject of this paper deals with the ways in which direct energy inputs, fossil fuels and electricity are used for agricultural production. Indirect energy inputs, such as the energy for manufacturing chemical fertilizers, farm machinery, and pesticides, are not included in this study. It is recognized that these indirect energy forms do constitute a significant portion of total farm energy use; Pimentel (1) calculated that the indirect energy forms made up 62% of the total energy use for growing corn. However, since the objective of

* Research Associate, Research Specialist, Professor, Associate Professor and Associate Professor, Agricultural Engineering Department, Cornell University, Ithaca, New York 14853.

the overall project is to determine the feasibility of directly sub-stituting methane for other energy forms, only the fossil fuels and electricity which have this potential for substitution will be discussed.

OBJECTIVES

The specific objectives of this paper include:

1. For a 40-cow dairy farm, a 100-cow dairy farm and a 1000-head beef feedlot, determine the quantities of diesel fuel, gasoline, heating oil, and electricity that are used annually for crop and livestock production, as well as in the farm home for family living.

2. To determine the yearly distribution of each of these direct energy inputs on a monthly basis.

3. To determine the photosynthetic energy flows on the farm situations in question, including total crop biomass production, crop residues, bedding, feed for the animals, animal food products, and manure production.

4. To compare the photosynthetic energy flows with the fossil fuel energy inputs, in order to determine the relative energy ef-ficiencies of the different farm situations. Also to assess the poten-tial for energy recovery from crop residues and animal manures as compared with the fossil fuel energy requirements.

PROCEDURE

It was decided to investigate two specific farm situations: an average sized dairy farm in New York State and a large beef feed-lot in the upper midwest U.S. These operations were defined in detail with respect to equipment, land, labor and capital require-ments. The crop and livestock cultural operations were detailed for a typical year to determine the time distribution of capital and labor use. Finally the direct energy use by both agricultural and personal home activities was determined, both on a yearly and on a monthly basis. This direct energy flow was then compared with the photosynthetic energy flow in order to determine the energy potential of the manures and crop residues.

DAIRY FARM DEFINITION

The various components of the dairy farms are listed in Table 1. These include land, machinery, animals, buildings and feed storage, and labor.

It was decided to investigate two specific dairy farm sizes, a 40-cow and a 100-cow dairy. The smaller 40-cow dairy used a stanchion type barn and was found to be the average size farm which can be successfully operated by a single family's labor. For larger farm sizes, it was found that about 100 cows were needed to sup-

Table 1. Components of dairy farms.*

Component	40-Cow Dairy			100-Cow Dairy		
	No.	Description	Value	No.	Description	Value
Land (hectares):	105		$56,300	237		$150,000
Corn Silage	10.5	30,300 kg/ha		30	35,300 kg/ha	
Corn Grain	2.5	5760 kg/ha		9	5930 kg/ha	
Hay	30	5300 kg/ha		64	5570 kg/ha	
Oats	6	1690 kg/ha		8	2220 kg/ha	
Other	56	Pasture, Woodland and unused		126	Pasture, Woodland and unused	
Machinery:			$28,000			$50,000
Tractors	3	2 gas 30-60 hp 1 diesel 60+ hp		5	3 gas 30-60 hp 2 diesel 60+hp	
Trucks	1	¾-ton		2	¾-ton 2-ton	
Materials Handling		Barn Cleaner Silo Unloader			Bunk Feeder Silo Unloaders	
Milk Handling		Storage Tank Vacuum Pump			Storage Tank Vacuum Pump	
Heating		Space and Water Heaters			Space and Water Heaters	
Animals:	67		$29,400	163		$70,000
Cows	40	473 kg/hd (1260 lb)		100	573 kg/hd (1260 lb)	
Young Stock	27	295 kg/hd (650 lb)		63	295 kg/hd (650 lb)	
Buildings:			included w/land			included w/land
Barns	2	Stanchion Barn Loose Housing		2	Free Stall Barn, Loose Housing	
Houses	1			1		
Storage		Silos—327 m. tons Hay—182 m. tons			Silos—1273 m. ton Hay—227 m. tons	
		Total Capital $113,700			Total Capital	$270,000
Labor:						
Operator	1	1 Man Equiv.		1	1 Man Equiv.	
Family	5	0.4 Man Equiv.		5	0.4 Man Equiv.	
Hired		0.3 Man Equiv.			1.7 Man Equiv.	

* Basic References: U.S. Census Bureau (3), Bratton (4, 5), Kearl and Snyder (6) and Personal Interviews, New York (7).

port an additional full-time hired man, and that those larger farms use a free stall barn type.

The machinery category in Table 1 included those components of the dairy farm which directly consume the fuel and electricity energy inputs. The larger farm was found to use more and larger

diesel tractors than the smaller farm, thus affecting the proportions of gasoline and diesel fuel use. The free stall barn used on
the 100-cow dairy has a separate milking center with automated
milk handling, thus more electrical equipment than the smaller,
stanchion type barn.

It was found that dairy farms with a given number of milking
cows have additional younger calves and heifers being raised as
future replacements for older cows. The number of young stock
was about 60% of the number of milking cows and their average
weight was about half the weight of the milking cows. The outputs
of the farm are the milk produced by the cows and the meat included in cull cows that are sold for slaughter. The gross energy
value of these outputs was calculated and this energy will be compared with the direct fuel and electrical energy inputs in a later
section of the paper. It was determined that a milk production
figure of 5682 kilograms (12,500 pounds) per year per cow was representative of the dairy sizes in this study, based on New York
State milk production averages, and data reported by the State
Dairy Herd Improvement Association (2).

The dollar values of the farm components show that the land
and buildings comprised the most significant share. The values
of the two farm sizes were roughly in proportion to the number of
cows found on each farm, about $2700 per cow.

Since dairy farming is such a sizeable industry in New York,
there were considerable data available from which to define the
"typical" dairy farms. The sources included the U.S. Census
Bureau (3), Bratton (4,5), Kearl and Snyder (6), and Personal Interviews, New York (7). The general approach used in arriving at
the figures in Table 1 was to average data from the various sources
and scale the numbers to the 40- and 100-cow farm sizes.

BEEF FEEDLOT DEFINITION

The components of the 1000-head beef feedlot are listed in Table 2.
The categories are similar to the dairy farms, and include land,
machinery, feed storage and processing, feedlot structures, animals
and labor.

The 1000-head feedlot includes not only the feedlot structure but
also land and machinery to grow almost all the feed. It was found
that corn is the principal component of the ration for fattening
cattle. The corn is harvested as both whole crop silage, and high
moisture shelled corn grain. This typical farm needs to grow only
this crop to feed the cattle; it was found, however, that some beef
feedlot farms grow additional cash crops such as wheat and other
small grains.

This typical feedlot could be located in the upper midwest states

Table 2. Components of 1000-head beef feedlot farm.*

Component	Description	No. of Units	Value $ (1973-74)
A. Land			653,838
Corn Grain	Harvest High Moisture Corn @ 7417 kg/ha	192 hectares	
Corn Silage	Harvest Whole Crop Silage @ 37,120 kg/ha	168 hectares	
B. Machinery			198,727
Tractors	Diesel—125 hp	2	
	Gasoline—60-80 hp	3	
Tillage & Harvesting	12-Row Conventional Tillage Equipment		
	6-Row Combine	1	
	2-Row Silage Chopper	2	
	Wagons	6	
Feedlot Equipment	Liquid Manure Spreader	2	
	Pump	1	
	Truck-mounted Feed Mixer-Wagon	1	
C. Feed Storage and Processing			46,079
	Bunker Silos:		
	Silage—6,240,000 kg	1	
	Corn—1,425,000 kg	1	
	Roller Mill	1	
	Bulk Supplement Bin	1	
D. Feedlot Structure	1000-head capacity Housed, confinement barn with slatted floor and manure pit. 12.2 x 230 meters (40 x 754 ft)	1	142,880
E. Animals			254,000
	Beef Cattle Initial Weight 295 kg Ending Weight 477 kg	1000	
		Total Capital:	1,295,524
F. Labor			
	3 Full-Time	4 Man-Equiv.	
	2 Part-Time	per year	

* Basic References: Hughes (9), Boehlje (10), Edwards and Stoneberg (13), and Personal Interviews, Ohio (11).

such as Iowa, Illinois, Michigan, and Ohio. Here the climate and soil conditions are favorable for corn production and markets are available for fattened beef cattle. This type of feedlot is different from southwestern U.S. feedlots where several thousand cattle may be fed in one location on open dirt lots. The upper midwest feedlot is smaller, grows almost all its own feed, and has a harsher environment where shelter and manure handling are major considerations. Thus, this feedlot has a good potential for utilizing anaerobic digestion since some type of waste treatment is necessary, no matter what system is used. The confined and sheltered nature of the lot makes it easier to collect the manure.

Large diesel-fueled tractors are required to do the soil tillage and crop harvesting activities associated with corn production. The smaller gasoline-powered tractors handle feedlot activities including feed and manure handling. A small quantity of electricity is also required for such uses as water pumping, feed grinding, and lights.

The feedlot structure is a housed, confinement barn. The complete feedlot is under a roof, with no outside areas. The space provided for the cattle is about 1.6-2.1 m^2/head (15-20 ft^2/head). Manure is stored in an underfloor pit, and spread every six months. Such a system is typical of new feedlots being constructed in the upper midwest. Certain modifications would be required if the manure were to be anaerobically digested; however, this type of "liquid manure" handling system most closely approximates the requirements of anaerobic digestion technology.

The animals are fattening beef cattle with an average weight of 386 kg (850 lb). The fattening period required for these animals was found to be about 160 days per animal. Thus, the feedlot would handle 2.25 times the 1000-head capacity per year, or 2,250 animals.

It was determined that the equivalent of four full-time men would be required for running the whole feedlot farm. The part-time help is needed in the spring and in the fall when crop and manure handling activities are particularly intense.

The capital value of the components was found to be high, about $1300 per head of feedlot capacity, but still not as high as the capital requirements per dairy cow on the previously defined dairy.

The approach for defining the beef feedlot involved the use of both statistical data and engineering calculations. Since the feedlots in the midwest are spread over several states, there was not a consistent data base to draw all the information needed. Also, most cornbelt farms are smaller and diversified, containing other crops and animals besides corn and beef. The 1000-head beef size was specified in the original research proposal. The procedure for detailing all the farm components was the following:

1. The weight gain and rate of gain per animal was chosen based on nutritional studies (8), and actual case studies (9-11).

2. Using the weight gain, the corn needed in ration as well as the land needed to grow this corn was determined. The rations and crop yields were based on the above references and the USDA Statistical Reporting Service (12).

3. The sizes and numbers of tractors and machinery to grow and harvest the corn were determined based on the above references and Edwards and Stoneberg (17) reporting of available days for crop activities in various months of the year in Iowa.

4. The feedlot size and value were based on data reported by Boehlje (10).

5. The labor requirements were based on the references listed in item 1 above.

DIRECT ENERGY UTILIZATION FOR CROP AND LIVESTOCK PRODUCTION

The quantities and distribution of energy used by dairy farms and beef feedlots are shown in Tables 3, 4, and 5 and Figures 1, 2, and 3. For ease of comparison all energy values have been ex-

Table 3. Annual energy use by type of fuel and by operation in a 40-cow dairy.

Operation	Gasoline 10^6kcal (gal)	Fuel Type Diesel 10^6kcal (gal)	Electricity 10^6kcal (kwh)	Heating Oil 10^6kcal (gal)	Total 10^6kcal
Crop					8.7
Establishment					
Plowing		3.5 (100)			
Seedbed Prep.	2.3 (74)				
Planting	1.3 (42)				
Truck	1.6 (50)				
Harvest					30.2
Hay	8.7 (280)	9.0 (255)			
Corn	3.6 (115)	4.2 (120)			
Truck	4.7 (150)				
Manure Handling	6.8 (219)	5.5 (155)	0.1 (130)		12.4
Barn & Farmstead					37.7
Misc. Tractor	10.6 (340)	3.9 (110)			
Truck	6.2 (200)				
Feed Handling			0.3 (310)		
Milk Handling			5.7 (6620)		
Space Heating			0.9 (1000)		
Water Heating			4.1 (4800)		
Lighting			1.7 (2000)		
Other Electricity			4.3 (5000)		
Personal Home					74.2
Cooking			1.1 (1300)		
Clothes Drying			0.9 (1100)		
Water Heater			5.2 (6000)		
Other Electricity			7.0 (8100)		
Space Heating				60 (1700)	
Total	45.8 (1470)	26.1 (740)	31.3 (36,360)	60 (1700)	163.2

pressed in millions of kilocalories. The common units, such as gallons of gasoline are also included in parenthesis. Again, it is stressed that the energy figures included are only the direct fuel and electrical consumption, not indirect energy inputs. Also the electrical energy is expressed as kilowatt hours directly converted to kilocalories. In order to generate this electricity, at least three times more fuel energy would be required, to make up for engine and generator efficiency.

The energy data listed in Tables 3 and 4 and Figures 1 and 2, for dairy farms, are based partially on calculations utilizing the equipment sizes, animal numbers, and crop acreages previously developed. There were several major literature sources which reported direct energy use on dairy farms, and which were used in developing

Table 4. Annual energy use by type of fuel and by operation in a 100-cow dairy.

Operation	Gasoline 10^6kcal (gal)	Diesel 10^6kcal (gal)	Electricity 10^6kcal (kwh)	Heating Oil 10^6kcal (gal)	Total 10^6kcal
Crop					20.0
Establishment					
Plowing		9.5 (270)			
Seedbed Prep.		3.2 (90)			
Planting	3.6 (115)				
Truck	3.7 (118)				
Harvest					77.8
Hay	21.7 (700)	22.6 (640)			
Corn	10.5 (340)	12.0 (340)			
Truck	11.0 (352)				
Manure Handling	21.7 (700)	17.6 (500)			39.3
Barn & Farmstead					92.6
Misc. Tractor	29.6 (955)	10.2 (290)			
Truck	14.7 (470)				
Feed Handling			1.2 (1350)		
Milk Handling			14.3 (16,600)		
Space Heating			3.1 (3650)		
Water Heating			12.6 (14,700)		
Lighting			2.2 (2600)		
Other Electricity			4.7 (5500)		
Personal Home					74.2
Cooking			1.1 (1300)		
Clothes Drying			0.9 (1100)		
Water Heating			5.2 (6000)		
Other Electricity			7.0 (8100)		
Space Heating				60 (1700)	
Totals	116.5 (3750)	75.1 (2130)	52.3 (60,900)	60 (1700)	303.9

Table 5. Annual energy use by type of fuel and by operation in a 1000-head beef feedlot.

Operation	Gasoline 10^6kcal (gal)	Fuel Type Diesel 10^6kcal (gal)	Electricity 10^6kcal (kwh)	Heating Oil 10^6kcal (gal)	Total 10^6kcal
Crop					169
Establishment					
Shred Cornstalks		13.7 (389)			
Soil Prep.		82.5 (2338)			
Fertilizing		27.6 (782)			
Planting		12.9 (365)			
Weed Control		18.5 (525)			
Truck	14.4 (462)				
Crop Harvest					144
Corn Silage Chop		64.4 (1826)			
Corn Silage Transport	23.4 (750)				
Corn Grain Harvest		39.0 (1105)			
Corn Grain Transport		4.0 (114)			
Truck	14.4 (462)				
Feedlot					136
Silo Unload	68.4 (2190)				
Feed Wagon	33.7 (1080)				
Feed Processing			1.7 (2000)		
Water Pumping			3.7 (4320)		
Lighting & Misc.			4.1 (4720)		
Truck	28.9 (924)				
Manure Handling					52
Pump Manure	6.7 (214)				
Spread Manure	47.1 (1508)				
Personal Home					162
(3 Households)					
Electricity			39.1 (45,480)		
Home Heating				122.7 (3480)	
Total	237 (7590)	262 (7444)	49 (56,520)	123 (3480)	671

the data in this paper. Those sources included Gunkel, *et al.* (14), U.S. Census Bureau (3), and Bratton (4,5). The personal home heating fuel quantities were developed from data reported by Agway (15), Kearl and Snyder (6), and Personal Interviews, New York (7). Price (16), Gunkel, *et al.* (14), and New York State Electric and Gas (17) were the sources of home and farm electrical usages.

The significant information in Table 3 is that the personal home use of electricity and fuel oil predominates the energy use on the

40-cow dairy farm, making up about 45% of the total energy use. The heating fuel use is also significant with respect to the yearly distribution as shown in Figure 1, where during the three winter months, over 50% of the heating fuel is required. The barn and farmstead energy use is high, requiring 23% of the total energy. More energy is required for manure handling, 12.4 million kilocalories than for all crop establishment activities, 8.7 million kilocalories. The relatively low crop establishment fuel use is due to the small proportion of land that is devoted to corn and small grains, the crops which require yearly tillage and planting. The energy distribution shown in Figure 1 illustrates how the spring, summer, and fall gasoline and diesel fuel peaks, counterbalance the winter peak use of fuel oil. Electricity usage is relatively constant year-round. The sum total of all the energy uses results in higher energy use per month in the winter months, about 80% higher than in August when the energy use is at the yearly low. This distribution is due to the fuel oil use in winter dominating all the other energy uses.

The 40-cow dairy uses the equivalent of 63 gallons of gasoline (both actual gasoline and diesel fuel) per cow for all farm operations. The electrical use for the farm operation alone is about 493 kWh per cow. Of this electrical demand, 58% is required for milk cooling and water heating alone. When the farm home is included,

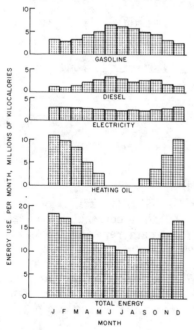

Figure 1. Energy distribution by month, and by type of fuel (40-cow dairy farm).

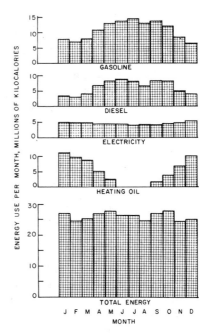

Figure 2. Energy distribution by month, and by type of fuel (100-cow dairy farm).

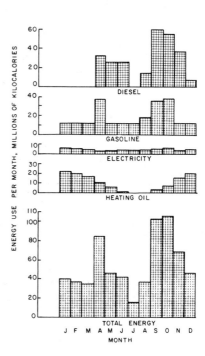

Figure 3. Energy distribution by month, and by type of fuel (1000-head beef feedlot).

the total energy use for the 40-cow dairy is 4.08 million kcal/cow. This energy does not include personal automobile use.

The 100-cow dairy was found to use the various energy types in roughly the same proportions as the 40-cow dairy, with the exception of fuel oil for heating. The personal home energy consumption was a smaller proportion of the total energy use due to the fact that one farm home was included in both sizes of farms, and the 100-cow dairy had a higher energy use for farm production than the 40-cow dairy.

Table 4 shows that gasoline and diesel fuel made up 38% and 25%, respectively, of the total energy use on a 100-cow dairy. The comparable figures for the 40-cow dairy were 34% and 16%, respectively for gasoline and diesel fuel. The larger farms thus used greater proportions of diesel fuel for tractor and truck operations than did the smaller farm. This was due to more and larger diesel tractors being present on the 100-cow dairy where larger acreages of crops were involved.

The yearly energy distribution shown on Figure 2 shows that the total farm energy use is a relatively constant amount per month, about 26 million kilocalories. This fact is the result of the high winter use of fuel oil balanced against the relatively high summer use of diesel and gasoline fuels. The crop tillage and planting activities in the spring, haying in the summer, and corn harvesting in the fall all contribute to the gasoline and diesel fuel distribution being high in these months. The heating fuel distribution was determined from degree-day data reported by Dethier and Pack (18). Agway (15) reported that they use this technique for basing fuel oil deliveries for home heating. The electrical consumption has a relatively even distribution year-round, for both dairy operations and the farm home.

The total gasoline and diesel fuel use for the 100-cow dairy was 67.3 gallons of gasoline equivalent per cow, or slightly higher than the per cow fuel use by the 40-cow dairy. The electrical consumption for the farm alone was found to be 444 kWh per cow, or slightly less than the per cow consumption on the smaller dairy. Thus, one of the original premises, that the 100-cow dairy is more energy intensive, was not found to be true; that is, both small and medium sized dairies use about the same amount of fuel and electrical energy per cow.

When the farm home energy use is included (excluding personal automobile use), the 100-cow dairy was found to use 3.04 million kilocalories per cow. This is lower than the energy use by the 40-cow dairy, 4.08 million kilocalories per cow, due mainly to the fact that only one farm home was included on both dairy farm sizes. If the personal home energy needs for the full-time hired man were included on the 100-cow dairy, its per cow energy requirements would be about the same as for the 40-cow dairy.

The energy data reported in Table 5 and Figure 3 for the 1000-head beef feedlot was determined primarily from averages of data reported by Hughes (9) for Michigan feedlots, Ayres (19), and Boehlje (10) for Iowa feedlot farms, and Personal Interviews (11) conducted on Ohio beef feedlots. Additionally, engineering information reported by the American Society of Agricultural Engineers (20) was used for calculating energy use where direct consumption figures were not available. The farm definition outlined in Table 2 was the basis for these calculations, where the tractor and machine sizes were specified, along with crop acreage and numbers of cattle. Where possible, independent estimates of energy usage for the various operations were also obtained. In almost all cases, the data reported by the various sources compared favorably.

Manure handling was the smallest energy-consuming category, using less than 7% of the total energy. This compared with the dairy farm case, where up to 13% of the total energy use was involved in manure handling. The beef feedlot used a liquid manure storage and handling system, which required comparatively less energy than the solid manure scraping and daily spreading system utilized by the 100-cow dairy.

Electricity was a very small proportion of the energy used for the farm operation. In the feedlot category, electricity was needed only for water pumping, feed grinding, lighting and farm shop use. The yearly electrical energy for these activities was only 11 kWh per head of feedlot capacity compared with over 400 kWh per cow for the dairy farms. This shows the difference in energy intensiveness between the two farm types, where the feedlot is more diesel fuel and gasoline energy-intensive.

Three farm homes were included with the 1000-head feedlot, these being for the three full-time men required to run the feedlot. The energy needs for these homes included electricity for appliances and lighting, and fuel oil for heating. The gasoline use for personal automobiles was not included in the total. The personal home energy use totaled 162 million kilocalories, or 24% of the total.

The monthly energy distribution shown in Figure 3 shows a radical departure from the distribution found for the dairy farms. Since corn is the only crop grown, the distribution of the diesel fuel energy used for crop production shows high peaks in the spring and fall. The spring peaks are due to crop tillage and planting activities which must be completed early in the summer for high corn yields. The fall peaks are due to the heavy energy requirement to harvest the corn silage, corn grain, and to plow the fields after harvest.

The gasoline requirement was found to have a regular monthly

energy requirement for the cattle feeding and maintenance activities. The April peak of energy use is due to manure spreading from the liquid manure tank which has a six-month storage capacity. The September peak is due to corn silage transportation for which the smaller gasoline tractors are used, and the October peak is due to the manure-spreading activities.

The electrical and heating oil distribution are roughly similar to the dairy farm case, since these are mostly for farm home use. The total energy distribution shows the same spring and fall peak usages as described above.

The tractor and truck fuel yearly requirements for the 1000-head beef feedlot farm was the equivalent of 18 gallons of gasoline per head capacity of the feedlot. The electrical requirement was 11 kWh per head capacity for the feedlot alone. The total energy requirement, including personal home use (excluding personal automobile use), was .67 million kilocalories per head capacity of the feedlot. One reason this figure is so much less than the dairy farm figures of 3 to 4 million kilocalories per farm is that the beef cattle weigh less and thus consume less feed per animal. Another reason is that a large proportion of the dairy cow's diet is alfalfa hay, which requires more energy per amount of feed produced. The high electrical requirement of the milking operation also adds to the total energy usage.

OVERALL ENERGY FLOWS IN MILK AND BEEF PRODUCTION

It was decided to compare the previously computed direct energy inputs with the photosynthetic energy associated with crop production and the animal production of the beef feedlot and dairy farm. This was done to 1) determine the relative energy efficiencies of the two agricultural systems, and 2) determine the energy included in the animal and crop waste products which could potentially be converted to a useful energy form, methane gas, via anaerobic digestion.

Figures 4 and 5 outline the energy flows on the two dairy farms. The energy numbers for the feed, crop residues, and manure are gross energy values of the total dry matter included in these products, as reported by several sources.* The labor energy input is

---

* Gross energy values of feedstuffs and milk as reported by the National Academy of Sciences (21). Numbers are kilocalories per kilogram-dry matter: Corn Grain, 4370; Corn Silage, 4390; Corn Stover, 3780; Corn Cobs, 4650; Alfalfa Hay, 4440; Grass Hay, 4490; Oat Grain, 4750; Oat Straw, 4600; Soybeans, 4630; Milk, 5160 kcal/kg-dry matter.

Gross energy values of manures as reported by Azevedo and Stout (22), kilo-

based on the numbers of full time man-equivalents per year, multiplied times the yearly calorie requirements for adult humans based on the assumption of 3000 kilocalories per day per person.

Both the 40-cow and 100-cow dairies have a feed and crop residue energy value which is about 10 times the value of the direct fuel and electrical energy inputs. However, by the time the feed has been fed to the dairy animals, the milk and meat outputs have an energy value roughly equal to the direct energy inputs, or one-tenth of the feed inputs. Milk contains food value in more areas than just calories, namely protein and vitamins.

Of importance on Figures 4 and 5 are the energy values of the waste products—manure and crop residues. The 40-cow dairy has a total of 723 million kilocalories of gross energy in these categories. This compares with 132 million kilocalories of petroleum fuel energy. Thus an anaerobic digester would have to be operated at $132/173 \times 100\% = 18\%$ overall energy efficiency to provide the petroleum fuel input. This is based on being able to collect all of the manure and crop residue produced. In order to provide electricity, considerably more than 31 million kilocalories of direct

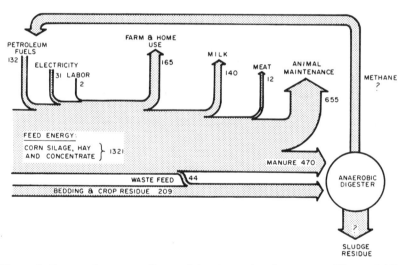

Figure 4. Energy flow on a 40-cow dairy farm (numbers are millions of kilocalories).

calories per kilogram-dry matter: Dairy Cow Manure, 3936; Beef Cattle Manure, 3887.

Gross energy values of animal meat: Dairy Cows, 2300 kilocalories per kilogram of whole body weight (23); Beef Cattle, 5060 kilocalories per kilogram of whole weight gain (24).

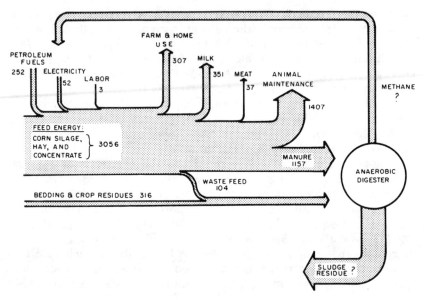

Figure 5. Energy flow on a 100-cow dairy farm (numbers are millions of kilo-
        calories).

electrical input would be required to generate the electricity. If
an engine-generator operated at 15% thermal-electrical efficiency,
the fuel input to generate the electricity for the 40-cow dairy would
have to be 207 million kilocalories. This would require the digester
to have an overall energy conversion efficiency of 29%.

The energies for the 100-cow dairy are in roughly the same
proportion as the 40-cow dairy. The crop residues, including corn
stalks, bedding, waste feed, and manure would provide a total of
1577 million kilocalories. In order to provide the energy to replace
petroleum fuels, 252 million kilocalories, the efficiency of the
anaerobic digester would have to be 16% of the gross input energy
value from organic matter. In order to provide 52 million kilocalories of electrical energy, a generator operating at 15% efficiency
would have to have a fuel input of 347 million kilocalories, or 22%
of the total biomass energy input.

These efficiencies appear to be reasonable, and thus there does
seem to be a possibility of anaerobic digestion providing a significant share of the input energies for the dairy farms. However,
because of the present day inefficiencies in converting fuel energy
to electrical energy, it does not seem possible to generate enough
energy from the agricultural wastes to provide *both* the electrical
and the petroleum fuel energy requirements of dairy farms.

Figure 6 shows the energy relationship for the 1000-head beef

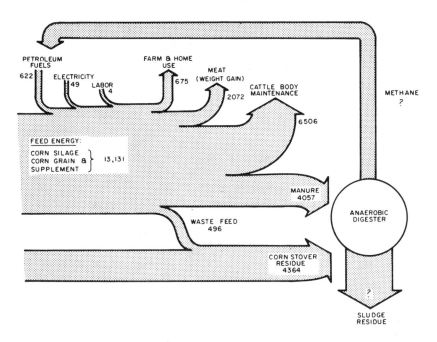

Figure 6. Energy flow on a 1000 head beef feedlot (numbers are millions of kilo-calories).

feedlot. Here there is a considerably higher gross energy value in the feed and crop residues, as compared with the petroleum and electrical energy inputs. The total energy value of the corn silage, corn grain, corn stover, and protein supplement is 26 times the direct energy inputs of petroleum fuels and electricity. Again, the inefficiencies of converting this feed to animal weight gain results in an energy output of the meat from the fat cattle being only about 12% of the gross crop energy value. The meat output still has an energy value of 3 times the value of the direct fossil energy inputs. Thus the beef feedlot has a higher energy efficiency than the dairy farm which produced about 1 kilocalorie of milk and meat per kilocalorie of fossil energy input. Again, it must be realized that the food outputs from beef and dairy farms contain more than calories. The protein is probably the most important nutritive components of these food products.

The energy values of the crop and animal wastes are high in comparison with the fossil fuel inputs. The corn stover residue includes as much energy as the manure produced by the cattle. This is because more than half the corn acreage is harvested for the corn grain alone, leaving stalks, leaves and cobs in the field.

The value shown in Figure 6 for the corn stover is the gross energy of all the material, and in actual practice, only a portion of this residue could actually be collected. Still, the potential energy availability is high. The total energy value of manure, waste feed, and corn stover residue is 8917 million kilocalories. If electricity could be generated at an efficiency of 15%, the fuel input for the 49 million kilocalories of electrical energy would be 327 million kilocalories. If this value is added to the petroleum fuel energy value, a total of 949 million kilocalories would be required. An anaerobic digester would have to convert energy from the biomass input at the rate of $949/8917 \times 100\% = 11\%$ to provide *both* the petroleum fuel energy *and* the electrical energy input. This efficiency seems within the range of feasibility, and leads to the conclusion that the potential for energy recovery from the waste of beef feedlots is much higher than the potential of dairy farms.

*   *   *

This study was supported by National Science Foundation Grant GI-43099.

REFERENCES

1. Pimentel, D., L. E. Hurd, A. C. Bellotti, M. J. Forster, I. N. Oka, O. D. Sholes, and R. J. Whitman. "Food Production and the Energy Crisis." *Science*, Vol. 182, 2 November 1973.
2. Taylor, R. "Management Factor Summary." Unpublished data, Dairy Records Processing Laboratory, Cornell University, Ithaca, NY. August 1974.
3. U.S. Census Bureau. 1969 Census of Agriculture, Vol. 1, Part 7, New York Section 1. pp. 202-217.
4. Bratton, C. A. "Dairy Farm Management Business Summary, New York, 1972." A.E. Res. 73-11, Dept. of Agr. Econ., Cornell Univ. Ithaca, NY. July 1973.
5. Bratton, C. A. "Dairy Farm Management Business Summary, New York, 1973." A.E. Res. 74-9. Dept. of Agr. Econ., Cornell Univ., Ithaca, NY. August 1974.
6. Kearl, C. D. and D. P. Snyder. "Field Crops Costs and Returns from Farm Cost Accounts." A. E. Res. 73-24. Dept. of Agr. Econ., Cornell Univ., Ithaca, NY. December 1973.
7. Personal Interviews, New York, with selected dairy farm operators, 1974.
8. National Research Council, "Nutrient Requirements of Beef Cattle," 4th Ed., 1970. Washington, DC.
9. Hughes, H. A. "Energy Consumption in Beef Production Systems as Influenced by Technology and Size." Unpublished Ph.D. Thesis, Department of Agricultural Engineering, Michigan State Univ., E. Lansing, Mich., 1973.
10. Boehlje, Mike. "An Evaluation of Iowa Beef Cattle Systems." Cooperative Extension Service, Iowa State University, Ames, IA. EC-900A, November 1973.
11. Personal Interviews, Ohio, with selected beef feedlot farm operators. 1974.
12. USDA Statistical Reporting Service. "Crop Production: 1974 Annual Summary of Acreage, Yield, and Production. CrPr 2-1(75), Washington, DC. 16. January 1975.
13. Edwards, William and E. G. Stoneberg. "Background Information for use with CROP-OPT System, FM1627. Cooperative Extension Service, Iowa State University, Ames, Iowa 50010. FM1628, November 1974 (6th Rev.).

14. Gunkel, W. W., D. R. Price, T. R. McCarty, S. McClintock, C. A. Sansted, G. L. Casler, and J. H. Erickson. "Energy Requirements for New York State Agriculture. Part I: Food Production." Ag. Eng. Ext. Bull. 405, Cornell Univ., Ithaca, NY. 1974.
15. Agway Petroleum Service, Personal Communication, Ithaca, NY. Dec. 10, 1974.
16. Price, D. R. "Electricity Use and Costs on Four New York Farms." Ag. Eng. Ext. Bull. 369, Dept. of Agr. Engr., Cornell Univ., Ithaca, NY. 1966.
17. New York State Electric and Gas Corp. Personal Communication, November 1974. Ithaca, NY.
18. Dethier, B. E. and A. B. Pack. "Climatological Summary, Rurban Climate Series No. 1." New York State College of Agriculture and Life Sciences, Cornell University, Ithaca, NY. 1963.
19. Ayres, G. E. "Fuel Required for Field Operations." A.E. 1079, Cooperative Ext. Service, Iowa State Univ., Ames, Iowa. 1974.
20. American Society of Agricultural Engineers. "Agricultural Engineer's Yearbook." 21st Edition. St. Joseph, Michigan. May 1974.
21. National Academy of Sciences. "Atlas of Nutritional Data on United States and Canadian Feeds." Washington, DC 1971.
22. Azevedo, J. and P. P. Stout. "Farm Animal Manures: An Overview of Their Role in the Agricultural Environment." California Agr. Exp. Station, Ext. Service, University of California, Davis, California. 95616. 1974.
23. Seinffendorfer, S. "Relationship of Body Type and Size, Sex and Energy Intake to Body Composition of Cattle." Unpublished Ph.D. Thesis, Animal Science Dept., Cornell University, Ithaca, NY 1974.
24. Reid, J. T. Personal Communication, Cornell University, Ithaca, NY, 1975.

4.

Assessment of Energy Inputs for Texas Agricultural Production

W. A. LePori and C. G. Coble*

Petroleum and Agriculture are the two most important economic sectors in Texas. Products from these two sectors in Texas also make an important impact nationally. Almost 40% of the oil and gas produced in the United States is produced in Texas (1). Fifty-two percent of the nation's natural gas is produced here and 27% of the nation's refining capacity is located in Texas.

Availability of energy resources in Texas has been responsible for a large number of related industries locating in the state and, as a result, it is the largest energy-consuming state as well as energy-producing state. It exceeds the second largest energy-consuming state, California, by over 30%. More than half of the natural gas produced in Texas and 10% of the hydrocarbon liquids are consumed in the state.

Texas produces a major portion of many petrochemicals used in the U.S. The percent of U.S. petrochemical capacity located here includes: Butadiene—87, Methanol—67, Toluene—65, Ethylene—60, Propylene—54, Benzene—51, and Carbon Black—36.

In addition to these industries, the availability of land and unique climate has allowed Texas to be a major producer of agricultural products. Texas generally ranks third in value of producing food

* Assistant Professors, Agricultural Engineering Department, Texas A&M University, College Station, Texas.

and fiber crops in the U.S. and is a major exporter of these items to the rest of the world. Texas ranks number one in production of cotton, sorghum, beef cattle, watermelons, cabbage, spinach, onions, sheep, lambs and wool. Texas also ranks high in production of rice, peanuts, hay, and other vegetable crops.

The energy crisis has created the need for understanding where nonrenewable energy resources are being used. This need includes use in food and fiber production. In Texas, the agricultural industry is concerned about the availability of adequate energy to meet its food and fiber production potential. Agriculture was fortunate to have priority in fuel allocations during the recent "energy crunch." However, other industries are pressuring for energy guarantees, and agriculture is dependent on some of these. Adequate energy supplies in the future may not be available for all the interdependent industries relied upon to produce our food and fiber needs.

Energy uses in agriculture have been criticized as being high by some and ignored by others because it is viewed as being only a small percent of the total. This chapter presents energy uses in Texas for several important phases of agricultural production. To maintain a viable Texas agricultural industry, it is essential that energy inputs be identified and reduced where possible.

ENERGY INPUT CATEGORIES

As in all states, the diversity of agriculture in Texas leads to problems in assessing energy used in agriculture. Many "gray areas" of energy assessments arise as well as many easily identified areas. An amazing interdependence exists between food and fiber production and many petroleum-related industries. Boundaries must be defined if energy assessments for different industries are to have meaning. In this study energy inputs were identified for different fuels needed in agriculture and converted to British Thermal Units (Btu) on the basis of standard heat values of the fuels.

In this study direct on-farm inputs of energy were assessed to agriculture, and certain other indirect inputs essential to a marketable product were also assessed to agriculture. Areas where energy was assessed were:

Direct Inputs
  Machine operations
  Irrigation
  First-line crop processing
  Forestry
  Transportation
  Animal production

Indirect Inputs
  Fertilizer
  Food and kindred products

Some of the above energy inputs are claimed by other industries, but it is important that the agricultural industry realize its interdependence and strive to conserve energy indirectly as well as directly.

ENERGY INPUTS

Energy inputs for Texas Agriculture for the areas given above are shown in Table 1. The total energy input of approximately

Table 1. Summary of energy inputs (Texas, 1973).

Input		Energy Btu $\times 10^{13}$
Machinery Operations		5.04
Irrigation		16.31
First-line Crop Processing		.4
Food and Kindred Products		3.5
Forestry		3.18
Transportation		7.6
Animal Production		.53
Fertilizer		5.1
	TOTAL	41.66

41.66×10^{13} Btu of energy represents about 7% of the 1972 gross energy input for Texas (2). The largest single use of energy in the state is for pumping irrigation water and represents about 39% of the total identified.

Table 2. Fuel purchases on Texas farms (July 1, 1973–June 30, 1974).*

Fuel	Energy Btu $\times 10^9$	Percent of Total
Diesel	37,800	13.8
Gasoline	47,120	17.2
LP Gas	27,744	10.2
Fuel Oil	648	.2
Natural Gas	148,680	54.4
Electricity	11,359	4.2
	273,351	100.0

* Texas Crop and Livestock Reporting Service, 1974 Texas Farm and Fuel Survey.

The Texas Crop and Livestock Reporting Service has identified fuel purchases and relative importance of various energy sources purchased by farmers and ranchers in Texas (3). The energy purchased July 1, 1973–June 30, 1974 by Texas farmers and ranchers is shown in Table 2. The totals of energy shown in Tables 1 and 2 cannot be directly compared because they do not include the same items. However, the relative importance of the different fuels shown in Table 2 are probably about the same for the total shown in Table 1.

Machinery Operations

Energy shown in Table 1 for machinery operations in crop production was determined by using data on machinery performance, cultural practices, and crop acreages. Fuel use in gallons of diesel equivalent was determined for 41 machine operations, 15 crops, and 19 districts in the state. Some of the 15 crops were subdivided into irrigated and nonirrigated acreages to reflect the different cultural practices used for each. Procedures used in combining this data are shown in Table 3.

Table 3. Method for estimating fuel use for machinery operations.

[1] $d_{ijk} = (a_{ijk})\ (b_{ijk})$

[2] $e_{jk} = \sum\limits_{i=1}^{N} (d_{ijk})$

[3] $f_{jk} = \sum\limits_{i=1}^{N} (d_{ijk})\ (C_{jk})$

[4] $TFU_k = \sum\limits_{j=1}^{M} f_{jk}$

[5] $TFUS = \sum\limits_{k=1}^{L} TFU_k$

Where:

i = subscript to identify machine operations
j = subscript to identify district
k = subscript to identify crop
L = No. of crops
M = No. of districts
N = No. of machine operations
a = No. of times over field for crop and district
b = Fuel use for one time over one acre
c = Acreage of crop for machine operation
d = Fuel use in gal/acre for crop and district
f = Fuel use in gallons for crop by districts
TFU = Fuel use for crop
$TFUS$ = Total fuel use for state

Harvested acreage of the different crops is usually less than planted acreage as some acreage is abandoned at some point in production, and in some cases the amounts can be significant. To incorporate energy used on this acreage, an assumption was made that the energy use per acre for the planted but unharvested acreage was 50% of that for the harvested acreage.

The Texas Crop and Livestock Reporting Service (TCLRS) also identified energy-consuming equipment on Texas Farms (3). They determined that the percentage of tractors of different fuel types were 43, 39 and 18 for diesel, gasoline, and LP gas, respectively. These percentages were used to convert the diesel equivalent to energy used for the State by machinery operations in production.

The 15 crops used represent about 90% of the crop production in Texas, and the total for machinery operations in Table 1 includes an estimate for the other 10%.

The cultural practices used in calculating energy used in machinery operations were examined to determine where energy conservation might occur. Effect of reduced tillage techniques was determined through elimination of excessive operations and using lower energy-consuming operations where considered possible. A reduction in energy use by 20% could be achieved by the reduced tillage practices selected. An additional 10% reduction might be achieved by converting entirely to diesel tractors since these engines produce 16-18% more work per Btu of fuel than gasoline engines.

Irrigation

Approximately 35% of the cropland in Texas is irrigated and most of the acreage utilizes ground water as shown in Table 4.

Table 4. Type of irrigation in Texas.

Surface Water	1,267,607 acres
Ground Water	6,648,553 acres
Combined Supplies	290,089 acres

Lifting water from deep underground aquifers consumes large amounts of energy as compared to use of surface water. Most of the irrigation wells are located in the High Plains of Texas where 90% of the wells have pumping depths greater than 125 feet and average pumping depth estimated to be 300 feet (4). Almost three-fourths of these wells yield less than 700 gallons per minute, requiring pumps to operate almost continuously during much of the growing season to provide enough water. The TCLRS survey showed that wells were pumped an average of about 2700 hours in the 1973-74 crop year.

The number of irrigation wells is shown in Table 5 according to type of fuel used as the power source. The amount of energy used in an average year is also estimated for the different fuel types. During a dry cropping season, these estimates could increase by as much as 35% due to increased hours of pump operation.

Table 5. Irrigation wells and energy used in Texas.

Fuel Type	No. of Units, 1974*	Btu $\times$ 10^{13}
Diesel	1,353	.19
Gasoline	6,454	1.12
LP Gas	6,113	.99
Natural Gas	59,804	12.39
Electricity	67,793	1.62
	TOTAL	16.31

* Texas Crop and Livestock Reporting Service, 1974 Texas Farm, Fuel and Fertilizer Survey.

Electricity is shown to account for slightly less than 10% of the energy for irrigation in Table 5, but this energy includes only that delivered to the motors. It does not include power plant efficiencies and transmission losses since they are not located on the farm. The other fuel types include losses due to inherent efficiency characteristics of the internal combustion engine since they are located on the farm. This helps give an illusion of low energy use for the large number of wells served by electricity. Also, wells powered by electric motors tend to be of smaller horsepower than other types.

Seventy-six percent of the total energy for pumping irrigation water is used in the form of natural gas. This is especially significant since natural gas reserves are being depleted faster than other petroleum fuels. Due to this, in the future it may be advantageous to convert more wells to electricity since electricity could be produced from alternate sources of energy.

The high energy demand for irrigation makes this an area in which a small improvement in efficiency could make a large reduction in energy input. Energy inputs for irrigation could be reduced by increasing engine efficiency, increasing pump efficiency, and increasing water use efficiency.

Ulich (5), in a sampling of pump and engine efficiencies for irrigation wells in Texas, found that overall pumping plant efficiencies were low. He found that the average internal combustion engine efficiency was 19% and that average pump efficiency was 52%.

Using this study as a basis for analyzing energy reduction in pumping irrigation water, a maximum reduction of 41% could be achieved. This would require pumps to be improved by 72% efficiency and engines to be improved to 24% efficiency. This would require upgrading about 90% of the wells in the state. By upgrading about 25% of the wells and achieving 62% pump efficiency and 21% engine efficiency, energy input could be reduced by approximately 25%. Since upgrading wells requires considerable capital investment, the latter case would be a more realistic savings to achieve.

The annual amount of irrigation water pumped is partly dependent upon rainfall. Amount of irrigation water needed in the High Plains area for three major crops was calculated using consumptive use and effective precipitation data (6). It was found that cotton, grain sorghum, and wheat require 13.3, 15.3 and 26.2 acre-inches of irrigation water per year, respectively. Assuming a linear increase in yield with application of irrigation water, the amount of yield increase attributable to irrigation water was 161 lb/acre-ft for cotton, 47.9 bu/acre-ft for grain sorghum, and 7.7 bu/acre-ft for wheat.

Present water use efficiencies are characteristically higher in Texas than in many irrigated areas. Water tables are dropping in many wells of the state and maximum water application efficiency is desired to minimize rate of drop. Energy savings can be achieved through individual management practices and these savings could approach that achieved by upgrading wells. However, accomplishing increased water application efficiencies would probably take longer to achieve and increase pressures on management.

First-Line Crop Processing

Most agricultural products require some form of processing before they can be used or stored. Failure to process or condition a crop for storage can cause deterioration due to infestation with insects, fungal diseases, or molds. Cotton ginning, corn, sorghum, rice and peanut drying are considered in first-line crop processing.

In Texas, all rice, 95% of corn, 80% of peanuts, and 10% of grain sorghum are mechanically dried to some extent during an average crop year. Commercial and cooperatively-owned grain dryers account for practically all of the crop drying in the state and only 1400 on-farm dryers and aeration units were reported in the TCLRS 1974 Farm Fuel and Fertilizer Survey (3). Energy used in first-line crop processing is shown in Table 6 along with the amount of each processed.

As shown in Table 6, ginning operations consume more energy than drying the other crops. Wilmot and Watson (7) show that for

Table 6. Summary of energy in first-line crop processing.

Operation	Harvested 1973*	Percentage Processed	Energy Level Btu/Unit	Energy Used Btu $\times$ 10^{11}
Cotton Ginning	4,698,800 bales	100	5.86 $\times$ 10^5	27.
Rice Drying	20,530,000 cwt	100	1.67 $\times$ 10^4	3.4
Sorghum Drying	417,000,000 bu	10	6.2 $\times$ 10^3	2.6
Corn Drying	60,800,000 bu	95	1 $\times$ 10^4	5.8
Peanut Drying	4,710,000 cwt	80	1.9 $\times$ 10^4	.72
			TOTAL	39.52

* TCLRS Field Crop Statistics, 1973 and Cotton Statistics, 1973

Texas high capacity gins from 50 to 65% of the total energy consumed in ginning is consumed in conveying seed cotton, trash, lint, and cottonseed through the different subsystems of the gin. He also showed that another 15 to 25% is used in drying seed cotton. These operations account for about 65 to 90% of total energy used in ginning operations and offer potential for energy reduction.

Other types of conveying systems which are more efficient could be substituted for the pneumatic systems presently used but would require considerable equipment modification and capital investment. Managing seed cotton drying to minimize operation time of mechanical dryers could also give significant energy savings in ginning. Energy used in first-line crop processing might be reduced by about 30% with most of this reduction obtained from ginning operations.

Forestry

Forestry is an often forgotten branch of agriculture, but in Texas it is a large revenue earner. Forest acreage in Texas totalled 11.5 million acres for 1972. Energy consumption estimates showed that of the 3.18 $\times$ 10^{13} Btu of energy attributed to forestry, about 20% was consumed in production (*i.e.,* to get the forest crop from the field to the plant). The rest was used in processing wood and pulp products (8, 9).

Transportation

Trucks are the major means of transporting agricultural products in Texas, and the estimate for transportation does not include energy used in transporting products by railroads. About 22% of all trucks in Texas were engaged in hauling farm products, forestry products, and processed foods during 1972 (10). Over 4.7 billion miles representing approximately 20% of total truck mileage were for these purposes.

Pickup and panel trucks represented about 83% of the agricultural related truck numbers; however, they represented only about 70% of the mileage. Miles of travel and energy estimates for truck transportation are shown in Table 7.

Table 7. Truck use for agriculture (Texas, 1972).*

Major Use	Miles $\times$ 10^6	Energy Use Btu $\times$ 10^{13}
Excluding Pickups and Panels		
Farm Products	805	2.4
Forest Products	77	.2
Processed Foods	536	1.6
Pickups and Panels		
Farm Products	3,016	3.1
Processed Foods	305	0.3
	4,739	7.6

* Census of Transportation, 1972. Truck Inventory and Use Survey, Texas. Bureau of Census, Washington, D.C.

Animal Production

Livestock products accounted for approximately 55% of the agricultural cash receipts for Texas in 1973. Cattle dominate the livestock industry in Texas, and there are large numbers in feedlots as well as large numbers on ranges. Energy estimates for livestock did not include energy value of feeds and only included direct fossil fuel inputs.

Of the energy purposes determined by the TCLRS (3) in the farm fuel and fertilizer survey, slightly over 10% of the total was attributed to livestock and poultry production. The approximately 2.8 $\times$ 10^{13} Btu did not include energy used in large feedlot operations, but includes transportation fuel purchased on-farm. In the spacious ranges of Texas transportation is a very significant input. The energy value of .53 $\times$ 10^{13} Btu shown in Table 1 does not include transportation and is based on energy used for other items related to maintaining animals.

Fertilizer

Nationally, Texas ranks fourth in total fertilizer consumption but third in consumption of primary plant nutrients. Fertilizer is dependent upon mineral reserves and fossil fuels, particularly natural gas. Although energy used in fertilizer manufacturing could be assessed to the chemical manufacturing industry, fertilizer is essential in present agricultural production techniques, and the energy for its manufacture needs to be considered as part of the input for agricultural production.

The 5.1 × 10¹³ Btu assessed to fertilizers in Texas represents an estimate of the energy required to manufacture the primary ingredients N, P_2O_5 and K used in the state. The tonnage of each ingredient used and energy requirements per ton were used to make the estimates on energy input for fertilizer.

In 1973, when feed cattle production reached a peak in Texas of 4.41 million head, most of the 4 million tons of manure produced, plus stockpiles from previous years, was used on cropland. Today, Texas feedlots are not operating at full capacity but feedlot manure is still being used on croplands in large quantities. Sweeten, *et al.* (11) have estimated that 26.2 × 10¹¹ Btu of energy is saved through using manures instead of manufactured fertilizer. If manures were not being used, this additional energy would be required as an input to agricultural production.

Food and Kindred Products

There are many processes using energy to prepare these products for consumption by humans and animals. The Standard Industrial Classification of the Directory of Texas Manufacturers lists 43 categories of processes for human food and 5 for animals.

Of the 3.5 × 10¹³ Btu of energy assessed to this area, meat packing and processing consumed almost 24% of the total (12). Although Texas is a large producer of fruits and vegetables, it does not have a large processing industry. Slightly over 7% of the total was consumed in processing fruits and vegetables in the state.

SUMMARY AND CONCLUSIONS

Energy inputs were assessed for several important categories of the Texas Agricultural Industry to assist in evaluating where major reductions could be made in using some of our rapidly depleting fossil fuels. Energy inputs totaling 41.66 × 10¹³ Btu were identified which represents about 7% of the 1972 gross energy input for the state. Pumping irrigation water accounted for about 39% of the total and has a large potential for savings.

Natural gas is the most important fuel type for agriculture and is also the fuel being depleted most rapidly. This must be viewed with concern by agricultural interests. The agricultural industry must show that it is making the best use of its energy inputs and reduce them where possible.

More research is needed in Texas to follow this study and the 1974 Farm Fuel and Fertilizer Survey (3). More information is needed on an individual crop basis to further identify these inputs. Complete assessment of energy inputs by crops would provide information on relative importance of different crops and practices related to conservation. It would also enable the agricultural industry to be compared to competitive industries on a benefit to energy input basis. Should energy curtailment occur to Texas farmers, recommendations concerning maximizing food production while minimizing energy inputs need to be available.

* * *

This research was supported by funds from the Texas Department of Agriculture, Honorable John C. White, Commissioner. Special acknowledgment is also given to a related study by the Texas Crop and Livestock Reporting Service, Charles E. Caudill, Director.

REFERENCES

1. Miloy, L. Texas Energy. 1:1 1974.
2. Coble, C. G. and W. A. LePori. Energy Consumption, Conservation and Projected Needs for Texas Agriculture, Unpublished Report S/D - 12, Special Project B, Governor's Energy Advisory Council. 1974.
3. Caudill, C. E., P. M. Williamson, M. D. Humphrey, and D. Adkisson. 1974 Texas Farm Fuel and Fertilizer Survey. Texas Department of Agriculture and USDA, Special Project A, Governor's Energy Advisory Council. 1974.
4. New, L. 1973 High Plains Irrigation Survey. Texas Agricultural Extension Service, Texas A&M Research and Extension Center, Lubbock, Texas. 1974.
5. Ulich, W. Power Requirements and Efficiency Studies of Irrigation Pumps and Power Units. Agricultural Engineering Department, Texas Technological University, Lubbock, Texas. 79 pp. 1968.
6. McDaniels, L. L. Consumptive Use of Water by Major Crops in Texas. Texas Board of Water Engineers. Bulletin 6019, 51 pp. 1960.
7. Wilmot, C. A. and H. Watson. Power Requirements and Costs for High-Capacity Cotton Gins. Marketing Research Report No. 763, 23 pp. USDA, Washington, D. C. 1966.
8. Barron, E. H. "Harvest Trends 1973." Texas Forest Service. Texas A&M University, College Station, Texas. 12 pp. 1974.
9. Wagoner, E. R. Personal Communication. Texas Forestry Association, Lufkin, Texas. 1974.
10. "Census of Transportation, 1972. Truck Inventory and Use Survey, Texas." Bureau of Census, Washington, D.C. 1973.
11. Sweeten, J. M., et al. "Feedlot Manure as an Energy Source." Texas Agricultural Extension Service, Texas A&M University, College Station, Texas. 1974.
12. Industrial Energy Study of Selected Food Industries. Development Planning and Research Associates, Inc. Manhattan, Kansas. 1974.

5.

Accounting of Energy Inputs for Agricultural Production in New York

D. R. Price,* W. W. Gunkel* and G. L. Casler**

During the most serious gasoline and diesel fuel shortages of 1974, agricultural production was fortunately included in the highest priority class for fuel allocations. The allocations process was hindered to some extent by not having adequate information to plan for allocations to all end users. Little was known to determine the fuel requirements for agricultural production in New York State. Gross estimates were made without the benefit of data.

The justification for doing energy accounting in the agricultural sector is based on the need for accurate data on which to base allocations, and the need to establish baseline information for future energy research. Agriculture and energy research programs will continue to be most important due to critical food supply needs in the world and a dwindling reserve of many fuels.

Agriculture is an energy-intensive industry that relies heavily on electricity and fossil fuel inputs. During the past decade, yield or production output per farm worker has drastically increased. However, this increase has been accompanied by a corresponding increase in energy consumption per farm worker. Today's agriculture is a very specialized and highly mechanized industry that requires extensive inputs of energy including electricity, fuel, chemicals,

* Associate Professor and Professor respectively in Agricultural Engineering, Cornell University, Ithaca, N.Y. 14853.
** Professor in Agricultural Economics, Cornell University, Ithaca, N.Y. 14853.

machinery, labor and management. As demands for food and fiber increase, greater inputs of energy will be required. Specific information on agricultural production requirements will be of paramount importance as allocation programs for fuel distribution are developed or modified.

According to a recent U.S. Government report, USDA (1), about one-third of the energy used on farms in the United States is for family living purposes. Fuel oil and L.P. (liquid petroleum) gas is used extensively for home heating. Liquified petroleum gas and electricity are major cooking fuels. Farm numbers are expected to decline by 21% by 1980, but energy use for family living is expected to decline only by about one-half that amount.

Farm production in the United States requires about eight billion gallons of fuel, or about 3% of the U.S. total. There have been shifts in the types of fuels used on farms, as diesel engines have replaced gasoline engines in tractors and combines. The U.S. Government estimates (USDA 1974) that by 1980, over 80% of agricultural tractors and 90% of self-propelled combines sold new will be diesel-powered.

The objectives of this energy accounting study were to: 1) determine the fuel and electricity requirements for agricultural production in New York State, and 2) establish an index or model to use in projecting energy requirements for agricultural production systems. If the fuel requirements on a per acre basis were available, the projections of fuel requirements could be made for future years by knowing the estimated acreages of each crop. Crop acreages are available annually from the Crop Reporting Service in the State Department of Agriculture and Markets.

PROCEDURES

A team of researchers was assigned to a summer project to obtain the needed energy documentation (acknowledgments of those participating is included at the end of this report). The documentation was developed using three different procedures and sources of data. One procedure was to evaluate data collected from cost account farms or those who maintain enterprise cost accounts in cooperation with the Department of Agricultural Economics at Cornell University. A second procedure involved engineering analysis of farm operations for each crop. Data from the Agricultural Engineering Yearbook was used to calculate fuel use rates for various operations. The lists of operations for each crop were developed by personal communication with many growers, Cornell University staff, extension field staff and others knowledgeable of farm operations. A third procedure involved the analysis of data collected from a New York State fuel survey conducted jointly by the New York State Extension Service and the State Department of Agriculture and Markets. Approximately 15,000 returns from 55,000 survey forms mailed to commercial farmers were received and the data was transferred to to magnetic tape for analysis with the digital computer.

Agricultural Fuel Survey Data

The survey was conducted jointly by the cooperative extension association and the ASCS (Agricultural Stabilization and Conservation Service). The information obtained included numbers of each animal and acres of each crop for 1973 and the expected number of animals and crops for 1974. The monthly consumption of gasoline, diesel fuel, propane, and electricity was obtained. A copy of the survey form is included in a comprehensive report of the results in a publication by Gunkel, *et al.* (2).

It was assumed that personal use of gasoline and electricity was included in the survey results since the cover letter with the form did not specifically ask farmers to subtract that portion attributed to personal use. To correct for this, 800 gallons of gasoline were subtracted from each farm. Also, 5500 kWh of electricity was subtracted from each farm for home use.

To obtain estimates of the energy requirements for each major crop, the returns were searched for farms having a substantial acreage of the crop of interest. In general, these farms also grew small acreages of other crops. In order to separate out the fuel required by these crops a system of simultaneous equations was used.

To use simultaneous equations it is necessary that the system be well conditioned. In the simple two dimensional case an ill-conditioned system is one in which the two equations have nearly parallel slopes. In this case round-off errors both in reading the problem into a computer and doing the calculations may cause a large change in the answer obtained. In a well-conditioned problem, the equations have much different slopes so that round-off and truncation errors do not affect the answer significantly. In order to obtain a well-conditioned problem, it is necessary to approach the identity matrix as closely as possible, *i.e.*, writing equations in the typical form:

$$A_{11}X_1 + A_{12}X_2 + \ldots + A_{1n}X_n = b_1$$
$$A_{21}X_1 + A_{22}X_2 + \ldots + A_{2n}X_n = b_2$$
$$\cdot$$
$$\cdot$$
$$\cdot$$
$$A_{m1}X_1 + A_{m2}X_2 + \ldots + A_{mn}X_n = b_m$$

Then A_{11} in the first equation must be large with respect to the rest of the coefficients in that equation, as must A_{22} in the second equation, A_{33} in the third, etc. With regard to the survey, this meant that a farm contributing information for a particular crop must have a large fraction of its total fuel consumption due to that crop. To insure this, all activities reported on the farm were weighed according to their estimated fuel consumption (estimated from cost account and engineering data in this report). The consumption due to the crop of interest was then compared with the total estimated fuel use on the farm. To use this technique for the survey information required an assumption. In the equations above, X_1 and X_2 have the same values in both equations, however, the example below is essentially two equations in four unknowns.

Equation	Potatoes	Other		Fuel
Potato	$A_{11} X_1$	$A_{12} X_2$	$=$	b_1
Other	$A_{21} X_1$	$A_{22} X_2$	$=$	b_2

The requirement for exact solutions then is that X_1, the fuel consumption per acre of potatoes on farm (1), be equal to X^1_1 on farm (2), and that X_2, fuel consumption due to other activities on farm (1) be equal to X^1_2 on farm (2). It was believed that if a well-conditioned system could be obtained from a large sample, the equality requirement, though met only approximately, would hold well enough so that a reasonable answer could be obtained.

The final problem in obtaining a system of equations was to find a method to handle minor crops and animals which did not appear as a variable in the system. The method used was to include the estimated fuel consumption (based on engineering and cost account data) due to other activities as the coefficient for the "other" variable in each equation.

Engineering Analysis

Fuel use per operation was determined in a low, average, and high range by using the range of draft requirements and speeds as presented in the Agricultural Engineers 1974 Yearbook published by the American Society of Agricultural Engineers, St. Joseph, Michigan. After calculating the power take-off horsepower required for each operation as outlined in the Yearbook, a tractor size most closely meeting this need was selected from the Nebraska Tractor Test data. The fuel efficiency from this tractor was then used to calculate the fuel requirements. Certainly a tractor with higher than optimum horsepower will require more fuel for the operation, but this method was selected to permit standardization of data. In the final compilation and comparison, the optimum fuel use was increased by 10% to account for mis-matching of tractor and operation. Fuel use was calculated in terms of gasoline. Diesel fuel use was converted to gasoline equivalency by dividing by a factor of 0.73.

An example of the engineering calculations is given in Tables 1 and 2. A complete analysis of all operations for crops is given in the comprehensive report by Gunkel *et al.* (2).

Table 1. Tillage operation.

| Description | Fuel Consumption (gal/acre) | | |
	Low	Average	High
Moldboard Plow (or Disk Plow)[1]	.50	2.94	3.76
Chisel Plow[2]			
1. Previously tilled	—	1.60	2.40
2. Previously untilled	1.00	1.74	2.75
Tandem Disk[3]	.49	.77	1.02
Spring Tooth Harrow[4]	.38	.70	1.24
Float[5]	.36	.67	1.18
Subsoiler			
1. Drainage (muck)[6]	.42	.45	.47
2. Tillage (muck)[7]	1.89	2.18	2.43

1. 4, 14″ bottoms, 4.0 mph.
2. 12′ tool, 4.5 mph.
3. 14′ tool, 4.0 mph.—fuel 80% as much for post-crop disking
4. 14′ tool, 4.0 mph. to 5.0 mph.
5. 14′ tool, 4.0 mph. to 5.0 mph.
6. 60 lb/inch depth; 3.0 mph; depth—24″, 27″, 30″; spacing—1 rod apart
7. 60 lb/inch depth; 2.0 mph; depth—18″, 21″, 24″; machine width—12′

Table 2. Fuel consumption—snapbeans.

| Operation | Fuel Consumption (gal/acre) | | |
	Low	Average	High
Moldboard Plow	.50	2.94	3.76
Disk (2)	.98	1.54	2.04
Spring Tooth Harrow	.38	.70	1.24
Plant (corn planter)	.64	.94	1.10
Herbicide	.23	.29	.38
Shallow Row Crop Cult. (2)	.86	1.18	1.70
Spray	.23	.29	.38
Harvest			
1. Chisholm—Ryder	3.02	4.30	5.79
2. Trucking	1.20	1.40	1.60
	8.04	13.58	17.99

Cost Account Analysis

The fuel consumption per acre and per head of livestock figures used in this study were derived from the records of cost account farms in New York State. For the most recently published cost account year (1972), there were 35 New York State farmers who completed detailed records on their businesses in cooperation with the Department of Agricultural Economics, Cornell University. The participating farms were located throughout the major farming areas of the state. They were generally well-run, full-time, commercial farms and were representative of the "better" farms in New York.

With reference to the present study, each of the individual farm enterprise records was evaluated for each of the field crops and livestock classes under consideration. Data was extracted as to the total acreages for each crop and the total hours of tractor use and truck miles attributed to growing and harvesting operations for the crop. Similar data was derived for livestock maintenance and production activities. In addition to tractor and truck use, and the total cost of other equipment used, *e.g.*, self-propelled equipment, irrigation pumps or custom-hired machinery was extracted from the records.

To derive a figure for average fuel consumption per hour of tractor use on each farm, tractor enterprise records were consulted. Each of these records gave full details on fuel consumption for each tractor in the farm inventory. These figures were reduced to a consumption per hour of use basis and then aggregated on a weighted average basis to yield a single fuel consumption per hour of use figure for all tractors in the farm business.

The tractor fuel consumption figures and the hours of tractor use per crop were then multiplied to derive the total fuel consumption figure for the crop under consideration. When the fuel consumption figure per crop acre for tractor use had been isolated for each farm enterprise, the individual farm figures were aggregated on a weighted average basis to give a single figure for fuel consumption for tractor use per crop acre for all farms having the same crop enterprise.

Exactly the same process was followed to derive a single figure for fuel consumption for truck use per crop acre for all farms having the same crop enterprise.

With regard to all other equipment used in the crop and/or livestock enterprises, it was conservatively estimated that 5% of the total costs in this category could be attributed directly to fuel costs for the machinery in question. It was assumed that, while the category obviously included some equipment for which there was no percentage of total costs attributable to fuel costs—*i.e.*, plows or harrows—and the percentage of total costs attributed to other equipment—*i.e.*, a self-propelled combine—might be considerably more than 5% of the total, the 5% of total costs for all other equipment combined was a reasonable approximation.

Greenhouse Energy Accounting

A survey form was mailed to all commercial growers asking for fuel use for heating their greenhouses. Approximately 200 forms were completed and returned and the results analyzed. To extrapolate the data to obtain total fuel use for New York State, census data was used to determine the total area.

Spot checks of the fuel use factors with several large operations showed good correlation with the survey data results.

The survey was conducted by the Department of Floriculture in cooperation with the College Task Force on Energy and Agriculture.

RESULTS OF ANALYSIS

Field Crops Gasoline and Diesel Fuel Use

The results of the analysis are given in Table 3. The fuel requirements for all major and most minor field crops are listed. A fuel factor (gallons per acre) developed from the analysis was used to determine total New York State fuel use per crop. Diesel fuel was converted to gasoline equivalent.* Diesel fuel represents 28% of total gasoline equivalent for field crops.

Fuel requirements were adjusted for crops planted but not harvested as indicated in Table 3. The total gasoline equivalent for field crops in New York State was calculated to be 54,690,900 gallons. This represents 39,377,443 gallons of gasoline and 10,938,180 gallons of diesel fuel. A number of explanations, assumptions, and qualifications are required and these are listed as references to Table 3.

Livestock Gasoline and Diesel Fuel Use

Table 4 provides a summary of fuel use for livestock and poultry. The fuel use factors for this table were determined from cost account data alone. The total gasoline equivalent for all operations was calculated to be 13,072,200 gallons annually.

Propane and Electrical Consumption for Field Crops and Livestock

Propane and electrical consumption were calculated from the fuel survey. Tables 5 and 6 represent the results from the analysis. In addition, total New York consumption of electrical power for farm use was calculated from data supplied by the state's major electric utilities. A total of 5500 kWh's (representing an estimated farm home use) were deducted from the total use per farm. The average consumption for residences in New York State was about

---

* Gasoline equivalent of diesel fuel equals 1.4 times gallon of diesel fuel.

Table 3. Gasoline equivalent for New York State crops.

Crop	Acres Harvested in 1973 (1000's of acres)	Average Yield (Units/acre)	Total Production (1000's of units)	Fuel Factor (gal/acre as gasoline equivalent)	Total Fuel Use (100's of gals as gasoline equivalent)
Field Crops					
Barley	12	40 bu/A	480 bu	9.8	117.6
Corn					
grain	360	77 bu/A	27,720 bu	9.7	3,492
high moist.	—			8.8	
silage	610	12.5 T/A	7,625 T	15.7	9,577
Dry Beans	39	0.48 T/A	18.6 T	15.8	616.2
Hay					
baled }	2,295	2.27 T/A	5,204 T	10.1	24,097.5
silage }				11.8	
Oats	325	55 bu/A	17,785 bu	9.8	3,185
Rye					
cover	114	—		1.2[1]	136.8
harvested	16	30 bu/A	480 bu	4.0	64.0
Soybeans	11	23 bu/A	253 bu	7.6[2]	83.6
Wheat	140	36 bu/A	5,040 bu	8.0	1,120
Requirement for acreage planted but not harvested[10]					592.2
Sub-total	3,922				43,081.9
Vegetable Crops					
Beets	4.4	14.5 T/A	63.8 T	29.8[3]	131.1
Cabbage	10.9	14.5 T/A	158.1 T	50.7[4]	552.6
Carrots	1.8	16.8 T/A	30.2 T	33.1[3]	59.6
Cauliflower	2.7	5.4 T/A	14.6 T	25.0	67.5
Celery	1.2	16.3 T/A	19.6 T	37.4[5]	44.9
Cucumbers	2.1	5.5 T/A	11.6 T	22.2	46.6
Lettuce	3.6	10.0 T/A	36.0 T	24.8[5]	89.3
Onions	13.6	11.0 T/A	149.6 T	57.6[5]	783.4
Peas	6.0			16.9	101.4
Potatoes	54.0	11.3 T/A	608.5 T	46.6[6]	2,516.4
Snapbeans	65.4	2.0 T/A	130.8 T	18.8	1,229.5
Sweet Corn	15.2	3.6 T/A	54.7 T	26.8[7]	407.4
Tomatoes	5.6	9.8 T/A	54.9 T	38.8	217.3

	(acreage)				
Other::	30.0[8]			29.6[9]	828.06
Asparagus				29.6[9]	828.06
Chinese Cabbage				20.1	
Eggplant				26.6	
Green Peppers				42.7	
Melons				37.2	
Spinach				39.8	
Squash				19.4	
				21.2	
Requirement for acreage planted but not harvested[10]					209.0
Sub-total	216.5				7,284.1
Fruit Crops					
(1970 acreages unless noted)					
Apples	72.6	6.5 T/A	471.9 T	38.1	2,766.1
Cherries	8.0	2.7 T/A	21.6 T	36.6	292.8
Grapes	36.9	4.1 T/A	151.3 T	27.5	1,014.8
Pears	5.0	2.7 T/A	13.5 T	23.0	115.0
Peaches	2.5	3.9 T/A	9.8 T	26.4	66.0
Plums	1.6	—	—	27.4	43.8
Raspberries	—	—	—	19.5	26.4
Strawberries	1.2	2.0 T/A	2.4 T	22.0	
Sub-total	127.8				4,324.9
Total	4,266.3				54,690.9

1. Excludes plowing under which is counted with the crop following.
2. Excludes any cultivation.
3. This is an average figure and may vary as much as 100% where it is necessary to transport over long distances to market.
4. 75% of crop for processing, 25% for fresh market.
5. Does not include fuel used for irrigation in dry periods to prevent wind erosion.
6. This is an average figure and may vary by 50-100% depending on soil type and distance hauled to market.
7. This figure varies depending on the amount of irrigation required.
8. Estimated at 15% of total vegetable acreage reported.
9. Average of fuel factors for other vegetables.
10. Based on estimate that 60% of total fuel consumption is required for crop establishment.

Table 4. Livestock and poultry operation fuel use.

Livestock Class	Total Numbers (thousands)	Avg. Production Per Unit	Total State Production (thousands)	Gallons/ Fuel Per Unit	Total Fuel For State
All Cattle and Calves	1,764				
Cows and Heifers in milk production	914	10,733 lb	9,809,962 lb	7.7	7,038,000
Cows and Heifers for beef	106			3.2	339,200
Heifers 500 lb and over for milk replacement	314			4.8	1,507,200
Heifers 500 lb and over for beef replacement	32			3.2	102,400
Other Heifers	37			3.2	118,400
Bulls 500 lb and over	23			3.2	73,600
Steers 500 lb and over	40			3.2	128,000
Calves less than 500 lb	298			3.2	953,600
Cattle and Calves on feed	(14)				
Calf crop	(959)				
			Sub-total		10,260,400
All Hogs and Pigs for breeding	89			2.6	231,400
Pigs crop	(126)			0.8	100,800
Pigs marketed	(105)				
All stock Sheep for breeding	85			1.6	136,000
Lamb crop	70			1.1	77,000
Chickens (Layers)	8,965	230 eggs	171,829 doz.	0.06	537,900
Chickens (Pullets)	3,270			0.02	65,400
Chickens (Other)	75			0.008	600
Turkeys	184			0.08	14,700
Ducks	2,000			0.02	40,000
Horses	500			3.2	1,600,000
Other (primarily fur bearers)	1,000			0.008	8,000
			Total		13,072,200

6,000 kWh's. A slightly lower consumption was assumed for the average farm home. The results of this analysis are listed in Table 5. Listed in Table 6 is the estimated total New York State propane use. The propane total use was estimated from the fuel survey data.

Farmers use of electricity has tripled since 1950 in the United States (USDA, 1974). In 1950 only 77% of the farms had electric service. In 1973, over 40 billion kilowatt hours were used on farms.

However, much of this was consumed in family living. In 1970, farm family living purposes used 64% of the electricity serving farms. Thus, only 13.4 billion kWh's were used for such production purposes as power milking machines, feed mills, elevators, augers, welders, heating lamps, and countless other labor-saving devices. New York State, being a large dairy state, has a higher percentage used for farmstead operations than for family living.

Table 5. Estimated yearly electricity requirements for selected crops and animals.

Activity	No. Units, Acres or Animals in Sample	Electricity[1] kWh/Unit
Potatoes	3,195	213
Onions	1,034	280
Snapbeans	925	90
Cabbage	435	234
Sweet Corn	2,000	52
Corn Grain	985	56
Dry Beans	390	21
Apples	4,077	583
Grapes	2,443	104
Milk Cows	36,260	603
Beef	969	295
Layers	325,880	4

1. An allowance of 5,500 kWh/farm for personal use was made and is not included in these results.

Table 6. Estimated yearly propane requirements for selected crops and animals.

Activity	No. Units, Acres or Animals in Sample	Propane[1] gal/unit
Potatoes	4,277	1.8
Onions	1,542	13.6
Snapbeans	625	4.2
Sweet Corn	3,517	4.0
Corn Grain	1,420	22.1
Dry Beans	153	9.8
Apples	10,989	16.6
Grapes	1,358	5.6
Milk Cows	8,543	24.1
Beef	1,250	7.4
Layers	417,780	0.11

1. Estimated usage on only those farms reporting propane use

Table 7. New York State total propane and electric use.

No. Farms	Average kWh/Farm	Total Electric (1000 kWh)	Total Propane (1000 Gallons)
50,000	14,500	725,000	8,567

Table 8. Greenhouse fuel use.

Type Fuel	Type Operation	N.Y. Total Area 1000 ft^2	% Area[1] Each Fuel	Fuel Factor	N.Y. Total Fuel Use
Heating Oil #2 Equivalent	Florist	13,285	85%	1.63 gal/ft^2	18,406,367 gal
	Vegetable	556	85%	0.81 gal/ft^2	382,806 gal
	Nursery	400	85%	0.95 gal/ft^2	323,000 gal
				Total	19,112,173 gal
Natural Gas	Florist	13,285	10%	278 ft^3/ft^2	364,323,000 ft^3
	Vegetable	556	10%	85 ft^3/ft^2	4,726,000 ft^3
	Nursery	400	10%	125 ft^3/ft^2	5,000,000 ft^3
				Total	374,049,000 ft^3
L.P. Gas	Florist	13,285	2.5%	1.78 gal/ft^2	591,182 gal
	Vegetable	556	2.5%	0.48 gal/ft^2	6,672 gal
	Nursery	400	2.5%	0.65 gal/ft^2	6,500 gal
				Total	604,354 gal
Coal	Florist	13,285	2.5%	236 tons/acre	1,799 tons
	Vegetable	556	2.5%	96.8 tons/acre	30 tons
	Nursery	400	2.5%	121 tons/acre	28 tons
					1,857 tons

1. From fuel survey, the percentage area heated by each fuel was determined.

Greenhouse and Nursery Crops Fuel Use

The gasoline requirement for the greenhouse and nursery operations was also determined from the fuel survey and is listed in Table 9.

Table 9. Gasoline use in greenhouse and nursery operations.

Type Operation	Total N.Y. Area 1000 ft²	Gasoline Fuel Factor gal/ft²	Total N.Y. Use gallons
Florist	13,285	0.11 gal/ft²	1,461,350
Vegetable	556	0.25 gal/ft²	139,000
Nursery	400	0.18 gal/ft²	72,000

The value of wholesale and retail commercial flower production in New York State is about $44 million. The United States production of flowers is reported to be about $961 million. Vegetable plants produced in New York State are worth nearly a half million dollars. United States production of vegetable plants is about $17 million. The New York State total value of nursery crops is about $12 million and the United States about $285 million.

The greenhouse industry is an energy-intensive operation that consumes fuel for space heating at a higher rate than about any other space heating application.

Miscellaneous Farm Uses of Fuel

Agricultural Aircraft: In a telephone survey conducted in January 1974 there were 24 fixed wing and 11 helicopters engaged in custom agricultural chemical application activities in New York State. In 1973, these aircraft treated 836,500 acres and consumed 225,400 gallons of fuel. In 1974 it is estimated that 1,000,000 acres will be treated and 270,800 gallons of fuel will be consumed. Of the fuel consumed, 80% is octane and the remaining 20% is 100 octane gasoline.

Maple Syrup Industry: In New York State, the estimated production of maple syrup is 450,000 gallons annually. Fuel oil and natural gas is used to process an estimated 300,000 gallons with 80% fuel oil and 20% natural gas. The remaining 150,000 gallons is mostly processed using wood as the fuel input. The summary is given in Table 10.

Honey Production: New York State apiarists annually produce 8 to 10 million pounds of honey. This honey is produced in 160,000 colonies averaging 50 to 65 pounds of honey per colony. Of these, 80,000 colonies are operated by commercial producers. Another 40,000 colonies are operated on a semicommercial basis while the

Table 10. Maple syrup processing in N.Y. State.

Type Fuel	Quantity Processed gallons	Total Fuel Use
Heating Oil #2 Equivalent	240,000	840,000 gallons
Natural Gas	60,000	36,000,000 ft³
Wood	150,000	——

remaining 40,000 colonies are operated by hobbyists. The primary fuel requirement for honey production is gasoline used to truck the colonies to various producing sites. Colonies are clustered in groups of 30 to 40 colonies, approximately 2 to 3 miles apart. These cluster colonies are visited about 10 times per year. Some of the commercial producers transport their colonies to Florida for winter production. It is estimated that approximately 15,000 colonies are transported annually in loads of 250 to 400 colonies. Using an average figure of 8 miles per gallon for the truck, the transport and site visits require 36,430 gallons of gasoline per year.

Family Living Fuel Use

According to a recent U.S. government report (USDA, 1974) in 1970 farm families in the United States used for family living purposes an estimated 36 billion cubic feet of natural gas from public utilities, 850 million gallons of fuel oil, 1 million tons of coal, 26 million barrels of L.P. gas, and 23 billion kilowatt hours of electricity. This totaled nearly 368 trillion Btu's.

For other than electricity, space heating consumed most of the fuels used for family living. In 1970, only 11% of farm household electricity was used in space heating. L.P. gas (propane) and natural gas are used for both space heating and cooking, but space heating consumed about three-fourths of these fuels.

Fuel oil, including kerosene, was the most widely used fuel to heat farm homes in 1970 with 36% of all homes so heated. Over 31% were heated by L.P. gas; 10% by natural gas; 8% by electricity; 8% by coal or coke; and 7% by wood and other sources.

The amount of fuel used in farm automobiles for family living is difficult to determine. In a 1970 ERS/SRS (Economic Research Service/Statistical Reporting Service of USDA) survey of fuel needs in farming in the United States, nearly 20,000 respondents reported using an average of 610 gallons per auto. Farmers reported that 25% of the automobile use was for business purposes. Therefore, an estimated 458 gallons of fuel per automobile was determined to be for family living purposes. Farm family members also use trucks —particularly pick-ups—for family living purposes. In surveys by the government, farmers estimated as much as 25% of truck fuel was for family use. This would indicate that an additional 355 million gallons is used in trucks for family living purposes.

In a later survey (1972) by ERS, fuel use in automobiles was reported sharply higher—about 1,000 gallons per auto. At 25% for business purposes, this leaves 750 gallons per auto for family living.

In this study, 800 gallons of gasoline was deducted from the state fuel survey for each farm. It is expected that the family living use exceeds 800 gallons; however, it is also expected that some gasoline is purchased away from the farm.

Gasoline and Fuel Use Per Month

The fuel use per month was tabulated from the fuel survey forms. Table 11 indicates the percentage distribution of fuel usage by months for the year 1973.

Table 11. Percent usage of fuel by months, 1973.

Fuel	Jan.	Feb.	Mar.	Apr.	May	June	July	Aug.	Sept.	Oct.	Nov.	Dec.
Gasoline	5.7	5.3	6.3	8.1	9.5	10.8	11.0	10.4	10.1	9.4	7.3	5.5
Diesel	5.2	4.5	5.9	8.9	11.0	11.6	10.5	8.5	10.7	9.9	7.5	5.2
Propane	8.4	7.3	7.5	6.3	4.9	4.8	4.3	5.3	7.5	16.0	17.8	9.2

ACKNOWLEDGMENTS

The authors acknowledge the encouragement and support of this energy accounting study by Dr. Noland L. VanDemark, Director of Research, David L. Call, Director of Cooperative Extension, and the Energy Task Force, New York State College of Agriculture and Life Sciences, Cornell University, Ithaca, N.Y., State and County Extension Services, Agricultural Stabilization and Conservation Service(ASCS) and the New York State Electric Utilities. Partial support of this study was obtained from the National Science Foundation—RANN Grant Number GI43099.

Special acknowledgment is given to T. R. McCarty, C. A. Sandsted, and S. McClintock, graduate students in Agricultural Engineering at Cornell University for their assistance in the analysis from the fuel survey and the engineering analysis and to J. H. Erickson, graduate student in Agricultural Economics at Cornell University for his assistance with the analysis from the cost account farm data.

Although it is impossible to list all the resource personnel contacted, the authors do appreciate the assistance of the following persons:

1. Engineering Data—Richard W. Guest, Everett D. Markwardt, Gerald E. Rehkugler, and Carl S. Winkelblech, Department of Agricultural Engineering; Phillip A. Minges, Roger F. Sandsted and Joseph B. Sieczka, Department of Vegetable Crops; William D. Pardee, Department of Plant Breeding, Cornell University; James Anderson, Robert Backer, Raymond Nichols, Horace A. Smith, Warren H. Smith, Leon Weber, and Dale Young, County Extension Agents.
2. Bee Keeping Data—Roger A. Morse, Department of Entomology, Cornell University;
3. Greenhouse Data—Carl F. Gortzig, Department of Floriculture, Cornell University;
4. Maple Syrup—Fred E. Winch, Jr., Natural Resources, Cornell University;
5. Agricultural Aircraft—Vic Mason, President N.E. Agricultural Aviation Association and Arthur A. Muka, Department of Entomology, Cornell University.

REFERENCES

1. USDA, 1974, The U.S. Food and Fiber Sector: Energy Use and Outlook, Economic Research Service, USDA, Government Printing Office No. 38-906, Washington, D.C.
2. Gunkel, W.W., D. R. Price, G. L. Casler, T. R. McCarty, S. McClintock, C. A. Sansted, J. H. Erickson, 1974. Energy Requirements for New York State Agriculture: Part I, Food Production, Agr. Engr. Extension Bulletin 405, Cornell University, Ithaca, N.Y.

6.

Energy Requirements for Agriculture in California

V. Cervinka,* W. J. Chancellor,** R. J. Coffelt,**
R. G. Curley** and J. B. Dobie**

Fuel and electrical energy are essential inputs to California agriculture in its present form. Recent appearance of shortages and irregularities of energy supply has caused concern among all persons connected with agriculture as well as among many others who might be affected by changes or decreases in agricultural production. In order that problems of energy supply to California agriculture be dealt with in the most appropriate way, it is necessary that quantitative information be available on fuel and electricity requirements of farms, agricultural suppliers and processors of agricultural products. To meet this need, the State of California Department of Food and Agriculture, working in conjunction with the Department of Agricultural Engineering, University of California at Davis, with financial support from the California Farm Bureau Federation and the Committee on Relation of Electricity to Agriculture, has conducted this study to determine energy requirements for California agriculture.

The objective of this study has been to develop a detailed estimate of the fuel and electricity requirements for California agriculture during 1972. The fuels considered are gasoline, diesel fuel, aviation fuel, LP gas, and natural gas. Requirements include those for:

* California Department of Food and Agriculture.
** Agricultural Engineering Department, University of California, Davis, California.

Field Crops, Vegetables and Fruits
Tillage and planting
Cultural practices
Harvesting
Transport to market or processing plants
Processing
Storage prior to entry into distribution channels

Livestock
Livestock production and feeding
Raising of young stock for replacement of production stock
Transportation of livestock to local markets by nonfarm vehicles
Transport from markets to processing plants or resale markets
Processing of livestock products
Storage of livestock products prior to entry into distribution channels
On-farm feed processing

Irrigation
On-farm water pumping by electric motors and internal combustion engines for purposes of crop irrigation and frost protection
Water pumping in major water supply projects that is used specifically to meet irrigation needs

Fertilizers
Fertilizer production
Fertilizer distribution
Fertilizer application and transport associated therewith

Agricultural Aircraft
Operation of all agricultural aircraft
Operation of ground support vehicles for aircraft loading, supply, etc.

Frost Protection
Operation of heaters
Operation of wind machines

Greenhouses
Heating of greenhouses by natural gas during periods of cold weather
Ventilating, air circulation and evaporative cooling of greenhouses

Vehicles for Farm Business
On-road and off-road operation of farm automobiles during the portion of time required for farm business
On-road and off-road operation of farm trucks (including pickups) during the portion of time required for farm business and transport

The information presented here has been developed with the co-operation of a great number of individuals and organizations who have generously contributed their knowledge and time. A list of categories and groups of contributors, bibliographic and institutional sources of information appear in the complete report.

METHODS

Information for on-farm use of energy was developed for three stages of crop production—crop establishment, cultural operations, and harvest—for the following major crops.

Field Crops	Vegetables	Fruits and Nuts
Alfalfa, hay	Asparagus	Almonds
Alfalfa, seed	Beans, green	Apples
Barley	Broccoli	Apricots
Beans, dry	Carrots	Grapefruits
Corn, grain	Cauliflower	Grapes
Cotton	Celery	Lemons
Oats	Lettuce	Oranges
Rice	Melons	Peaches
Safflower	Onions	Pears
Sorghum, grain	Potatoes	Plums
Sugar Beets	Strawberries	Prunes
Wheat	Tomatoes	Walnuts

The energy requirements for transport and processing associated with each of the common methods of processing these crops were also determined. In addition, energy use in feed transport, husbandry, market transport, and processing each of the following major livestock product groups was calculated.

Dairy (milk)	Hogs	Eggs
Cow-calf	Sheep and lambs	
Feedlots	Broilers and fryers	

Energy requirements for crops and types of livestock not listed were included, according to their tonnage or value, in the totals for each commodity group. Fuel and electricity requirements for irrigation pumping, fertilizer manufacture, frost protection, agricultural aircraft, and farm vehicles were separated from those for crop and livestock production, transport and processing, and were presented on a total state-wide basis.

The basic approach used for crop and livestock categories was to determine energy-use values for each ton of product in that category and to then multiply those values by the tons of product reported for California in 1972 by the California Crop and Livestock Reporting Service.

The units used for the various fuels and electricity were the ones most commonly used in the field, *i.e.,* gasoline—gallons, diesel fuel—gallons, LP gas—gallons, electricity—kilowatt hours, natural gas—therms.

When appropriate to consolidate all energy sources into a single dimension, the unit of barrels of crude oil was used.

A broad range of methods were used to collect the information presented. Methods were selected and developed individually for each category of information sought.

Fertilizer production and distribution. Fuel and electricity use data for fertilizers were obtained by sending questionnaires to all California fertilizer producers and to a representative sample of fertilizer distributors and commercial applicators.

Irrigation pumping. Personal communication with each of the major electricity suppliers serving rural areas and of major water project agencies was used to obtain either data or estimates of electrical energy used for irrigation pumping. One of these suppliers also furnished information on the number and size of internal combustion engines used for irrigation pumping in its service area (a major portion of rural California). This firm also furnished data by district on electricity consumption per installed horsepower.

On-farm crop husbandry. For each crop three or more representative "sample cost of production data sheets" representing typical production practices were obtained. These sheets prepared by Agricultural Extension Farm Advisors, in cooperation with Extension Agricultural Economists, list for each field operation a value for "fuel and repairs." This value was converted to gallons of fuel based on the data given for each specific operation in a standard reference, which was frequently used in the formulation of the cost-of-production-data-sheets themselves.

Using this method, three or more estimates were obtained for each crop, thereby providing gallons of fuel required per acre for operations falling into each of three groups—crop establishment, cultural practices, and harvest. Estimates were then averaged for each group.

In order that energy used in on-farm crop husbandry could be conveniently related to that used in transport and processing, the values of gallons of fuel used per acre for each operation group in each crop were divided by the average per acre yield of the respective crop as reported for the period of 1968 through 1972.

On-farm livestock production. Information on energy use in livestock production was obtained from questionnaires distributed by Agricultural Extension Farm Advisors to livestock producers whose operations were more or less limited to one type of livestock production. Additional questionnaires were completed by poultry and egg producers who received the questionnaire through the Pacific Egg and Poultry Association.

Energy use values for animal husbandry were adjusted to exclude energy use for irrigation, crop production and farm vehicle use on livestock farms.

Fuel for feed transport was computed using standard per-head feed consumption data for hay and concentrates. Transport distances were reported in the questionnaires.

Fuel for transport of livestock products to market was computed similarly to that for feed transport, with the exception that the number of tons transported per farm reporting was based on the average weight per unit marketed.

Transport of crop products to processing plants. Questionnaires were distributed by the California Trucking Association to members commonly hauling agricultural products. Data reported allowed determination of average round-trip haul, average pay load and average miles per gallon for each commodity. Trucks used were almost exclusively diesel-powered.

Processing of agricultural products. Questionnaires were sent to numerous processors selected to represent various products and processes.

Each one usually reported on several products. Division of fuel and electricity used among the various products was done according to processing schedules throughout the year and according to respective tonnages handled. All values were put in terms of fuel or electricity required per ton of incoming product.

Vehicles for farm business. Census data was used to estimate the number of farm trucks and autos in 1972. Per-vehicle gallons of gas used for farm business were obtained from a study reporting analyses of gasoline tax refund claims during 1969.

Agricultural aircraft. Questionnaires were prepared and distributed by the California Agricultural Aircraft Association. These provided a basis for estimating the average fuel consumption per aircraft and by associated ground equipment. The number of operating aircraft and auxiliary units used in 1972 was estimated and used to compute the quantities of gasoline, diesel fuel and aircraft gasoline required for their operation.

Frost protection. Questionnaires were sent to Agricultural Extension Farm Advisors specializing in fruit crops in counties where frost protection is practiced. Citrus, deciduous fruit and nuts, and grapes were considered separately and the extent of each type of frost protection (heaters, wind machines, heaters plus wind machines, and sprinklers) in each crop category was estimated. Energy consumption by various frost protection devices was determined from references.

Energy required by sprinkler or furrow irrigation when used for frost protection was included with that for irrigation and is not reported separately.

Greenhouse heating and ventilation. Data on greenhouse heating energy requirements per square feet of house were obtained from gas company representatives for one Central Coast county and one South Coast county. These were compared with climatological data for these counties and were found to correlate well with the percentage of days during the year when the minimum temperature fell below 50°F. Using this as a basis, energy use requirements per square foot were determined for all counties having major greenhouse square footages. Actual square footage in each county was projected from census data.

Electricity required for air movement, etc., in greenhouses was based on data obtained from a power company representative for one Central Coast county.

Confirmation and adjustment of data. After all data were assembled and computed, information pertaining to energy use for each crop or livestock category and for each processing operation were presented to industry and extension specialists in each respective field.

These specialists indicated whether the values seemed generally appropriate or if there appeared irregularities in the values shown. In some cases, specific changes or values were suggested. Upon receipt of the comments of these specialists, values requiring further examination were investigated, additional data was sought, and adjustments were made to reflect the new information developed.

The values presented in the following tables are those which have been developed in the above processes.

RESULTS

The distribution of requirements for each of the agricultural energy use categories among electricity and fuel sources is shown in Table 1. A comparison between agricultural energy use and total energy use for California is shown by energy source in Table 2.

Data on the use of diesel fuel and gasoline in various field operations, and for product transport associated with the production of field crops, vegetables, and fruit and nuts, are given in Table 3.

The energy requirements for five selected crops are shown in Table 4. These figures include the fuel and electricity for several categories of operations that were not broken down by commodity

Table 1. Energy requirements for agriculture in California (1972).

Category	Energy Source (1,000,000 Barrels of Crude Oil)*						
	Natural Gas	Electricity	Diesel Fuel	Gasoline	LP Gas Propane Butane	Aviation Fuel	Total
Field crops	6.289	0.273	2.327	0.416	0.039	—	9.344
Vegetables	2.862	0.211	0.936	0.535	0.072	—	4.616
Fruit and nuts	2.192	0.242	0.631	0.269	0.054	—	3.388
Livestock	1.847	0.859	1.121	0.167	0.199	—	4.193
Irrigation	0.700	4.220	0.158	0.010	0.072	—	5.160
Fertilizers	5.271	0.341	0.163	0.075	0.018	—	5.868
Frost protection	—	0.024	1.448	0.147	0.015	—	1.634
Greenhouses	1.771	0.049	—	—	—	—	1.820
Agr. aircraft	—	—	0.026	0.034	—	0.192	0.252
Vehicles (farm use)	—	—	0.252	2.518	—	—	2.770
Other use	—	—	—	—	0.387	—	0.387
Total	20.932	6.219	7.062	4.171	0.856	0.192	39.432
Totals in Millions of Normal Units	1,214.128 Therms**	10,575.128 kWh	292.584 Gal.	195.198 Gal.	52.629 Gal.	8.994 Gal.	—

* One barrel of crude oil is equal to 5,800,000 Btu.
** One therm equals 100,000 Btu.

Table 2. Proportion of energy consumed by agricultural industry in California.

Energy Source and Units		State of California[1]	California's Agricultural Industry	
		1,000,000 Units		%
Natural Gas	Therms	23,588.537	1,214.218	5.15
Electricity	kWh	135,241.711	10,575.340	7.82
Diesel Fuel	Gal	2,659.356	292.584	11.00
Gasoline	Gal	10,037.916	195.198	1.94
LP Gas	Gal	458.933	52.629	11.47
Aviation Gasoline	Gal	42.738	8.994	21.04

	1,000,000 Barrels of Crude Oil Equivalent[2]
Total direct energy use in agriculture	39.432
Heat energy rejected in electricity generation[3]	11.815
Total energy associated with agriculture	51.247
Total energy use in California (1972)[1,4]	1,010.247
Percent of California energy used in agriculture	5.072

1. Sources of information:
   "Electric Power Statistics," Federal Power Commission, January 1972 through December 1972 issues.
   U.S. Dept. of the Interior, Bureau of Mines, Mineral Industry Surveys, Liquefied Petroleum Gas/Annual, October 25, 1973.
   Personal Communications, California Department of Conservation, Division of Oil and Gas, January 10 and 24, 1974.
   U.S. Department of Interior, Bureau of Mines, Mineral Industry Surveys, Natural Gas/Annual, November 8, 1973.
2. One barrel of crude oil is equal to 5,800,000 Btu or to 58 therms, 1699 kWh, 41.43 gal. diesel fuel, 46.77 gal. gasoline, 61.37 gal. LP gas, and 47.15 gal. aviation gasoline.
3. The oil equivalent of heat energy rejected (assumed efficiency=34.5%) upon generation of electrical energy used in agriculture.
4. Includes energy from coal, jet fuel, and residual fuel oil, as well as the input oil equivalent (at 34.5% assumed efficiency) of electrical energy generated by hydroelectric, nuclear, and geothermal installations.

in the overall study. These are irrigation pumping, fertilizer manufacture, agricultural aircraft, and farm vehicles.

Figures 1, 2, and 3 show the distribution of energy requirements for the major commodities within the crop and livestock categories. It should be emphasized that the distribution shown in these three figures does not include the fuel and electricity for irrigation pumping, fertilizer manufacture, frost protection, agricultural aircraft, and farm vehicles.

The major findings from this study are:

1. Farm operations and transportation and processing of agricultural products used 5.07% of the total energy consumed in California in 1972.

2. Agricultural production is an annual sequence of interde-

Table 3. Consumption of fuels in field operations and product transport (California, 1972).

Field Operations and Transport	Crop Categories and Fuel Types						Total Consumption			
	Field Crops[1]		Vegetables[2]		Fruit and Nuts[3]				Percent	
	Diesel	Gasoline	Diesel	Gasoline	Diesel	Gasoline			D	G
	1,000,000 Gallons									
Crop Establishment	37.720	2.173	12.049	1.151	2.228	.268	D	51.977	37.1	
							G	3.592		8.6
Cultural Practices	7.932	4.529	4.335	2.825	12.534	6.119	D	24.801	17.7	
							G	13.473		32.2
Harvest	15.535	8.960	4.166	8.642	3.575	5.254	D	23.276	16.6	
							G	22.856		54.6
Transport	19.601	.852	13.862	.848	6.658	.221	D	40.121	28.6	
							G	1.921		4.6
Total	80.788	16.514	34.412	13.466	24.995	11.862	D	140.195	100.0	
							G	41.842		100.0
Percent—D	57.7		24.5		17.8		D	100.0		
—G		39.5		32.2		28.3	G	100.0		

NOTE: 140,195,000 gallons equals 47.5 percent of all diesel fuel used in California's agriculture. 41,842,000 gallons equals 21.4 percent of all gasoline used in California's agriculture.

1. In 1972, 25,711,000 tons of field crops were produced on 5,859,000 acres.
2. In 1972, 11,300,000 tons of vegetables were produced on 842,000 acres.
3. In 1972, 6,703,000 tons of fruits and nuts were produced on 1,397,000 acres.

Table 4. Energy requirements for selected crops in California.

| Operation | Gallons of Diesel Fuel Equivalent Per Acre | | | | |
	Melons	Tomatoes	Alfalfa	Cotton	Rice
Crop Establishment	14.0	18.4	3.9	13.6	11.2
Cultural Practices	4.1	9.5	.8	7.7	0.0
Harvest	14.5	24.2	6.3	4.0	9.0
TOTAL	32.6	52.1	11.0	25.3	20.2
Percent Gasoline	(45)	(43)	(31)	(17)	(5)
Transport	7.4	39.4	3.4	3.1	2.4
Processing*	7.7	488.0	—	20.6	13.5
Fertilizer	40.2	36.2	4.6	25.2	25.1
Irrigation	29.6	29.6	29.6	29.6	29.5
Aircraft & Farm Vehicles	25.3	30.7	5.2	12.1	9.4
GRAND TOTAL	142.8	676.0	53.8	145.9	100.1

* Melons—Cooling and packing
Tomatoes—Canning
Alfalfa—Baled hay only; does not include processing
Cotton—Ginning and seed processing
Rice—Drying and milling

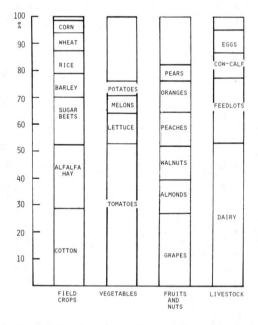

Figure 1. Distribution of diesel and gasoline requirements for production and processing of crop and livestock commodities (Calif., 1972).*

* This distribution does not include diesel and gasoline requirements for irrigation pumping, frost protection, agricultural aircraft, and farm vehicles.

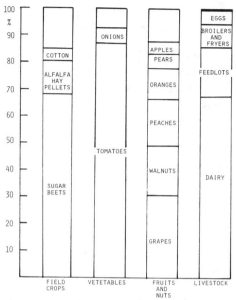

Figure 2. Distribution of natural gas requirements for processing of agricultural products and in livestock production (Calif., 1972).

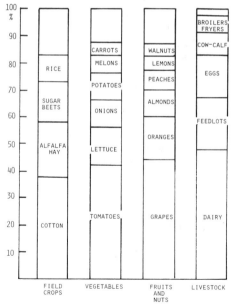

Figure 3. Distribution of electricity requirements for processing of agricultural products and in livestock production (Calif., 1972).

pendent energy-using activities, starting with fertilizer manufacture and continuing through the processing of agricultural products. Restricted energy flow during certain portions of the year, or in certain segments of this overall system, would likely cause major reductions in food production with only minor savings of energy.

3. Natural gas represents 35.1% of all energy consumed in California's agriculture.

4. Diesel fuel is the second major energy source for agriculture in California (17.8% of all energy used by agriculture). Diesel and other middle distillate fuels are primarily used for field operations, transportation, and frost protection.

5. Fertilizers, their production, distribution, and application, consume 14.9% of all energy supplied to agriculture. Of this amount, 89.9% comes from natural gas, most of which is used for production of fertilizers (13.4% of the total energy used by agriculture in California).

6. Water pumps for irrigation require 13.2% of the energy used by California agriculture. About 68% of all electricity used (*i.e.*, 10.7% of all energy) goes to irrigate crops.

7. Because of the size of California and the diversity and degree of specialization of its agriculture, farm products must be hauled distances between field and processor. One avenue for potential energy conservation is the development of systems, such as truck-rail combinations, for more efficient farm-to-processor transport.

8. Energy inputs per unit of final product are considerably higher for vegetables and fruits than for field crops. Thus, the costs of production for vegetables and fruits can be expected to be more directly linked to energy costs.

This study is an initial attempt to analyze the comprehensive system of agricultural production and processing practices from an energy standpoint. It is expected that continued efforts along this line will bring to light new opportunities for energy conservation and energy-efficient methods of production. A detailed report of the study is available from the Department of Agricultural Engineering, University of California, Davis.

7.

Energy and Monetary Requirements for Fed Beef Production

H. A. Hughes, J. B. Holtman, L. J. Connor*

To study the tradeoffs associated with technology and size of operation in beef production, capital outlays, annual costs, and energy usage were evaluated over a range of firm sizes employing various forms of technology. Capital outlays for various size-technology combinations under consideration and the annual costs per hundredweight of feedlot gain (measured as the sum of variable costs and annual charges assessed fixed factors of production) were measured in dollar terms. Energy requirements calculated were labor, fossil energy and land (as a proxy for solar energy).

The beef production technologies considered are summarized in Table 1. In all evaluations it was assumed that all the required high-moisture corn and corn silage were produced on-farm. Capital outlays, fossil energy and labor requirements include all of that which is required for feed production, materials transport and within the feedlot. However, the capital, fossil energy, and labor expended off-farm in the production of the inputs for on-farm use were not included. Furthermore, the fossil energy requirements do not include the electricity consumed on-farm, which was found to be very minute. Annual costs include the usage costs of capital as well as the costs of all material, energy and labor inputs.

---

* The authors are respectively: Assistant Professor of Agricultural Engineering, VPI & SU, Blacksburg, VA; Associate Professor of Agricultural Engineering; and Professor of Agricultural Economics, Michigan State University, East Lansing, Michigan.

Table 1. Specification of feedlot technologies considered.

Component	Feedlot Identification Number									
	1	2	3	4	5	6	7	8	9	10
A. Type of Housing										
Partial shelter, unpaved lot	X									
Partial shelter, paved lot		X								
Open, unpaved lot			X							
Completely covered				X	X	X	X	X	X	X
B. Type of Feed Storage										
Moist corn storage, tower silos for silage					X					
Moist corn storage, bunker silos for silage	X	X	X	X				X	X	X
Tower silos for silage						X				
Bunker silo for silage							X			
C. Type of Ratio										
All silage						X	X			
Corn and silage	X	X	X	X	X			X	X	X
D. Type of Waste Handling										
Liquid								X		
Solid	X	X	X	X	X	X	X		X	X
E. Type of Animal										
Calves									X	X
Yearlings	X	X	X	X	X	X	X	X		
F. Sex of Animal										
Steer	X	X	X	X	X	X	X	X	X	
Heifers										X

Size of feedlot was expressed as head capacity (the number of fully grown animals accommodated at one time in the lot). It was assumed that year round feeding was practiced (for yearlings the annual volume was nearly twice the feedlot capacity).

Machinery and field labor costs were evaluated based upon results from a field machinery selection algorithm (1). Transport costs and energy requirements were evaluated by a procedure based upon the following assumptions:

1. A relationship exists between average effective transport distance (field to farmstead) and farm area which is the same for harvested crops and the wastes, *i.e.*, it was assumed that over a period of years the wastes are uniformly distributed over the land.

2. Transport system component parameters are fixed in size. The only computation made by the model was to determine the number of units required.

3. An average transport speed exists which is independent of transport distance.

The overall model structure and the waste handling systems comparison results are described by Hughes (2) and Hughes and Holtman (3). Computation details and parameter values used are given by Hughes (4).

LIMITATIONS OF STUDY

The results and conclusions drawn from this study are specifically applicable to southern Michigan, although they are also applicable to larger areas in the Corn Belt. However, it should be noted that individual feedlot situations may depart from the technology and production practices assumptions specified in this study. Thus, caution should be exercised in applying these results to specific feedlot situations. Also, recent inflationary trends have resulted in larger prices for some inputs than were assumed for this study. The models and procedures used in this study can easily incorporate different assumptions relative to input and product prices and various technologies.

EFFECT OF HOUSING TYPE AND FEEDLOT CAPACITY ON CAPITAL OUTLAYS, ANNUAL COST AND ENERGY USE

The effect of housing type and feedlot capacity on capital outlays, annual cost, and energy usage is shown in Table 2 and in Figures 1 through 4. The results are summarized by each monetary and energy category. Feedlots 1-4 are compared in this analysis (see Table 1). Feedlot 4 is used as the base feedlot for all comparisons.

Land

Under the assumptions of this study, no measurable economies of size for land were found for any of the feedlots analyzed. Differences were observed, however, in the land requirements for the various housing types for a given capacity size (Table 2). The open, unpaved lot (Feedlot No. 3) required the largest acreage and land per hundredweight of gain. As total land requirements are determined mainly by feed requirements, crop yields and feed efficiencies determined land requirements.

Fuel

Fuel consumption per hundredweight of gain for the four basic types is shown in Figure 1. Diseconomies of size were noted for all housing types for fuel consumption.

Fuel is consumed for the field operations, in the feedlot, and for transportation. Variation in energy consumption with system capac-

Table 2. Land requirements for 1000-head capacity feedlots, by alternative housing technology.

Feedlot Identification Number[1]	Total land (acres)	Hundredweight of gain (cwt.)	Land/cwt. of gain (acres/cwt.)
1 (partial shelter, unpaved lot)	1,051	8,395	.125
2 (partial shelter, paved lot)	1,051	8,395	.125
3 (open, unpaved lot)	1,069	7,847	.136
4 (completely covered)	1,051	8,577	.123

1. See Table 1 for specification of feedlot components.

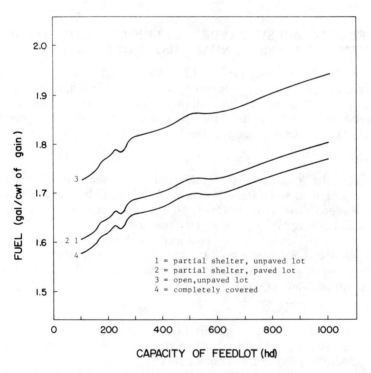

Figure 1. Fuel requirements per hundredweight of gain by housing type and capacity of feedlot.

ity is mostly attributable to transportation. Fuel consumption per hundredweight of gain in the feedlot is a small percentage of the total and is nearly constant. Fuel per hundredweight of gain for field operations is also independent of capacity. Transport energy increases monotonically, but at an ever decreasing rate, because the required fractional increase in average transport distance is smaller than the fractional increase in system capacity. The major differences in fuel consumption between various housing types is attributable to differences in feeding efficiency. The open, unpaved lot had the highest consumption and the completely covered feedlot had the lowest consumption per hundredweight of gain.

Labor

Economies of size for labor (Figure 2) were noted for the major housing types, but were mostly realized at low capacities (approximately 200-head capacity). Labor requirements were lowest for the fully sheltered and highest for the open systems because of feeding efficiency of the animals in the systems.

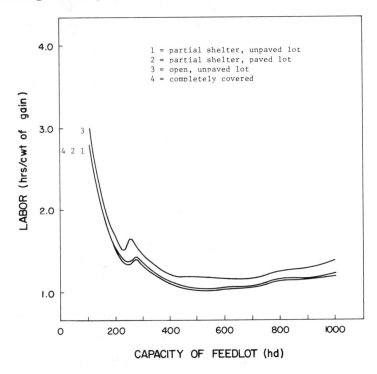

Figure 2. Labor requirements per hundredweight of gain by housing type and capacity of feedlot.

The shape of the labor requirement curves results from the use of labor by various parts of the system. For systems with low capacity, the administration time per hundredweight of gain is high. Field crop production labor per hundredweight of gain is high for small systems because small machines are used which require more labor per unit of cropland. At the upper end of the capacity range, administrative labor per unit gain is small. In this region, the upper limit on machinery size has been reached and there are no economies of size in field labor. The total labor increases after reaching a minimum as shown in Figure 2, because of increasing time required for transportation.

Capital

Economies of size were present for all major housing types, but were largely attained with 300-head capacity feedlots (Figure 3). Small systems used small field equipment and small feed storages which tend to have higher initial costs per bushel or per ton of feed handled than the larger units. In some instances, the smaller feed-lot does not fully utilize the capacity of the smallest item available.

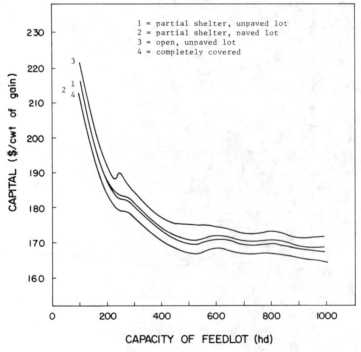

Figure 3. Capital requirements per hundredweight of gain by housing type and capacity of feedlot.

This high initial capital cost accounts for the high capital cost per hundredweight of gain for the small systems.

The highest and lowest capital per hundredweight of gain were required for the open feedlot and the completely covered feedlot, respectively. The partial shelter lots had capital requirements midway between the other two systems. Although the open lot had the lowest total capital requirement, the lower feed efficiency brought about by lack of shelter resulted in the lowest hundredweight of gain. This surprisingly resulted in a higher capital cost per hundredweight of gain than was the case for the other housing types.

Annual Cost

Annual cost includes all nominal costs of production. The lowest annual cost was for the confined, completely covered feedlot and the highest cost was for the open, unpaved lot (Figure 4). The economies of size were largely exhausted at approximately 300-head capacity levels.

Prices assumed for this analysis were approximate 1972-73 values. Based upon these price assumptions, the annual cost per hundredweight of animals sold would be approximately $36.00. This would be the break-even price required to cover the producers' costs.

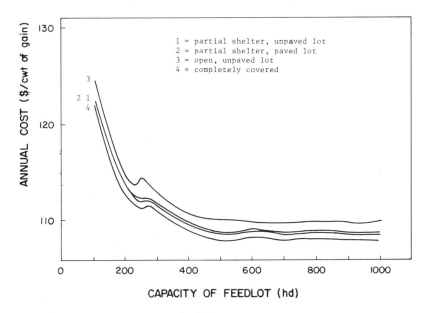

Figure 4. Annual cost per hundredweight of gain by housing type and capacity of feedlot.

EFFECT OF RATION, FEED STORAGE SYSTEM, AND CAPACITY OF FEEDLOT ON CAPITAL OUTLAYS, ANNUAL COST, AND ENERGY USAGE

The effect of ration, feed storage system, and feedlot capacities on capital outlays, annual cost, and energy usage is shown in Figures 5 through 8.

Total land and land per hundredweight of gain were highest for Feedlot 4 and lowest for Feedlot 6 (Table 3). Feedlots utilizing corn

Table 3. Land requirements for 1000-head capacity feedlots, by ration and feed storage system.

Feedlot Identification Number[1]	Total land (acres)	Hundredweight of gain (cwt.)	Land/cwt. of gain (acres/cwt.)
4 (moist corn storage, bunker silos for silage)	1,051	8,577	.123
5 (moist corn storage, bunker silos for silage)	1,022	8,577	.119
6 (tower silos for silage)	801	7,482	.107
7 (bunker silo for silage)	839	7,482	.112

1. See Table 1 for specification of feedlot components.

and silage rations required more acreage than feedlots using all silage rations. Because of higher silage losses in bunker silos, systems using bunker silos required more land than systems with lower silos.

Fuel

Fuel consumption is shown in Figure 5. Again, diseconomies of size were noted for all systems. Systems with bunker silos require less fuel than tower silo systems. The packing operation consumes less energy than blowing silage into tower silos (the difference depends on silo height), but part of the advantage is lost in the energy expended to produce and transport the extra silage needed to compensate for high bunker silo losses.

The mixed rations require more fuel/cwt gain than the all silage ration. This result is not immediately obvious as there are several factors involved. More fuel is used in tillage and planting for the

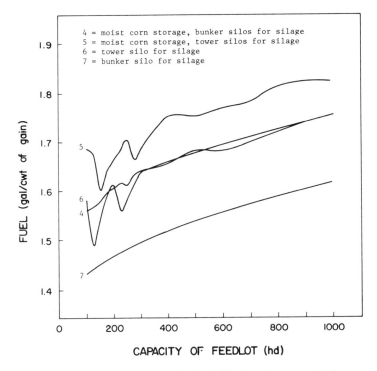

Figure 5. Fuel requirements per hundredweight of gain, completely covered feedlot, by alternative rations and feed storage systems, and capacity of feedlot.

mixed ration as approximately 9% more crop acreage is involved. More transport energy is required for the silage ration. We also have a tradeoff of silage chopping versus corn combining. In the bunker systems, shelled corn must be elevated whereas the silage is not. The net result is a fuel advantage for the all silage rations.

Labor

The plots of labor requirements show that for small systems the corn and silage ration requires the same or more labor as for all silage rations, and for systems larger than 250-head capacity, the all silage systems require more labor (Figure 6). The higher labor requirement for the large all silage systems results from two factors. More labor is required in the field to complete the harvest and more transport labor is consumed. The all silage system requires more loads of silage than total loads of both feeds when the mixed ration is used.

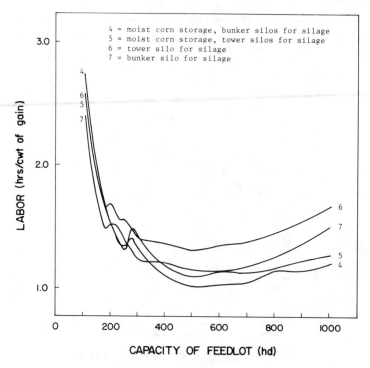

Figure 6. Labor requirements per hundredweight of gain, completely covered feedlot, by alternative rations and feed storage systems, and capacity of feedlot.

Again, economies of size were noted but were largely utilized at 250-head capacities. However, diseconomies of size also eventually occurred because of transportation demands for labor.

Capital

Capital requirements are shown in Figure 7. The all silage system had the lowest capital requirements for all sizes if bunker silos were used. Because of the great crop acreage required, the mixed ration requires more capital than the all silage ration.

Annual Cost

The annual cost of beef production is lower for systems using bunker silos than for systems using tower silos for all capacities evaluated (Figure 8). For most capacities, the mixed ration had a lower cost than the all silage ration.

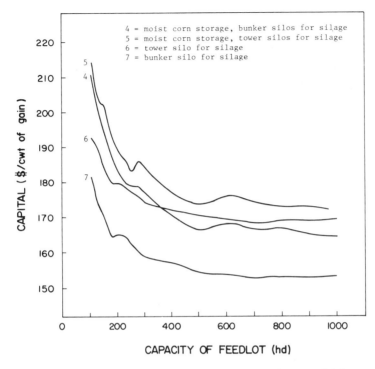

Figure 7. Capital requirements per hundredweight of gain, completely covered feedlot, by alternative rations and feed storage systems, and capacity of feedlot.

EFFECT OF TYPE AND SEX OF ANIMAL AND FEEDLOT CAPACITY ON CAPITAL OUTLAYS, ANNUAL COST, AND ENERGY USAGE

Two alternative systems were compared with Feedlot No. 4 to determine the effect of type and sex of animal (Feedlots 9 and 10). Feedlot No. 9 resembles Feedlot No. 4 except that calves are fed out in place of yearlings. Feedlot No. 10 is identical to Feedlot No. 4 except that heifer calves are fed out in place of yearling steers.

Fuel

Fuel requirements are higher for yearlings for all capacity levels (primarily due to feeding efficiencies). The lowest fuel requirements per hundredweight of gain are achieved by the system utilizing steer calves. Diseconomies of size are again shown for all types and sexes of animals (Figure 9).

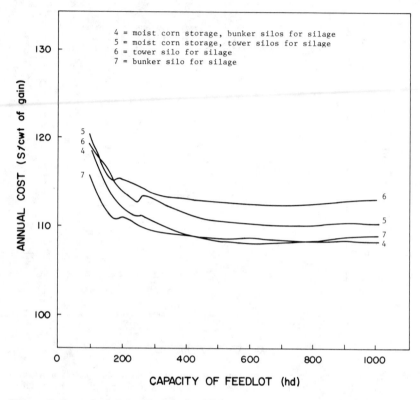

Figure 8. Annual cost per hundredweight of gain, completely covered feedlot, by alternative rations and feed storage systems, and capacity of feedlot.

Table 4. Land requirements for 1000-head capacity feedlots, by type and sex of animal.

Feedlot Identification Number[1]	Total land (acres)	Hundredweight of gain (cwt.)	Land/cwt. of gain (acres/cwt.)
4 (yearling steers)	1,051	8,577	.123
9 (steer calves)	746	7,482	.0997
10 (heifer calves)	685	6,610	.103

1. See Table 1 for specification of feedlot components.

Labor

The technology requiring the least labor varies with the specific capacity level. Yearlings have the lowest labor requirements up to approximately 275 head, but have the highest labor requirements at capacity levels of 600 head or greater. Steer calves have the

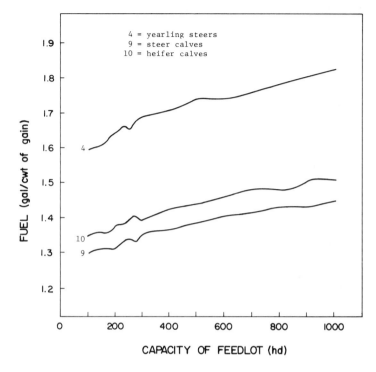

Figure 9. Fuel requirements per hundredweight of gain, completely covered feedlot, by yearlings, calves, heifer calves, and capacity of feedlot.

highest labor requirements for low capacity levels, and have the lowest labor requirements for capacity levels of 500 head or greater (Figure 10). These results occurred because yearlings are the most efficient utilizers of management labor, whereas the calves have higher labor efficiency.

Capital

Yearlings have the highest capital requirements for all capacity levels because of the capital tied up with animals (Figure 11). The lowest capital requirements for all capacity levels are for the system with steer calves.

Annual Cost

The inclusion of the cost of input animals in annual costs naturally led to a much higher cost per hundredweight of gain for the yearling system (Figure 12). However, these annual costs do not necessarily favor the production of calves in the feedlot. At feedlot capacities

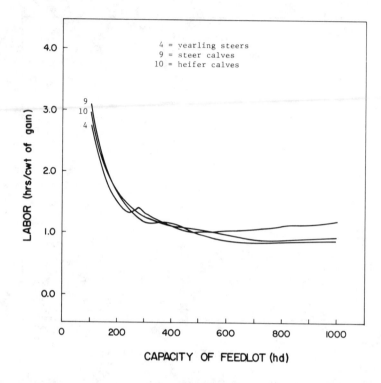

Figure 10. Labor requirement per hundredweight of gain, completely covered feedlot, by yearlings, calves, heifer calves, and capacity of feedlot.

where the economies of size have been virtually exhausted (300 head and greater), the average total cost per hundredweight of animals sold is approximately $36.00 per yearlings and $34.00 for steer calves, respectively. Thus, the cost differences per hundredweight of animals sold are not far apart for yearlings versus calves.

SUMMARY

Analysis of energy and monetary requirements for feedlots with varying technologies indicated that economies of size are present for all energy usage and monetary cost items except fuel. Fuel consumption increases with the capacity of the feedlot for all technology combinations studied. Fuel consumed in field and materials transport operations was approximately 15% higher for the 1000-head capacity lot than for the 100-head lot. All of the increase was in materials transport. Economies of size were realized in capital and in labor. The capital economies of size can be attributed in

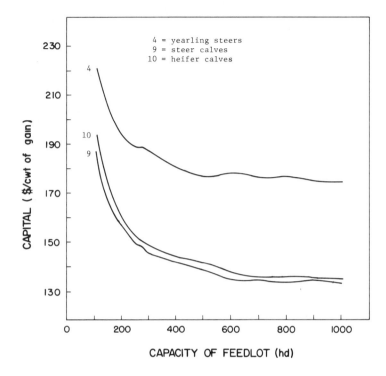

Figure 11. Capital requirements per hundredweight of gain, completely covered feedlot, by yearlings, calves, heifer calves, and capacity of feedlot.

large part to the unavailability of system components sufficiently small so that they can be fully utilized on the smallest lots. Labor economies of size can be attributed to two factors: 1) larger equipment, and 2) spreading management time over larger volumes. Most economies of size were realized at relatively low capacity levels (250 to 300 head).

Technological variations considered were alternative housing, feed storage, ratio, animal type, and sex. Many of the cost and energy usage variations among alternative technologies can be attributed to variations in feed efficiency.

* * *

This paper was prepared specifically for the Seventh Annual Cornell University Conference on Energy, Agriculture and Waste Management. Syracuse, N.Y., April 16, 17 and 18, 1975. Michigan Agricultural Experiment Station, Journal Article No. 7194. This project was supported in part by the NSF Grant GI-20, "Ecosystem Design and Management," directed by Herman Koenig.

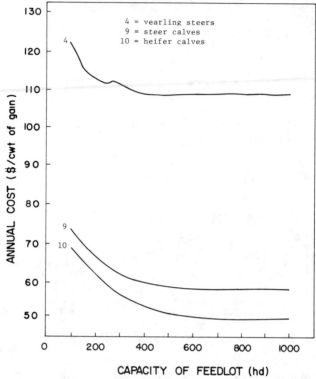

Figure 12. Annual cost per hundredweight of gain, completely covered feedlot, by yearlings, calves, heifer calves, and capacity of feedlot.

REFERENCES

1. Hughes, H. A. and J. B. Holtman, 1974. Machinery Complement Selection Based on Time Constraints. ASAE Paper No. 74-1541, ASAE, St. Joseph, Michigan.
2. Hughes, H. A., J. B. Holtman and L. J. Connor, 1974. Energy and Economic Costs for Two Beef Cattle Waste Disposal Systems, Proceedings of Sixth National Agricultural Waste Management Conference, Rochester, New York. March 25-27.
3. Hughes, H. A. and J. B. Holtman, 1975. Systems Analysis of Energy Consumption in Beef Production, Transactions of ASAE, 17:6. Pgs. 1161-1166.
4. Hughes, H. A., 1973. Energy Consumption in Beef Production Systems as Influenced by Technology and Size. Unpublished Ph.D. Dissertation, Michigan State University, East Lansing, Michigan.

8.

Economic Restraints on the Reallocation of Energy for Agriculture

Lowell D. Hill and Steve Erickson*

Recent concern over dwindling supplies of fossil fuels and inadequate supplies of food has focused attention on the relative efficiency of different systems of food production and marketing in meeting the nutritional requirements of a growing population. Although many of the popular proposals for increasing the "energy efficiency"[1] in food production fail to recognize important economic and social effects and incentives, the international importance of food supplies makes it imperative that agricultural scientists face the question "are there alternative systems of food production that will increase the quantity of food obtained from a given quantity of fossil fuel?" The answer to this question is complex and requires an understanding of the economic and physical relationships between energy and food, and the role of nutrition in consumer choice. This paper discusses these relationships and will also examine some of the social and economic restrictions on several proposals for improving energy efficiency. Finally, the criteria which should be useful in selecting economic policies for a rational system of food production will be examined.

FOOD AND ENERGY

The production of food is a process of energy transformation—changing energy from an inedible form into an edible form. A loss

* Professor of Agricultural Economics, and Graduate Student, respectively, University of Illinois, Urbana-Champaign.

of total energy is inevitable in any transformation, however agriculture is often viewed as unique because one of its major energy sources is the sun. Even though the resources of sun, wind, and rain are considered "free," agriculture is still a transformation activity whether primitive or mechanized. The goal of this transformation is to find the least expensive (*i.e.*, the most plentiful) supplies of energy to convert into a dearer (*i.e.*, higher valued) form. The choice of energy sources is not technological but economic and the choices are guided by relative values of inputs and outputs. Conversion of soil, water, and labor into food is not motivated by the ratio of energy produced to the energy used but by the value of the food, established by its basic demand for satisfaction of human needs and wants.

But food can be produced in many forms, by any of several processes using different combinations of energy inputs. The choice among these is based on relative value and costs. Consumer choices among food items are based on their concept of value relative to the price. Farmers' choices of energy sources and enterprise combinations are explained by the returns of the resources which they control. If the producer can purchase a quantity of a certain input (*e.g.*, fertilizer) for $1.00 and transform it into a food product worth $3.00 (*e.g.*, a bushel of corn) by combining it with his supply of land and labor, he will do so within the limits of his physical and capital capability. Furthermore, he will select that combination of energy sources that provides the greatest returns for each dollar invested.

This economic principle governing decisions at the micro level for an individual entrepreneur also provides an explanation of observed aggregate changes in resource use. The substitution of chemical fertilizers and fossil fuels for land and labor in agricultural production have been a response to the rate of return per dollar invested in a unit of each of these resources. For example, returns per hour to operator and family labor on corn belt grain farms range from a low of $0.24 per hour in 1931 to a high of $2.62 per hour in 1946. Between 1930 and 1957 labor returns were above $2.00 per hour in only four years. Returns to land have also been low. Farm record data for Illinois show a return of about 5% on the farm owners' capital (primarily land) in 1964,[2] and a return of 5.6% on capital in 1971.[3] Other studies indicate returns to land from agricultural production to be in the range of 4 to 6% on investment. In contrast, returns to fossil fuel energy were 2 to 3 times the value of the input from 1959 through 1973 (3). Pimentel (4) shows that each additional kilocalorie of input resulted in an increase of 2.6 kilocalories of corn production between 1964 and 1970. The price ratio between inputs and output results in an economic return even more favorable than this physical ratio indicates. For example, dividing the changes in the value of output (corn) by the changes in the cost of inputs (all energy) between 4 periods of 1959, 1964, 1970, and 1973 suggests a marginal value product of $3.34 for each dollar of energy in 1959-64, $1.91 in 1964-70, and $3.33 in 1970-73.[4] Even the lowest of these returns provides a strong economic incentive for increased use of energy rather than less. At a more micro level the application of an additional dollar's worth of nitrogen fertilizer on central Illinois land returns $7.23 in increased corn yield at 1964 prices, $13.36 at 1971 prices and $11.00 at 1974 prices.[5] The changes in farming practices requiring less labor, more fossil fuels, more fertilizers, more chemicals, more steel, and more of our non-renewable resources may be deplored by energy conservationists but they cannot be labeled irrational or inefficient decisions in economic terms.

Efficiency comparisons that exclude price relationships ignore the primary function of a market—to allocate scarce resources to their most valuable uses. While 1920 agriculture used much less fossil fuel to produce a bushel of corn than 1970 agriculture (5), 1920 techniques used much more harness leather, more steel wheeled wagons, and more husking pegs per bushel of corn production. Measuring efficiency by bushels of corn per unit of input, 1920 agriculture appears very "inefficient" relative to 1970 on the basis of these latter inputs, just as 1970 agriculture appears "inefficient" in the use of fertilizer relative to 1920. Both time periods, however, were economically rational in their allocation of the resources given the prices that existed at each point in time.

The fallacy in this exaggerated example is present in comparisons of "efficiency" between cultures as well as between time periods. Chinese wet rice agriculture may use very little fossil fuel per unit of output (6), but by U.S. standards it is very "inefficient" in its use of labor. The choice in both countries is based on relative costs of resources and products, and decision makers in both cultures tend to select the least cost techniques for production.

Similarly, critics of modern agriculture who point out that "Labor costs are now so high that farmers prefer to let their dairy cows eat as much as they want, rather than pay someone to monitor the feeding to be certain that each cow gets only the amount of food designed to yield the greatest amount of milk per unit of feed," (7) indicates a lack of understanding of the basic principles of economics and the economic incentives that direct resource allocation.

These economic principles are well known and will be restated only to illustrate their applicability to the statements on relative inefficiencies in agriculture currently being circulated even in pro-

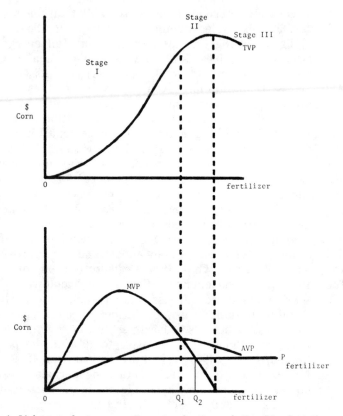

Figure 1. Value product curves for a production relationship depicting a single output from a single factor.

fessional journals. As shown in Figure 1, if it is profitable to use any of the input (*e.g.,* fertilizer) it will be profitable to use at least OQ_1 since each successive unit of fertilizer would return a yield even greater than the first unit. The decision to purchase OQ_2 units of input equates the marginal value of the last unit to the marginal cost of that unit of input, thereby maximizing returns to the input. In general, the quantity of OQ_2 lies beyond the maximum point of the average product curve. The suggestions that a reduction in the use of fossil fuels, or fertilizer will increase "efficiency"[6] (*i.e.,* average physical product) is not evidence of inappropriate resource use but evidence that farmers have in fact been using resources in the most economical fashion. If reducing input decreases the returns per unit, the farmer would be operating in the economically irrational stage I. Only if the price of the input is exactly equal to maximum average value product will OQ_1 be an economically rational level of input use. At all prices below this level, input use will exceed that which yields maximum physical product per unit of input. It is also evident from Figure 1 that any reduction in the use of inputs will reduce total output—a fact generally ignored by proponents of maximizing output per unit of energy.

Fertilizer provides an excellent illustration of the principles just discussed. Figures 2 and 3 show a typical response for nitrogen applied to corn in central Illinois. The maximum return per unit of energy input occurred with the first 50-pound increment of applied nitrogen. The yield response per unit of nitrogen decreased with each additional increment of nitrogen, reducing the energy ratio

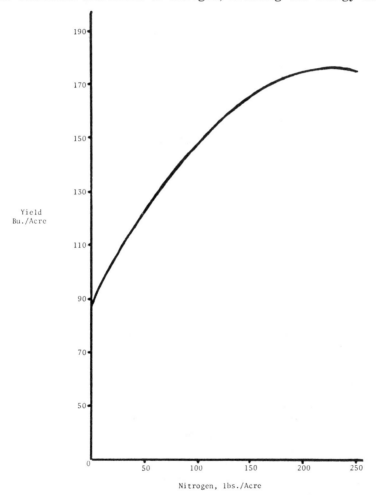

Figure 2. Effect of nitrogen fertilization on corn yield, Urbana, Illinois, 1968-1970.

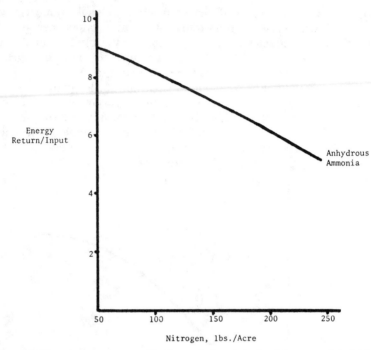

Figure 3. Energy return per unit of energy input derived from applying nitrogen fertilizer to corn.

from 8.86 to 1 for the first 50 pounds to 5.95 to 1 at application rates of 200 pounds per acre. Although the "energy efficiency" is decreased by nearly 1/3, the higher application rates still produce nearly 6 calories of corn for every calorie of energy used in production and transportation of the nitrogen fertilizer (8). As indicated previously, attaching 1974 prices to the inputs and output results in an even higher ratio of returns to fertilizer. In a world of food shortages it is difficult to argue that society should save 1 calorie in fossil fuel by a 6-calorie reduction in food supplies.

NUTRITION AND CONSUMER CHOICE

Most comparisons of dietary levels among groups in the United States or among countries are based on caloric intake. This measure provides a quick and convenient way to contrast the nutritional levels of countries such as India and Africa reporting diets of 1200 to 2000 calories per person per day with the nutritional levels in the United States where the average diet includes 3000 calories. However, comparisons on the basis of calories can be misleading because nutritional requirements include more than just calories.

Consequently, attempts to compare food products on the basis of the calories produced per calorie of energy used in production may lead to errors from a nutritional point of view. Corn meal or wheat flour cannot be substituted for fresh fruits and vegetables even though it requires much less energy to produce a calorie in grain than a calorie in vegetables. Although production of a calorie in the form of meat requires many times the energy for production of a calorie in grain, the comparison is not valid since meat is seldom if ever consumed for its caloric content. Malnutrition in countries where diets are based primarily on corn or cereal crops is often the result of protein deficiency rather than (or in addition to) a shortage of calories.

Comparisons among foods to encourage production of energy efficient forms also overlooks the convenience factor that heavily influences consumer preferences. A raw potato, a live chicken, and a basket of unshelled peas are not good substitutes for a TV dinner in the mind of a working housewife. The energy in the form of fossil fuels required to produce, process and market the two meals is irrelevant to most consumers.

Comparison of caloric levels can also be misleading because of differences in palatability, dietary habits, or culture. These and similar factors are often lumped together under the heading of "consumer preference." Thus a baked potato is not an acceptable substitute for a potato chip even though the cost per calorie is much less when measured in dollars or in the fossil fuel needed for production and processing. The unwillingness of consumers to purchase food on the basis of nutritional value has been an issue in the food stamp plan, or other welfare programs, and in the widely publicized breakfast food investigation.

In 1945 an economist published a minimum cost diet, meeting all nutritional requirements. This diet would also use much less production and processing energy than the present diets of most Americans. However, housewives have not been very enthusiastic about the suggested diet of 535 pounds of wheat flour, 107 pounds of cabbage, 13 pounds of spinach, 134 pounds of pancake flour and 25 pounds of pork liver (9).

The fallacy of comparing countries, diets, or food items on the basis of calories illustrates a fact that everyone recognizes with a moment's reflection. In developed countries, food is a form of entertainment, a business function, a social amenity, a psychological escape or a form of peer acceptance. Only rarely is it used *primarily* as a source of nourishment to meet biological requirements. Caloric content, or even the more general criterion of nutritional content, is seldom the basis on which food is purchased. There is a very limited market for nutrition and no market for calories as such. In many cases the price for food calories is negative as evidenced by the volume of "weight watcher" products and low calorie foods in the super markets.

Even in countries where malnutrition is important, traditional diets are difficult to change. The development of high lysine corn was thought to be a panacea for the lysine-deficient diets of South America where corn was a major source of calories. However, 10 years after its introduction it is not widely accepted in the country.

INCREASING THE EFFICIENCY OF FOOD PRODUCTION

World food supplies will not increase as rapidly as the desire for better diets and they are unlikely to equal the minimum subsistence requirements for an increasing world population. The need to increase food production in the face of shrinking supplies of fossil fuels requires that we examine alternative systems for agricultural production with lower energy requirements and the consequences of these alternatives. Inefficient use of our fossil fuels in food production will diminish our ability to feed the present population and will increase the speed of our headlong sprint on a collision course between food and population.

Several opportunities exist for increasing energy efficiency in agriculture. All of these involve an economic, social, or political cost. For purposes of evaluating their potential and assessing some of their impacts, these suggestions have been categorized under four headings: 1) Reduced use of inputs, 2) Input substitution, 3) Development of new energy sources and, 4) Changes in consumption habits.

Reduced Use of Inputs

Implicit in much of the writing on energy efficiency is the thought that a reduction in the quantity of fossil fuels relative to other inputs would result in a more "efficient" use of our scarce fossil fuels. As indicated previously this type of resource adjustment will increase the average physical productivity of the fossil fuel input but, in most cases, at a significant cost in terms of total output. For example, 1970 corn production used 133% more machinery, 47% more gasoline, 1500% more nitrogen, and 344% more phosphorus than 1945 corn production, when measured in energy equivalents (4). However, yields during the latter time period were 2.38 times the yields of 1945. In order to produce the 1972 corn crop using the 1949-51 input-mix, farmers would have required 2 times the acreage and 5 times as much labor. Production of the 1972 corn crop with high energy techniques rather than the 1945 technology released about 50 million acres of land for other grain production—mostly soybeans (10).

An increase in production per unit of fossil fuel can be obtained by a reduction in total output, but given the present balance of world food and population this choice does not appear to be politically or economically viable. There is a sound theoretical basis as well as strong empirical evidence to support Steinhart's statement: "It is likely that further increases in food production from increasing energy inputs will be harder and harder to come by" (5). This is the economic law of diminishing returns stated in its simplest form; but without costs and prices it does not provide a basis for recommending a reduction of total input in order to increase efficiency of the fossil fuel input.

Input Substitution

Most recommendations for conservation of fossil fuels in agricultural production indicate an awareness of the need to substitute other forms of energy for the fossil fuels to be withdrawn from production. Unfortunately, many of these recommendations have ignored differences in the cost of alternative energy sources for the production of an additional unit of food. The substitution of labor for fossil fuels seems a simple suggestion when considering incremental isolated substitutions such as hand spraying on a small plot. The magnitude of the problem is more visible however in a comparison of the cost of a complete substitution. Dr. Earl Cook has estimated the cost of fossil energy on U.S. farms to be about $15 per million Btu. In contrast a farm laborer earning $3.00 per hour and producing 500 Btu per hour of work costs $6000 per million Btu (11). The economic relationships necessary to justify such substitution requires that human labor must be valued at nearly zero or the price of fossil fuels increased by a multiple of 400. The increase in energy efficiency is unequivocal in Steinhart's observation that "Hand application of pesticides requires more labor than machine or aircraft application, but the reduction of energy for application is from 18,000 kcal/acre to 300 kcal/acre" (5). This information, however, is not an adequate basis on which to recommend substitution of labor for equipment because it ignores the costs of hand spraying, the health dangers, and the magnitude of the task when spraying several hundred acres of corn with a knapsack sprayer.

In other examples, the recommended substitution not only ignores the relative direct costs of the inputs, but the changes in the total system and the increased costs that these changes imply. Examine more carefully the conclusion that "more than 1.1 million kcal per acre could be saved by substituting manure for manufactured fertilizer" (5). The manure required to provide this rate of application is approximately 10 tons per acre, and would require 4 head of yearling steers on feed for 180 days for each acre of crop production. During the feeding period these cattle would consume 162 bushels of corn or essentially all the production obtained from the acre fertilized (12). Since livestock and grain production are specialized among regions and among forms within regions the cost of application (in dollars and in energy) must include transportation and storage.[7] Heichel (13) states that Connecticut farmers can afford to transport dairy manure only two miles at present price relationships. The increased cost of dispersion of livestock feedlots makes it highly improbable that animal wastes will become a significant source of fertilizer in the future.

Crop rotation can be used to reduce requirements for commercial fertilizer, but the total grain production cannot be maintained without an increase in land in crops. The use of the simple corn, small grain, legume rotation as a substitute for continuous corn would require 3 acres of land to maintain every 1 acre of corn. The total supply of tillable land in Illinois has remained fairly constant over the past decade, at about 30 million acres. In 1974 about 10 million acres were planted in corn. The use of the 3-year rotation would require another 10 million in small grain and 10 million in legumes to support current levels of production. Assuming that all the present small grain, hay and rotation pastures would be worked into the rotation scheme, Illinois would require an additional 15.1 million acres just to maintain the same total corn acreage. Differences in productivity would still reduce the total supply of corn from present levels. An expansion of the magnitude required for this type of substitution would require expensive and extensive land reclamation, clearing, drainage, and irrigation—clearly not an alternative for saving energy.

It has been suggested that the substitution of mechanization for animal power in the United States between 1940 and 1973 has reduced the efficiency of agriculture and conversely a return to draft animals would solve some of our energy shortage. Pimentel estimates that the caloric output ratio of corn in Mexican agriculture using bullocks for power is 3.4 kcal per kcal of energy input. In contrast he shows mechanized agriculture in the United States produces only 2.5 kcal of corn for each kcal of energy input (14). The amount of feed required to maintain the draft animals is indicated by USDA statistics showing that 42 million acres of crop land was used for feed for draft animals in 1940 (15). This feed requirement was necessary in addition to fossil fuel used in the 1.5 million tractors (16) on farms in 1940. Even at a modest corn yield of 50 bushels per acre, over 2 billion bushels of corn would be withdrawn from food production in order to provide the hay and oats needed for animal power. Such a substitution among inputs is obviously unacceptable on any criteria except minimum fuel use. The emphasis on crop yield per acre or per energy unit detracts from the important goal of increased total food production.

There are opportunities for substituting replaceable forms of energy for the scarce fossil fuels. The most rapid response is obtained from changes in price relationships. Increased prices of fertilizer lower the relative profitability of corn compared to soybeans. Additional expenditures on engine tune up are substituted for diesel fuel as the price of fuel increases. The lower cost of diesel fuel compared to gasoline provides much of the incentive for the rapid increase in the number of diesel tractors with the accompanying increase in energy efficiency. Higher prices for grain-drying fuel has also encouraged farmers to delay corn harvest to decrease the fuel needed for drying despite increased field losses and delays in fall plowing. Additional investments in building design and insulation can be substituted for fossil fuel for heating or cooling livestock buildings. These are only a few examples of the many opportunities for substitution among inputs to reduce the consumption of fossil fuels. All are dependent on an economic stimulus or at least the removal of a negative stimulus for their adoption. Present price relationships encourage continuation of the trend toward a more energy-intensive agriculture and the inescapable result of lower energy efficiency.

Development of New Energy Sources

The idea of substitution among existing inputs can be extrapolated to inputs and technologies not yet known or available. Given developmental research, several substitutes for fossil fuel in agriculture exist. Solar energy for heating livestock buildings (17) and for grain drying (18) are already in limited use and will undoubtedly expand in the next decade. Use of plants with the ability to utilize nitrogen from the air and transfer it to the roots and soil can reduce fertilizer requirements. There is some evidence that this ability is not limited to legumes and future research may find ways of genetically incorporating the ability into major crops in addition to soybeans. Yield research also holds promise for improved energy utilization. An increase in corn yield through genetics has only a small effect on total energy required to grow the crop. Genetic control of the chemical composition of plants such as corn is a proven capability although changes are not independent of yields and costs. High lysine corn and Triticale already are offering promise of producing greater nutrition per acre. Short season hybrids will mature at lower moisture contents, reducing fuel for drying. Since the total yield of corn per acre is a function of the length of the total growing season this requires a trade-off between total production and drying fuel.

The use of crop residues as a source of fuel or energy holds increased potential as price relationships become more favorable to the exploitation of less readily available sources. The energy available from the refuse (cobs and stalks) of a 130-bushel corn crop is nearly 70 million Btu/acre. If all of this energy could be used in some form of heat engine, with an efficiency of even 2% it would provide the mechanical power of about 1.4 million Btu required to produce the crop. However, the materials handling and storing problems are significant.

The efficiency of plants in converting sunlight into food is quite low. Although incident radiation energy on Illinois during a year averages 1250 Btu per square foot per day, a 130 bushel per acre corn crop extracts only .6% of this energy. At the peak of the growing season photosynthetic efficiency is only about 3-4%. Research is needed to develop techniques for capturing a higher proportion of the tremendous quantities of energy delivered by the sun and converting this into food (19).

Changes in Consumption Habits

The calorie content of an average diet in the United States has been estimated to be nearly 50% greater than the calories consumed in the less developed countries. Even worse, the 3000-calorie diet in the U.S. requires another 3000 calories to produce the high proportion of meat. These widely publicized comparisons have encouraged numerous suggestions for sharing U.S. food supplies with needy nations. The emphasis on meatless diets, increased grain shipments overseas, and selective export restrictions are inadequate solutions to the short-run problem of food shortages because they ignore the difficulty of transportation and distribution of large quantities of food grains; they ignore the effect that a drastic decline in meat consumption would have on the economic incentive for production of grain; they incorrectly assume that the corn not fed to livestock would be readily consumed by people who have eaten rice or wheat for generations; they consider only calories as a measure of nutrition ignoring the fact that consumption of U.S. corn would not solve protein deficiencies.

In the longer run, a recommended shift in U.S. consumer diets has two additional fallacies. 1) The population increase in response to improved diets (lower child mortality rates, longer life span, and increased fertility rates) will quickly reverse any short-run gains in the per capita food supplies and 2) current food production is only a minute part of the total question of energy allocation. Food cannot be separated from other choices of energy use. Forcing a choice between meat and grain consumption while ignoring other important uses of energy ignores the relationship between food and fuel. Adequate food supplies are possible on a world-wide basis if sufficient fossil fuels are provided to use the available land and solar energy. Energy—not land—is the limiting factor in food production. Current world food shortages could be eliminated if present supplies of fossil fuels were used for food production rather than for production of other consumer goods.

Conservation of fossil fuels in transportation, in households, or in industry provides the same potential for increased grain supplies for less developed countries as a reduction in beef consumption. In addition, the transportation and distribution costs of fertilizer and fossil fuels would be less than for equivalent yield of grain. Reduction of meat consumption has a popular appeal, and it provides an element of self sacrifice among an over-fed population but it is not the most effective approach to a goal of relieving widespread famine on either a long-run or short-run basis. The solution lies in an economically rational use of our total energy supply and an allocation of this supply among all choices available to consumers.

THE ROLE OF CONSUMER SOVEREIGNTY IN ENERGY USE

Economists have long debated the issue of consumer sovereignty. For the most part it is still unresolved whether the consumer's willingness to purchase stimulates production, or if the manufacturer's efforts to move his supply of goods results in manipulation of the consumer's willingness to purchase. Recent consumerism activity suggests that the consumer still has considerable "free will" in his decisions and he is not adverse to making his wishes known through boycotts or shifts in purchase patterns.

As a basic philosophy the market system has allowed consumers to choose between beer and coke, chicken and pork, and cake mixes and "bake it from scratch." Recommendations that agriculture shift from products with a high ratio of energy input per calorie, to products with a low energy input, either visualizes a new social order in which consumer preferences are legislated or incorrectly assumes that consumer preferences do not influence the allocation of resources into their highest value use.

If the intent is to legislate consumer preferences in order to reduce energy consumption, then careful thought should be given as to whose value judgments are to provide the basis for the "acceptable" market basket of goods.

The criterion on which certain food items such as meat have been selected for elimination in the interest of energy conservation is not clear, As shown in Table 1, there are several foods whose ratios of energy output per unit of input are lower than poultry and not significantly different from beef—especially when one considers that some of the energy input to beef is not suitable for direct human consumption.

Table 1. Ratios of crop calorie content to fuel and electrical energy input in crop production.*

Commodity	Ratio
Barley	6.609
Corn	3.250
Wheat Flour	5.363
Green Beans (raw)	.545
Green Beans (canned)	.288
Strawberries (raw)	.461
Broccoli (raw)	.246
Tomatoes (canned)	.167
Cauliflower (frozen)	.123

* Values for all commodities include energy requirements up through the first stage of processing. A more thorough list of these commodity ratios is contained in (20), Table 88, p. 113.

On the basis of caloric efficiency, corn production under mechanized agriculture ranks very low. However, a comparison with other food products quickly demonstrates that calories produced per unit of energy in unprocessed green beans are much lower and frozen cauliflower is even less "efficient" than hogs (20). The comparisons of Table 2, showing total energy consumed in processing and manufacturing, suggest other candidates where reduced consumption would have as great an effect on energy consumed with less effect on nutritional levels. For example the processing of sugar utilizes almost as much energy as the processing of meat. Coffee and cigarettes, offering no nutritional value, together require almost as much energy as the entire frozen fruits and vegetable industry. Beverages utilize more energy than the manufacture of agricultural chemicals and farm machinery combined. Before central planners can optimally allocate energy resources among alternative uses in agricultural production, they must develop allocation criteria that encompass more than just calories and allow for a range of con-

Table 2. Total energy used for heat and power in processing and manufacturing of selected agricultural products (millions of kilowatt equivalent).

Commodity	Energy Use
Meat products	32,308
Frozen fruits and vegetables	7,657
Sugar	31,899
Beverages	29,151
Roasted coffee	2,511
Cigarettes	3,335
Agricultural chemicals	17,617
Farm machinery	11,320

Source: (30)

sumer preferences broader than the traditional definition of food. To restrict the use of fuel for drying grains while permitting the production and drying of tobacco is inconsistent with a goal of maximum nutrition and health, or caloric efficiency. Returning to field drying of corn could save the equivalent of 476 million gallons of L.P. gas in United States agriculture[8] but, depending on weather conditions, could result in as much as a 10% reduction in bushels harvested. Considerable fuel is also used in drying other crops including 20 million gallons for soybeans, 11 million for rice, 10 million for peanuts, and 7 million for grain sorghum (21). However, 121.5 gallons of L.P. gas are used for drying tobacco.[9] Whose value judgments are to be used in determining that feed and food should be reduced in order to save the energy used to dry grain while continuing to use more fuel in drying tobacco than for rice, peanuts, sorghum, and soybeans combined?

The soft drink and beer industries use an estimated 66 billion kWh equivalent of energy just in producing containers, assuming ½ of the containers are returnable (22). In addition the beverage industry as classified by the U.S. Census of Manufacturers uses 29 billion kWh equivalent in manufacturing (13), while the distillers alone consume approximately 40 million bushels of grain per year. Reallocation of energy in food production must include non-food alternatives as well.

Steinhart and Steinhart (5) raise questions as to the appropriateness of frost-free freezers and kitchen appliances in an energy conscious society. It is not clear why these are less desirable than TV's and electric golf carts. These examples are given only to illustrate the danger of legislating consumer preferences. Necessities are often defined to be the products I buy—luxuries are the products that *other people* buy. This is especially true when we recognize that food choice cannot be made independently of all uses of land and energy. For example, good agricultural land is often diverted to recreational purposes or industrial use. Fertilizer—the basic ingredient for increasing food supplies in the U.S. and the world—is in short supply due to lack of natural gas. At the same time millions of gallons of gas are used to dry our tobacco crop, to manufacture alcoholic beverages, etc. We cannot simply choose between beef and bread. We must extend this choice to "instant on" T.V. units, electric golf carts, recreational vehicles and every use that is made of land, labor, and fossil fuels. They all compete for energy supplies capable of providing food to alleviate starvation no less than the beef animal. There is no satisfactory system for allocating products among people, and resources to the production of different products, except a market price—permitting people to express their preferences by the dollars they are willing to spend. A rationing system that substitutes coupons for dollars reflects dissatisfaction with distribution of wealth—it does not indicate a failure of the market system.

An extensive search of economic literature and observation of numerous market transactions has failed to identify a single instance of a market for calories. Without a market or a price, analysis of optimum combination of inputs and outputs to maximize calories out per calories in becomes little more than an academic exercise. The important decisions have to do with pricing of resources in such a way as to reflect future value as well as present value of those resources in alternative uses including conservation for future consumption.

CRITERIA FOR ECONOMIC POLICIES

The public policy choices affecting food and energy are complex and second and third order effects must be included in the analysis of alternatives. A carefully developed set of criteria are needed against which to measure each. Some of the criteria suggested by the relationships discussed in this paper are listed below as questions to be answered for any proposed solution to the food and energy problem.

1. What will be the effect on total food supply in the short and long run?

2. What mechanism will allocate energy sources among alternative end use products?

3. How will consumer preferences for alternative kinds of food and for all consumer goods influence the allocation?

4. What will be the total cost of implementing the change?

5. How will priorities for energy use be established?

6. What will be the economic and social effects on different groups and countries?

7. How can each individual producer be motivated to use the available energy in the production of goods most valued by society?

FOOTNOTES

1. The term "efficiency" is subject to many interpretations. To avoid confusion, the term "energy efficiency" will be used in this paper to refer to the ratio between food output and fossil fuel input. Inputs and outputs will be measured either in calories or Btu. "Economic efficiency" will be used to designate the degree to which the 2nd order marginal conditions of economics have been fulfilled.

2. These returns are the Landlords' returns to capital and management on 260- to 339-acre farms as calculated from Illinois Farm Business Farm Management records for 1964 (23).

3. These returns are the Landlord's returns to capital and management on farms over 660 acres as calculated from Illinois Farm Business Farm Management records for 1971 (24).

4. These MVP values were calculated from table 37, as reported in (15).
5. A yield increase of 12 bushel for a 25-pound increment of nitrogen was obtained at an average yield of 104 bushel per acre on Central Illinois soils. This yield response varies with soil type and level of fertilizer application (25).
6. Pimentel uses a nitrogen response curve for corn to illustrate a potential increase in efficiency from a reduction of nitrogen application rates. The curve is a simple factor-product diagram of APP. Obviously, without prices no conclusion as to optimum economic levels can be drawn (26).
7. Heavy applications during periods when the ground is frozen results in runoff that would exceed pollution standards and losses of nutrients of about 7% of nitrogen, 3.9% of phosphorus, and 23.1% of the potash (27). Application during the growing season is limited by crop development and requires storage during the remainder of the year. Storage losses with a liquid storage and spreading system have been estimated at 70-80% for N, 50% for P_2O_5 and 30% for K_2O (28).
8. Calculated on the basis of 1974 production of 4.6 billion bushel, 67% of production artificially dried and 6.5 bushels dried per gallon of L.P. gas.
9. Fuel for drying tobacco was estimated on data published by USDA (21) and 1974 production (29). The Task Force report by CAST (21) estimated a much greater volume of tobacco artificially dried.

REFERENCES

1. U.S. Department of Agriculture. "Costs and Returns on Commercial Farms, 1930-1957," ERS Statistical Bulletin No. 297, 1961.
2. U.S. Department of Agriculture. "Farms Costs and Returns," ERS Agriculture Information Bulletin No. 230, Revised 1968.
3. Ruttan, V.W. "Food Production and the Energy Crisis: A Comment," submitted for publication in *Science.*
4. Pimentel, D., L. E. Hurd, A. C. Bellotti, M. J. Forster, I. N. Oka, O. D. Sholes and R. J. Whitman, "Food Production and the Energy Crisis," *Science,* Vol. 182, November 2, 1973.
5. Steinhart, J. S. and C. E. Steinhart. "Energy Use in the U.S. Food System," in Energy and Agriculture: Research Implications, Report No. 2, NCRS Committee on Natural Resource Development, October 1973.
6. Perelman, M. "Mechanization and the Division of Labor in Agriculture," *Am. J. Agr. Econ.,* Vol. 55, No. 3, August 1973.
7. Science and Government Report. Newsletter, April 1, 1973, Washington, D.C.
8. Hoeft, R. G. and J. C. Siemens. "Energy Consumption and Return from Using Nitrogen Fertilizer on Corn," unpublished paper, University of Illinois 1974.
9. Stigler, G. "The Cost of Subsistence," *JFE,* Vol. 27, No. 2, May 1945.
10. Williams, S. W., S. Strong, and C. B. Baker. "Benefits Stem from Research on Corn Production," *Illinois Research,* University of Illinois, Agricultural Experiment Station, Vol. 15, No. 3, 1973.
11. Cook, E. "Saving the Environment Goat and the Energy Cabbage" in *Exxon* Magazine, Spring 1974. (Dr. Cook is Dean of the College of Geosciences, Texas A & M University.)
12. VanArsdall, R. N. "Resource Requirements, Investments, Costs and Expected Returns from Selected Beef Feeding and Beef Raising Enterprises," AE 4075, Illinois Agricultural Experiment Station, 1965.
13. Heichel, G. H. "Auxiliary Energy Requirements and Food Energy Yields of Selected Food Groups," unpublished paper delivered at the AAAS Symposium, Energy and Agriculture, San Francisco, California, February 1974.
14. Pimentel, D., W. R. Lynn, W. K. MacReynolds, M. T. Hewes, S. Rush. "Workshop on Research Methodologies for Studies of Energy, Food, Man

and Environment, Phase I," Cornell University, Center for Environmental Quality Management, June 18-20, 1974.
15. U.S. Department of Agriculture. "The U.S. Food and Fiber Sector: Energy Use and Outlook" ERS, U.S. Senate Document 38-906, September 20, 1974.
16. U.S. Department of Agriculture. "Changes in Farm Production and Efficiency, A Summary Report," Farm Economics Division, Statistical Bulletin No. 233, August 1958.
17. Hall, M. D. "Solar Heated Ventilation System for Confinement Livestock Buildings," Farm Structures Day Proceedings, Agricultural Engineering Department, University of Illinois, November 18, 1965.
18. Peterson, W. H. "Solar Heat for Crop Drying," in Grain Conditioning Conference Proceedings, Agricultural Engineering Department, University of Illinois, January 15-16, 1975.
19. Hunt, D. "Energy and Agriculture," unpublished paper, Department of Agricultural Engineering, University of Illinois, 1974.
20. Cervinko, V., W. J. Chancellor, R. J. Coffelt, R. G. Curley, and J. B. Dobie. "Energy Requirements for Agriculture in California," California Department of Food and Agriculture in California, Davis, January 1974.
21. Council for Agricultural Science and Technology, *Potential for Energy Conservation in Agricultural Production*, unpublished Task Force paper prepared January 24, 1975, D. R. Price, Department of Agricultural Engineering, Cornell University, Chairman.
22. Hannon, B. "System Energy and Recycling—A Study of the Beverage Industry," CAC Document No. 23, University of Illinois, January 5, 1972.
23. Reiss, F. J. "Landlord and Tenant Shares 1964," Department of Agricultural Economics, Agricultural Experiment Station, University of Illinois, AERR 79, November 1965.
24. Reiss, F. J. "Landlord and Tenant Shares 1971," Department of Agricultural Economics, Agricultural Experiment Station, University of Illinois, AERR 120, November 1972.
25. Taylor, C. R. and E. R. Swanson. "Experimental Nitrogen Response Functions, Actual Farm Experience and Policy Analysis," *Illinois Agricultural Economics*, Vol. 13, No. 2, University of Illinois, July 1973.
26. Pimentel, D. "Energy Crisis and Crop Production," in Energy and Agriculture: Research Implications, Report No. 2, NCRS Committee on Natural Resource Development, October 1973.
27. Miner, J. R. "Farm Animal-Waste Management," NCR Publication 206, Iowa State University of Science and Technology, Ames, Iowa, Special Report 67, May 1971.
28. Vanderholm, D. "Land Application of Manure from Modern Livestock Production Facilities," Soil and Water Conservation No. 14, Department of Agricultural Engineering, University of Illinois, December 1974.
29. U.S. Department of Agriculture. "Tobacco Situation," ERS, September 1974.
30. U.S. Department of Commerce. Survey of Current Business, Census of Manufacturers, "Fuels and Electric Energy Consumed," June 1971.

9.

Machinery Energy Requirements for Crop Production in Delaware

N. E. Collins, T. H. Williams and L. J. Kemble*

With the demand for farm commodities increasing and the availability of willing labor decreasing, agricultural production has been increased through mechanization. Mechanization has resulted in field production practices using large, hi-speed equipment. As size and speed increase, so do energy requirements. The energy shortage caused by an increasing demand for fuel and the oil boycott necessitated that the amount of machine energy required for crop production be reviewed.

A preliminary investigation of machine energy was conducted during the 1973 growing season. The tractors and implements (see Table 1) used in the study were those available at the Delaware Agricultural Experimental Station (Newark) and the Substation (Georgetown). As a result, the implements and tractors were not matched for maximum efficiency. Since the experimental plots were small in relation to actual fields, the field efficiencies were estimated (1).

As an outgrowth of the preliminary investigation, a project entitled "Implement and Power Unit Energy Requirements for Agricultural Production in Delaware and Coastal Plains Soils" has evolved. The objectives of the project are:

* Assistant Professor, Extension Agricultural Engineer and Senior Field Technician, respectively, Agricultural Engineering Department, University of Delaware, Newark, Delaware 19711.

Table 1. Description of tractors and implements used in preliminary investigation (1973).

	Georgetown	Newark
Tractor	IH 656 gas, gear	IH 656 gas, hydrostatic
Moldboard Plow	3-14″ trail type	3-14″ trail type
Chisel Plow	7 chisels on 12″ spacing	6 chisels on 12″ spacing
Disc	12 foot	10.5 foot
Spring Tooth	12 foot	16.3 foot
Pulvimulcher	—	10.3 foot
Planter*	3-30″ rows, AC 600	3-30″ rows, AC 600

* Dry fertilizer, insecticide attachments and No-til fluted coulters.

1. To quantify basic energy requirement parameters for agricultural implements on coastal plain soils.

2. To determine soil strength relationships and tractive efficiency of power units on coastal plain soils.

3. To correlate energy requirements of implements and tractive requirements and capabilities of power units to evaluate management practices for agricultural production.

The collection of additional field data was justified for several reasons. First, soil parameters had not been measured when most of the information regarding traction devices was obtained (2). Second, the effect of specific parameters on certain tools is frequently determined under laboratory conditions (3-6). Third, soil types, climatic conditions and management practices in Delaware differ from those areas for which data is available (7). Fourth, no relationship exists which reasonably predicts implement draft and traction coefficients based on soil strength. Fifth, field data was needed for some alternative tillage operations in order to evaluate management practices as to labor and energy requirements.

PRELIMINARY INVESTIGATION

The horsepower and fuel consumption measurements were made by calibrating a vacuum gage on the intake manifold. The calibration curves were developed using a PTO dynamometer as the loading device. Tractor horsepower and fuel requirements for a particular operation were then determined by noting the average vacuum under load and referring to the calibration curves. Ground speed was determined by measuring the time required to travel a given distance. The draft of pull type implements was determined with a drawbar dynamometer.

The tractor and implement data for the preliminary investigation are shown in Tables 2 and 3. As expected, draft and power requirements for specific operations were influenced by the soil type. When primary tillage practices were considered, chisel plowing required less energy per acre (see Table 4) than did moldboard plowing. It

Table 2. Field machinery data for corn production on Evesboro loamy sand, Georgetown, Delaware 1973.

Quantity	Operation							
	moldboard plow	chisel plow	1st disking moldboard grd.	1st disking chiseled grd.	springtooth harrow	planter	planter with 2 ammonia chisels	anhydrous ammonia applicator (5 chisels)
Indicated speed (mph)	5.1	3.5	4.8	5.3	—	—	—	5.0
Actual speed (mph)	4.70	3.09	4.14	4.55	5.14	4.13	4.01	4.80
Percent slip	8.50	13.30	14.70	15.4	—	—	—	4.00
Tillage depth (inches)	10.0	12.0	8.0	8.0	4.5	—	10.0	10.0
Machine width (ft)	3.5	7.5	12.0	12.0	12.0	7.5	7.5	7.5
Draft (lb)	2500	4200	—	—	1200	900	1500	—
Unit draft (lb/ft)	714	560	—	—	100	—	—	—
Drawbar HP	31.3	34.6	—	—	16.5	9.9	16.0	—
PTO HP	54.29	49.84	60.97	55.63	29.82	20.92	32.04	38.27
Fuel Consumption (gph)	4.70	4.53	4.97	4.77	3.47	3.00	3.60	3.95
Field efficiency	0.8	0.8	0.8	0.8	0.8	0.65	0.50	0.60
Effective field capacity (acres/hr)	1.60	2.24	4.81	5.29	5.98	2.44	1.82	2.62

also was found that energy required for initial secondary tillage operation (disking) was influenced by the primary tillage tool used. On the loamy sand, the energy required for disking was less after chiseling. On the heavier soil, however, disking after the chisel plow consumed more energy than was required after the moldboard plow. Energy requirements did not appear to be influenced by the type of primary tillage after the first disking.

When primary tillage is required, a farmer can realize a savings in energy (Table 4), fuel (Table 5) and labor (Table 6) by using the chisel plow. This advantage, however, appears to be a function of local conditions as the data for the silt loam soil indicates.

The use of no tillage practices will result in a large reduction of on-farm energy (Table 4) and fuel (Table 5) use where conditions permit. As expected, no tillage and its reduced number of operations required the minimum amount of labor (Table 6) for the systems studied. In considering the planting operation, however, no-tillage

Table 3. Field machinery data for corn production on Matapeake silt loam, Newark, Delaware 1973.

Quantity	moldboard plow	chisel plow	1st disking moldboard grd.	1st disking chiseled grd.	2nd disking	springtooth harrow	pulvimulcher	planter	spraying residual herbicide	spraying contact herbicide
Indicated speed (mph)	5.0	3.0	4.5	4.2	4.5	4.8	5.3	3.2	3	2
Actual speed (mph)	4.55	2.27	3.86	3.57	4.09	4.29	4.93	2.90	—	—
Percent slip	9.9	32	16.5	18.4	9.0	10.4	6.0	5.0	—	—
Tillage depth (mph)	8.5	9.0	4.0	4.0	—	4.5	2.0	—	—	—
Machine width (ft)	3.5	7.5	10.5	10.5	10.5	16.3	10.3	7.5	30	30
Draft (lb)	2200	3500	1350	2250	1175	1980	928	1230	—	—
Unit draft (lb/ft)	686	493	128	214	112	121	90	164	—	—
Drawbar HP	26.7	21.2	13.9	21.4	12.8	22.6	12.2	9.5	—	—
PTO HP	53.40	48.06	45.39	51.62	36.05	42.72	40.05	23.14	17.36	13.35
Fuel Consumption (gph)	5.23	4.95	4.75	5.20	4.27	4.65	4.50	3.70	3.25	2.90
Field Efficiency	0.8	0.8	0.8	0.8	0.8	0.8	0.8	0.65	0.6	0.6
Effective field capacity (acres/hr)	1.51	1.65	3.93	3.63	4.16	6.78	4.92	1.70	6.54	4.36

Table 4. Energy requirements for corn production.

	Evesboro loamy sand Georgetown, Del.			Matapeake silt loam Newark, Del.		
	Moldboard Plow	Chisel Plow	No Tillage	Moldboard Plow	Chisel Plow	No Tillage
Operation	Horsepower Hours Per Acre					
Moldboard plow	33.93	—	—	35.36	—	—
Chisel plow	—	22.25	—	—	29.13	—
Disk	12.68	10.52	—	11.55	14.22	—
Disk	—	—	—	8.67	8.67	—
Springtooth	4.99	4.99	—	6.30	6.30	—
Pulvimulch	—	—	—	8.14	8.14	—
Plant	8.57	8.57	8.57	13.61	13.61	13.61
Spray	—	—	—	2.65	2.65	3.06
TOTAL:	60.17	46.33	8.57	86.28	82.72	16.67

Table 5. Fuel requirements for corn production.

Operation	Evesboro loamy sand Georgetown, Del.			Matapeake silt loam Newark, Del.		
	Moldboard Plow	Chisel Plow	No Tillage	Moldboard Plow	Chisel Plow	No Tillage
	Gallons of Gasoline Per Acre					
Moldboard plow	3.0	—	—	3.5	—	—
Chisel plow	—	2.0	—	—	3.0	—
Disk	1.0	0.9	—	1.2	1.5	—
Disk	—	—	—	1.0	1.0	—
Springtooth	0.6	0.6	—	0.7	0.7	—
Pulvimulch	—	—	—	0.9	0.9	—
Plant	1.0	1.0	1.0	1.9	1.9	1.9
Spray	—	—	—	0.4	0.4	0.6
TOTAL:	5.6	4.5	1.0	9.6	9.4	2.5

Table 6. Labor requirements for corn production.

Operation	Evesboro loamy sand Georgetown, Del.			Matapeake silt loam Newark, Del.		
	Moldboard Plow	Chisel Plow	No Tillage	Moldboard Plow	Chisel Plow	No Tillage
	Manhours per Acre					
Moldboard plow	0.63	—	—	0.65	—	—
Chisel plow	—	0.44	—	—	0.61	—
Disk	0.21	0.19	—	0.25	0.28	—
Disk	—	—	—	0.24	0.24	—
Springtooth	0.17	0.17	—	0.15	0.15	—
Pulvimulch	—	—	—	0.20	0.20	—
Plant	0.41	0.41	0.41	0.59	0.59	0.59
Spray	—	—	—	0.15	0.15	0.23
TOTAL:	1.42	1.21	0.41	2.23	2.22	0.82

does not appear to offer any advantage in labor savings. Additional acreage with no-tillage will require a second planter or the acceptance of a timeliness penalty for early or late planting.

In summary, the no-tillage system took only 1/7 to 1/5 the horsepower hours, 1/6 to 1/4 the fuel and 1/3 the labor required for the conventional tillage systems.

NEW PROJECT

A 97 PTO horsepower diesel tractor with dual-wheel option has been instrumented so that fuel consumption, draft forces and wheel slippage can be recorded on an oscillograph for this investigation. The instrument package also can monitor implement depth and forces on selected components. During field operation, soil moisture and density samples are taken. In addition, the cone index is measured using a penetrometer (8). The penetrometer is operated by a hydraulic cylinder mounted on the front of a tractor. Output from strain gages on the penetrometer shaft is recorded on the oscillograph.

The collection of field data with the fully instrumented tractor started in the fall of 1974. Tractor and machinery data for two operations, disking corn stalks and moldboard plowing, have been collected on a Sassafras sandy loam soil. The tractor and machinery data for these operations are shown in Tables 7 and 8. Typical cone index and moisture profiles are shown in Figure 1. Due to adverse

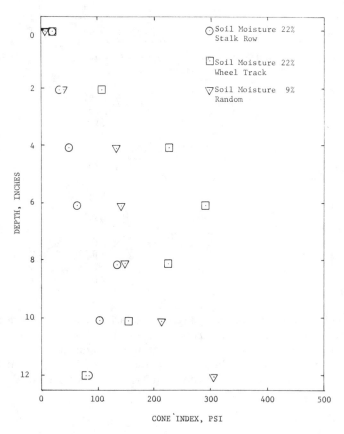

Figure 1. The cone index profile for Sassafras sandy loam with surface cover of corn stalks.

Table 7. Tractor and machinery data for disking* a Sassafras sandy loam corn stalks. Soil moisture (W.B.) 9-11%.

Disk Gang Angle	Speed (MPH)	Draft (#)	Depth (in.)	Unit Draft (PSI)	Machine HP	PTO HP	TTE **	Gallons per Hour	Avg. Rear Wheel Slip %
15°	3.2	2704	1.97	5.84	23.00	81.60	.28	4.57	.12
	4.3	2856	2.81	4.32	32.00	94.04	.34	5.67	.12
	5.5	2856	2.56	4.75	41.0	94.04	.44	6.69	.12
	5.0	2588	2.70	4.08	34.0	75.63	.45	5.22	.145
	4.6	1606	1.70	4.02	19.00	49.88	.38	4.90	.08
	5.8	2142	1.65	5.52	33.00	81.81	.40	5.86	.10
	7.6	2767	1.60	7.36	56.00	93.28	.60	7.40	.12
18°	4.4	2410	3.08	3.33	28.00	57.86	.48	5.42	.11
	5.9	2410	2.75	3.73	37.00	81.81	.45	6.83	.11
	6.4	2677	2.63	4.33	45.00	86.79	.52	6.44	.13
	5.1	2677	2.70	4.22	36.00	74.82	.48	5.22	.14
20°	2.8	3882	4.28	3.86	29.00	93.55	.31	6.02	.245
	3.9	4150	3.62	4.88	42.00	94.04	.45	7.28	.21
	4.6	3949	3.36	5.00	48.00	89.07	.54	7.10	.19
	4.1	3213	3.60	3.80	35.00	81.81	.43	5.79	.17
	5.3	3748	3.34	4.78	52.00	81.81	.64	7.40	.17

* 21-ft transportable disc with wings manufactured by John Deere.
** TTE = Machine Horsepower/PTO Horsepower.

Table 8. Tractor and machinery data for plowing* a Sassafras sandy loam with corn stalks. Soil Moisture (W.B.) 22%.

Surface Condition	Speed (MPH)	Draft (#)*	Depth (in.)	Unit Draft (PSI)	Machine HP	PTO HP	TTE **	Gallons per Hour	Ave. Rear Wheel Slip %
Disked	4.4	4750	8.25	7.20	55.00	89.17	.62	6.15	.105
	5.0	5333	9.00	7.41	70.00	77.28	.91	7.48	.085
	4.3	3979	9.00	5.53	45.00	77.28	.58	6.10	.07
	3.5	3573	8.25	5.41	33.00	89.17	.37	5.30	.105
	2.7	3410	8.25	5.17	24.00	77.28	.31	4.06	.075
	4.6	4547	8.25	6.89	56.00	89.17	.63	6.30	.08
	5.5	4609	8.00	7.20	67.00	93.62	.72	7.32	.065
	1.8	3086	8.00	4.82	14.00	65.39	.21	3.29	.065
Undisked	4.9	4249	9.00	5.90	55.00	93.13	.59	7.40	.125
	4.2	5160	9.00	7.17	58.00	57.46	1.01	6.53	.105
	4.1	5115	9.00	7.10	55.00	93.62	.59	6.02	.095
	3.5	3979	9.00	5.53	37.00	81.24	.46	5.42	.095
	2.7	3573	9.25	4.83	26.00	77.28	.34	4.19	.07
	4.6	3654	8.25	5.54	45.00	89.17	.50	6.30	.085
	5.1	4628	8.00	7.23	63.00	92.14	.68	6.90	.105
	1.8	3573	9.00	4.96	16.00	77.28	.21	3.29	.06

* 5-16"s manufactured by Oliver.
** TTE = Machine Horsepower/PTO Horsepower.

weather conditions, no additional data have been collected this spring as of March 28.

With only limited data available, conclusions would be premature. However, a couple of comments appear in order. First, the ratio of machine horsepower to power take-off horsepower has been lower than expected. The ratio appears to have a maximum value between 0.6 and 0.7 which is lower than expected on a firm soil (9). Second, the unit draft required to pull the disc increases while the depth of operation decreases with speed. The interaction, for this soil, does not result in a constant power requirement (10). Third, the unit draft for the moldboard plow increases with speed. As can be seen in Figure 2, the speed effect for the Sassafras sandy loam appears to be similar to that published by ASAE (9) for the general soil classification. The difference in magnitude may be due to soil moisture content.

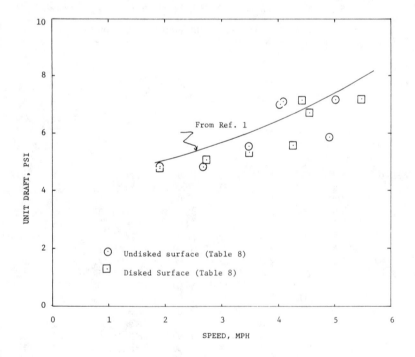

Figure 2. A comparison of unit draft for moldboard plow in Sassafras sandy loam with published data (1).

ACKNOWLEDGMENT

The authors express their thanks to Vance Morris and Sons of Bowers Beach, Delaware for providing both land and equipment used to collect a portion of data presented in this paper.

This paper is published with the approval of the Director of the Agricultural Experiment Station as Miscellaneous Paper No. 271.

REFERENCES

1. Smith, W., W. Frisby, and E. Constien. Machinery Management I — Field Machinery Capacity. University of Missouri. 1969.
2. Barger, E. L., J. B. Liljedahl, W. M. Carleton, and E. G. McKibben. Tractors and Their Power Units. John Wiley and Sons. p. 283. 1963.
3. Gordon, E. D. Physical Reactions of Soil on Plow Disks. Agricultural Engineering. 22:205-308. 1941.
4. Telischi, B., H. E. McColley, and E. Erickson. Draft Measurements for Tillage Tools. Agricultural Engineering. 37(9):605-608, 617. 1956.
5. Wang, J. K., K. Lo, and T. Liang. Predicting Tillage Tool Draft Using Four Soil Parameters. Trans. ASAE. 15(1):19-23. 1972.
6. Wang, J. K., and K. Lo. Predicting Moldboard Plow Draft in Different Processed Soils. Trans. ASAE. 16(5)851-854. 1973.
7. Promersberger, W. J. and G. L. Pratt. Power Requirements of Tillage Implements. North Dakota Experiment Station Bulletin No. 415 (Tech) 1958.
8. Waterways Experiment Station, Vicksburg, Miss. Trafficability of Soil-Development of Testing Methods. Tech. Memo 3-240. Third Supply. 1948.
9. American Society of Agricultural Engineers, Agricultural Engineers Yearbook, Section D 230.2, 1974.
10. Hunt, D. Farm Power and Machinery Management. Sixth Edition. Iowa State University Press. Ames, Iowa. 1973.

10.

The Real and Fictitious Economics of Agriculture and Energy

Michael Perelman*

The basic argument of this chapter follows the approach of David Pimentel and others who have pointed out the energy cost of modern agriculture. Such statistics are calculated by aggregating societal totals; however, individual farmers or businessmen are indifferent to these numbers. Instead their calculations are based on prices. Until now most economists have accepted the assumption that these prices will conform to the requirements of society; that is, the outcome of millions of decisions of firms, each of which is adjusting to their individual expectations of the future actions of the others, will lead to a socially desirable outcome. The laws of supply and demand, it would seem, generate prices as if the market were a giant computer. Like any other computer this price mechanism suffers from the universal weakness of garbage in-garbage out. Besides not knowing the future course of the economy from year to year, there is no means to provide for the long term maintenance of the resource base upon which the economy depends.

Recent literature demonstrates the theoretical difficulty of rationally pricing scarce stocks of natural resources (1, 2), although any analysis of environmental questions is dominated by considerations of price. Even ignoring these theoretical questions, prices give very conflicting signals. For example, while 1970 prices indicated that one Btu equivalent of labor energy was 300 times more expensive than

* Department of Economics, California State University, Chico, California 95926.

one Btu of Gasoline energy (3, 4), other writers, most notably Calvin, argues that agriculture is a viable source of energy production (5). Yet all of these analyses are incomplete. By relying on prices, they ignore most of the complex interactions which occur in the real interdependent world which they purport to study. Questions of environmental damage, threats to human health, and changes in the very structure of society as well as the potential destruction of the resource base upon which life itself depends are all left out of the price equations.

Furthermore, price considerations are made within a context of a given technological structure. The evolution of this technology is not an arbitrary movement; it depends on various socio-political forces. Thus, while alternative modes of production might not be profitable at today's prices, had research been directed toward developing them they might very well be superior to what exists at present. Also, the actual experience in producing and utilizing alternative technologies would have furthered their efficiency through the so-called learning-by-doing effect (6).

The generation of technical change is a matter of great import for students of agriculture. Yet much of the analysis of technical change is restricted to calculating the costs and benefits of technical change in the terms of traditional welfare economics—the wilingness to pay —and the theoretical understanding of the process underlying the creation of new technology is rather primitive.

The theory of induced innovations makes an attempt to explain the origins of technical change in terms of cost minimization. The work of Hayami and Ruttan is an outstanding example of this approach (7). One difficulty with the application of this theory to agriculture concerns the rapidly falling shares of labor in agriculture (8). According to some work on the theory of induced innovation, factor shares should tend to be fixed *ceterus paribus* (9).

Some economists argue on the other hand that the flow of technology is autonomously determined (10). This 'serendipity' theory of technical change is consistent with any pattern of labor shares; however, such intellectual agnosticism offers little in the way of guidance. But the pattern of rapidly dropping labor shares in agriculture seems too consistent to accept without some sort of attempt to explain it.

Looking at particular changes in U.S. agricultural technology does give us a clearer picture of the forces which determine technology. The development of the tomato harvester is a case in point (11). This development, hardly the fortuitous result of an individual brainstorm, is indicative of very highly directed agricultural technology. Moreover, Hightower's analysis suggests that the direction of this technology was not only foreseen in advance, but intentionally created for the benefit of the large corporate farmers (7). Gardner's statistical analysis suggests a similar conclusion (12).

Since profitability is determined by the farmers' ability to adopt the appropriate technology and since each successive innovation seems to imply a large scale of operation, new technology confronts the individual farmer as an imperative: grow or go. However, the corporations which develop much of the technology have good reason to consciously skew the development of their products in order to foster the trend toward larger units since profit margins are highest on the largest size equipment just as in the automotive industry (13, 14); see also the relevant experience of one industry pioneer (15).

As a result, a picture emerges of an optimality of increasingly large farm sizes; however, this determination is made only with respect to profitability. If the measuring rod of money is inaccurate, then the question scale remains open. For example, in light of the so-called energy crisis, the search for new sources of energy is supposed to be a national priority. As mentioned previously, some writers have suggested that photosynthesis offers a potentially efficient technology for harnessing solar energy (16, 17). Yet in terms of energy, U.S. agriculture uses several times more calories than it produces in the form of food (10). In terms of energy, it is highly inefficient.

Of course, energetic efficiency is not totally satisfactory as a guide to economic action; it does not encompass many social, political or environmental considerations which would normally be reflected in one way or another in our value systems. On the other hand, many such values are not accurately reflected in our price system either making calculations solely based on profitability or willingness to pay as open to criticism as those based entirely on energetic efficiency.

Working with prices gives the analysis of technology a peculiar slant. Much of the analysis of technology reads as if it were a panacea that magically produces something from nothing. Griliches' work on hybrid corn is an example of this type of presentation (18). While stressing the increases in quantity of corn produced by hybrid seeds, Griliches neglects to explain that the quality of the corn was reduced and that the new seeds necessitated massive increases in specialized inputs such as chemical fertilizer. Thus, while the new technology did give a farm more returns for a given cost of production, it did NOT give produce of the same output with the same inputs; but rather different amounts of different outputs for different amounts of different inputs in the same sense as if the farm had been converted into a golf course or a factory for producing tennis rackets. In short, comparability is extremely difficult and with rapidly changing prices such attempts at comparability may be quite misleading. Thus, although hybrid technology may be superior from the standpoint of a given price situation in a given environment, one can by no means say that it is absolutely superior. Regarding technical change in a partial equilibrium framework such as is usually done in studies such as Griliches' creates another significant theoretical problem. A new technique might not appear profitable because it requires expensive alterations in previously installed equipment.

This blocking of one sort of technical advance by the adoption of another is a familiar problem in economics although its implication is not as widely recognized; that is, that an efficient program of investment requires economic planning rather than the haphazard functionings of the marketplace.

More importantly, as was suggested before the misconception of technology lured the analysis of technological change away from the mark by forgetting that technological change is a change in the organization of society, as in the case of the example of the tomato harvester, and that the process by which new technologies are developed must be optimal before the trajectory of the economy adapting to new technologies can be judged to be optimal or even acceptable.

How then are the enormous increases in U.S. agricultural yields to be explained if not by technological advancement? Some of the increased productivity can be exchanged by a changing geographical distribution of plantings (19, 20). The weather has also played a major role in increasing yields over the past several decades. One study estimates that as much as 50% of the yield increases since 1950 can be attributed to favorable weather (21, 22, 23). Also part of the increases in yields have been bought by decreases in the protein content (24, 25). But most importantly, the use of purchased inputs on the farm has freed resources to produce marketable crops instead of farm inputs such as hay (10).

While plant breeding and other innovations have brought about some increases in yield, the extent of these increases is unclear and the data at hand at least should make economists hesitant to discuss technical change in agriculture as the simple outward shifting of production possibility curves.

The foregoing does not imply that modern technology does not increase the average product of labor at any period of time above what it would be using a less modern technology. In agriculture, the substitution of fossil fuel for human energy has allowed for an enormous savings in human labor power. However, in this case, any production possibility curve would have to include extra dimensions to account for the depletion of stocks of natural resources thus disallowing the simple picture of outward shifting production possibility curves seen in the textbooks.

The import of this discussion of technical changes is not so much to prove that one form of technology is superior or inferior to another. So long as analysis proceeds in terms of prices, technologies will be adopted which maximize profits; extending the results beyond the firm or industry, even in as simple a study as that of Seckler and Schmitz, leads to very unspecific results as the wide range of estimates of rates of returns to the tomato harvester imply.

Rather a far more thoroughgoing analysis of the implications of changes in technology that can be made in any single study is needed. A few examples of some of the questions that should be answered could include the following:

1. The world is facing a long period of increased goods scarcity, how much emphasis should maximation of yield be given? Since this objective generally implies a more labor-intensive agriculture while the direction of technological change in the U.S. has been away from labor-intensive techniques, what might be predicted about the future technology of U.S. agriculture. Further, would not a high yielding agriculture suggest a reversing of the trends of expanding farm size, since farm size is highly inversely correlated with increased value of harvests (26)?

2. One possible exception to the line of reasoning in the last paragraph might come from the so-called Green Revolution, although serious questions have been raised about the viability of this new technology. Pinmentel and Dovring, for example, question the ability of Third World Agriculture to command the massive resources required as inputs for the new High Yielding Varieties (27, 28). Their work points out the need for more study of the possibility of technologies which minimize the cost of agriculture in terms of resources since today's prices may not reflect tomorrow's scarcities.

3. A final point concerns future climatic change. The question remains as to how much modern technology has been adapting to unusual climatic conditions since, as has been mentioned before, the previous decades have exhibited unusually favorable climatic patterns (23). As an example, Indian plant breeders had directed much of their attention to producing drought resistance (29); the so-called miracle seeds give up the advantages of resistance for high yields just as the time when meteorologists worry about the future of the monsoons. How much of U.S. yield increases came at the expense of resistance and how much has U.S. agricultural technology been adapting to exceptional conditions? What sort of policy do such considerations suggest for the future?

The main thrust of research in agriculture has been to develop methods of utilizing inputs produced by agribusiness.

Farmers reacted to the relatively low price of these inputs; their utilization increased while the additional benefit from each increment fell. Farm output per unit of nitrogen for the nation as a whole was five times higher in 1950 than in 1971 (30). Output per horsepower of tractors fell by about 31% during the same period (30). Equipment expenditure/hour of labor has risen from $.03 in 1940 to $1.14 in 1973 (3).

With the expected continual rise in the price of fossil fuels, a far-sighted agricultural research establishment might be expected to set a high priority on research which could reduce the dependence on inputs which are becoming expensive in terms of prices as well as energy. Continued investment in research dictated by the fictitious economics of cheap energy will not only increase the use of these expensive inputs, but it will help to lock agriculture into the existing track which we can no longer afford to follow. On the other hand, research directed toward a less fossil fuel-dependent technology will do well to learn from the mistakes of the fossil fuel-oriented re-searchers and to consider the social, economic and environmental repercussions of their work thereby avoiding the creation of a new fictitious economics.

REFERENCES

1. Nuti, D. M. "On the Rates of Return of Investment, Kyklos, 27 (1974): 345-369.
2. Solow, R. M. "The Economics of Resources or the Resources of Economics," American Economic Review, 44 (1974): 1-14.
3. U.S. Department of Agriculture, Economic Research Service. The U.S. Food and Fiber Sector: Energy Use and Outlook," prepared for and published by the Subcommittee on Agricultural Credit and Rural Electrification of the Com-mittee on Agriculture and Forestry of the U.S. Senate, 93rd Congress, 2nd Session (1974).
4. Pimentel, D., et. al. "Food Production and the Energy Crisis," Science, 182 (1973); 443-449.
5. Calvin, M. "Solar Energy by Photosynthesis," Science, 184 (1974): 375-381.
6. Rausser, G. C. "Technical Change, Production, and Investment in Natural Resource Industries," American Economic Review, 44 (1974): 1049-59.
7. Hayami, Y. and V. W. Ruttan. Agricultural Development: An International Perspective, Baltimore: John Hopkins Press, 1971.
8. Lianos, T. P. "The Relative Share of Labor in the United States Agriculture: 1949-1968," American Journal of Agricultural Economics, 53 (1971): 411-422.
9. Kennedy, C. "Induced Bias in Innovation and the Theory of Distribution," Economic Journal, 74 (1964): 541-547.
10. ————. "Mechanization and the Division of Labor in Agriculture," Amer-ican Journal of Agricultural Economics, 55: 523-6, August, 1973.
11. Schmitz, A. and D. Seckler. "Mechanized Agriculture and Social Welfare: The Case of the Tomato Harvester," American Journal of Agricultural Eco-nomics, 52 (1970): 569-577.
12. Gardener, B. L. "Determinants of Farm Family Income Inequality," Amer-ican Journal of Agricultural Economics, 51 (1969): 753-69.
13. Barber, C. "The Farm Machinery Industry: Reconciling the Interest of the Farmer, the Industry and the General Public," American Journal of Agri-cultural Economics, 55 (1973): 820-828.

14. "The Small Car Blues at General Motors," Business Week 16 March 1974, pp. 76-8.
15. Fraser, C. Tractor Pioneer, The Life of Harry Ferguson, Athens: Ohio University Press, 1973.
16. Forthomme, P. A. "Can Rice Replace Petroleum?", Ceres (1968)L 50-51.
17. Kramer, M., T. Baker and R. H. Williams. "Solar Energy," Electric Power Consumption and Human Welfare, The Social Consequences of the Environmental Effects of Electric Power Use, eds. Howard Bokesenbaum et al. New York: Macmillan Technical service, forthcoming.
18. Griliches. "Research Costs and Social Returns: Hybrid Corn and Related Innovations," Journal of Political Economy, 66 (1958): 419-431.
19. Bray, J. O. and P. Watkins. "Technical Change in Corn Production in the United States, 1870-1960," Journal of Farm Economics, 46 (1964): 751-765.
20. Johnson, D. G. and R. L. Gustafson. Grain Yields and the American Food Supply, Chicago: University of Chicago Press, 1962.
21. Thompson, L. M. "Foreward," L. M. Thompson et. al., Weather and Our Food Supply, Iowa State Agricultural and Economic Research Report No. 20, 1964.
22. ————. "The Impact of Agricultural Drought in the Corn Belt and High Plains of the United States," A Report to the National Oceanic and Atmospheric Administration, 9 November 1973.
23. Bryson, R. A. "Climatic Change and Drought in the Monsoon Lands," presented at the 140th meeting of the American Association for the Advancement of Science, San Francisco, 27 February 1974.
24. ————. "Farming with Petroleum," Environment, 14 (1972): 8-13.
25. Perelman, M. Against the Grain, The Political Economy of Food and Farming, forthcoming.
26. ————. "Negative Production Functions and Agricultural Efficiency," Unpublished manuscript.
27. ————. Workshop on Research Methodologies for Studies of Energy, Food, Man and Environment. Center for Environmental Quality Management, Cornell University, Ithaca, 18-20 June 1974.
28. Dovring, F. "Macro Constraints on Agricultural Development," Indian Journal of Agricultural Economics, 27 (1972) 46-66.
29. Hopper, W. D. "The Mainsprings of Agricultural Growth," Rajendra Presad Memorial Lecture, delivered at the 18th Conference of the Indian Society of Agricultural Statistics, January 1965 cited by K. N. Raj, "Some Questions Concerning Growth, Transformation and Planning of Agriculture in the Developing Countries," Journal of Development Planning, (1969): 15-38.
30. National Academy of Science. Committee on Agricultural Production, Agricultural Production Efficiency, Washington, D.C., N.A.S., 1975.

Section II

Energy Consumption
For Controlling Wastes

11.

Energy Impacts of Water Pollution Control

Allen Cywin*

One of the major events affecting the pursuit of environmental quality over the past year has been the energy crisis.

Energy conservation, which had been primarily of interest to environmentalists and certain energy-intensive industries in the past, emerged this year as a matter of national imperative.

A rigid linkage between energy growth and economic growth is no longer accepted as completely necessary, and the importance of energy demand management in future energy planning is now broadly recognized. This ethic is in harmony with environmental and conservation programs.

Water pollution abatement programs affecting national energy needs can be divided into two major parts—municipal wastewater treatment and the control of industrial wastewater pollution. This paper assesses the energy requirements of both programs. The assumptions about energy requirements which are used throughout this paper are designed to be conservative. In reality, the Environmental Protection Agency (EPA) expects the energy penalty to be somewhat lower than the following predictions.

Underlying the analysis are three conservative assumptions:

1. No new technologies are used which would be more energy-efficient than those currently in use.

2. Energy prices are low (pre-embargo level) so that the incentive to save money by reducing energy consumption is minimal.

3. There is no explicit Federal energy conservation program.

---

* Director, Effluent Guidelines Division, U.S. Environmental Protection Agency.

A change in the assumed conditions can significantly decrease the predicted energy requirements.

In addition, there is a clear national trend, especially in the industrial sector, toward water conservation and water resource recovery. Since wastewater treatment costs and energy use are strongly related to the quantity and quality of water treated, any movement toward water conservaton, waste segregation, and/or water reuse will tend to reduce treatment costs and energy consumption. EPA therefore feels that the following estimates of energy demands are about at the upper level of future energy consumption.

MUNICIPAL WASTEWATER TREATMENT

Sewage treatment is not an energy-intensive process. However, because there are large quantities of sewage to be treated, demands for fuel or electricity can become significant on a national scale. The following assumptions were used to generate estimates of national requirements for energy by wastewater treatment plants:

1. Secondary treatment will be required at all plants by 1980.

2. No more than 10% of all sludge is incinerated. The balance is land-filled or used for fertilizer (84% of all present plants are land disposal).

3. Activated sludge treatment is utilized to attain secondary standards.

4. Advanced waste treatment is required for about half the plants which are on heavily polluted streams or lakes.

In accordance with the assumptions, the energy consumed by all municipal wastewater treatment plants and the amount of energy expected to be needed to meet the levels of treatment required by the Federal Water Pollution Control Act (FWPCA) are summarized below.

Table 1 differentiates between energy required to meet the "best practicable treatment" guidelines (secondary treatment) and advanced treatment which will be required in certain cases.

Table 1

Year	Level of Treatment	Energy Use (MBD)*	Total Energy (MBD)
1968	All levels of treatment	13	13
1974	All levels of treatment	20	20
1977	Best practicable treatment	20	
	Advanced treatment	6	26
1980	Best practicable treatment	24	
	Advanced treatment	20	44

* MBD = 1000 barrels per day of oil (18 million barrels per day consumed in U.S. in 1974)

Not included in this tabulation are energy requirements for 1) construction of treatment plants, 2) manufacturing of new equipment, 3) nonprocess related energy demand such as space heating for the laboratory buildings, 4) collection system pumping requirements or 5) manufacturing of chemicals used in processes, especially for advanced treatment processes.

On the other hand, this analysis also did not include energy recovery by collection of methane during the treatment process. Methane collection has not been extensive in recent years because alternative sources of energy were very cheap. However, current energy prices should again make methane collection cost-effective, and it can supply a significant percentage of energy needs for these facilities.

Energy demand for treatment increases very rapidly as effluent standards become more stringent. This estimate assumes that the solids are processed by digestion, and dewatered by vacuum filtration followed by landfill or agricultural disposal.

Energy demand can vary significantly from one plant to the next depending on specific plant types and designs. Table 2 indicates the energy demand of various size plants using different levels of technology. The demand was evaluated using examples of typical plants which have demonstrated the capability of meeting the minimum effluent limitations.

Table 2. Energy requirements for representative treatment.

Treatment Process	Flow (MGD)	Energy Demand (in kWh/Day)
Primary	10	1,500
	100	4,500
Secondary	10	5,000
	100	22,000
Advanced	1	4-10,000
	10	12-28,000
	100	40-150,000

(Source: Battelle Memorial Institute)

The 1968 inventory of all municipal waste treatment facilities which was used as a basis for the estimates is summarized in Table 3. The energy requirements for primary and secondary facilities were multiplied by the number of plants and the total flow for plants in each size category.

The 1971 estimate (Table 1) was computed by adding all new projects to the 1968 inventory. The estimated energy demand in 1974 is 11.4 x 10^6 kWh/day or 20 MBD.

The estimate obtained for 1974 has been adjusted upwards by 11% to account for growth in sewered population between 1974 and 1977 and the impact of the requirement for secondary treatment has been added. From these assumptions, the estimated energy demand for achieving secondary treatment is 16.8 x 10^6 kWh/day (26 MBD). Because of delays in funding, much of the actual energy impact of total secondary treatment is assumed in this analysis to be delayed until 1978 instead of 1977. The difference between the predicted 1978 energy use (26 MBD) and 1980 energy use (44 MBD) will be almost entirely for advanced waste treatment.

Table 3. Distribution of sewage treatment.

	Population Served	kWh/day capita	total kWh/day
Minor Treatment	1,360,870	.0185	25,175
Primary Treatment	36,947,397	.0286	1,056,656
Intermediate Treatment	5,857,690	.0286	167,530
Activated Sludge	41,264,036	.113	4,662,636
Trickling Filters	29,617,136	.043	1,273,537
Ponds	6,123,078	.0135	82,662
Other and Unknown	8,636,514	.0135	116,593
Tertiary Treatment	325,530	.226	73,570
TOTALS (1968)	130,132,251		7,458,359 (13 MBD)

EPA's best estimate is that by the mid 1980's between 78 and 80% of the energy requirement for municipal treatment will be for advanced treatment. The total then required for advanced plus secondary treatment will be about 105 MBD if water quality goals are met on all streams.

EFFLUENT LIMITATIONS FOR POWER PLANTS

The FWPCA requires that all industrial discharges provide the "Best Practicable Treatment" by 1977 and "Best Available Treatment" by 1983. These technology-based standards are being promulgated by EPA as effluent limitations for all major industrial categories. Table 4 is the estimated 1980 energy requirements of meeting these limitations. As noted, the requirements of one industry, the electrical power industry, are more than those of all the other industries combined and are primarily for thermal discharge control rather than for biological or chemical wastewater treatment.

The FWCPA requires EPA to promulgate effluent limitations for steam-electric power plants. These limitations will require many existing and proposed power plants to provide off-stream cooling, with an attendant energy penalty due to reduced efficiency and increased operating requirements. A conservative estimate of the 1980 energy penalty is 50 MBD.

Table 4. 1980 energy penalty of industrial effluent limitations.

Electrical Power Plants	
Cooling of Thermal Discharge	50 MBD
Chemical Treatment	negligible
Other Industries	
Cooling of Thermal Discharge	negligible
Wastewater Treatment	40 MBD
TOTAL	90 MBD

Assuming that the total steam-electric generating capacity, including nuclear power plants, in 1980 is about 530,000 MW, EPA estimates that about 70,000 MW will require closed-cycle cooling to meet the effluent limitations after consideration of exemptions under Section 316(a) of the FWPCA. Assumptions include:

—Thermal limitations will cover units larger than 500,000 kilowatts that were placed in operation after January 1, 1970, and all units larger than 25,000 kilowatts placed in operation after January 1, 1974. The affected units must comply by 1981, with extensions available up to 1983 for reliability considerations (except for those receiving a Section 316 exemption for 10 years).

—A 3% annual fuel penalty.

—50% of future units for which utilities are planning to install cooling towers are doing so for economic reasons and therefore are not included in the energy penalty.

—Energy penalties will be divided between coal and oil in the 1979-1983 period—80% coal, 20% oil.

EFFLUENT LIMITATIONS FOR ALL OTHER INDUSTRIES

These regulations will be in effect by 1977 but the total energy impact will not be felt until 1985. The energy requirements of the effluent limitations for industries other than electric power have been tentatively estimated, based on projected requirements for heating fuel and electricity which have been developed by consultants during their examination of alternate control technologies to meet the effluent limitations. These estimates are preliminary, but existing data indicate that the effluent limitations will conservatively require an energy penalty of approximately 70 MBD in 1983. It is estimated that in 1980 the penalty will be 40 MBD. These estimates are based on flow rates and level of treatment required for each of the industries for which effluent limitations have been promulgated. This represents about 50% of the pending permit applications for industry point sources but includes most of the major discharges.

FUTURE

Future effluent guidelines for additional industrial categories will of course increase this estimate commensurately. However, future energy requirements can be reduced significantly. To this extent water pollution treatment and control can be based upon alternatives that are less energy intensive. For example, industry and municipalities should use less water for processing (thereby pumping less water), recover and recycle materials of production (wastes) and consider multiple uses of the same water before treatment.

These practices will not only reduce the energy but also the costs expended for wastewater treatment.

With respect to such treatment, there are alternative techniques to consider, some less energy intensive than others. This is a challenge for consultants, researchers and plant operators. We can discontinue the flaring of sewage gas for example; check on the needed frequency of operating traveling screens; better control the generation and release of oxygen in accordance with demand in aeration processes through instrument-activated controls; instrument and automate sludge pumping (now largely done by hand); convert from sludge incineration to less energy intensive methods of disposal and convert pumps where possible to variable speeds rather than constant speeds—all controlled by advanced computerized systems.

A conventional wastewater treatment facility has a great many points of energy use. In a simple activated sludge plant, there are a minimum of nine unit processes in which electricity is applied directly or indirectly. A recent study noted that a 10-mgd, conventional activated sludge plant requires 5,000 kWh of electricity daily. At a rate of $0.036/kWh, the annual electrical costs would be $65,700. Proper controls such as those noted above could reduce this use of electricity significantly.

These controls would include advanced and computerized control systems that are just beginning to be used in substantial numbers in this field. One large water authority has estimated that its new system will save 10% in power costs let alone other savings.

In summation, a study by EPA has indicated that the energy impact of EPA regulations and programs is estimated to be about the equivalent of 525,000 barrels of oil per day (525 MBD) in 1980, or about 1.1% of forecasted total national consumption of energy. A summary of these energy impacts is shown in Table 5. To facilitate comparisons, all energy is reported in units of thousands of barrels of crude oil per day, regardless of the fuel source. Energy savings are presented in parentheses.

Table 5. Summary of energy impacts of EPA's Program in 1980.

Air Programs	(Thousands of bbls per day)	
Electric Power Plants	145	
All other	125	
Stationary Sources (Subtotal)		270
Auto Emission Controls	160	
Lease Free Regulations	60	
Low Lead Regulations	35	
Transportation Controls	(135)	
Mobile Sources (Subtotal)		120
Water Programs		
Municipal Wastewater		45
Electric Power Plant	50	
All other	40	
Industrial Effluent Guidelines (Subtotal)		90
Solid Waste Programs*		
Combustion of Solid Waste	(65)	
Recyling of Materials	(35)	
TOTAL ALL EPA PROGRAMS		525

( ) represent positive impacts.

* Energy benefits from solid waste programs have not been included in the total above because they primarily result indirectly from EPA's research and educational programs rather than from direct regulations. If included, this potential energy savings of 100 MBD would result in a net energy penalty of 425 MBD.

We believe these estimates are conservative—*i.e.,* for the most part they are close to the upper bound of the range of estimates which EPA and outside consultants have calculated.

12.

Energy Consumption in Wastewater Treatment

Richard A. Mills* and George Tchobanoglous**

Concern over the rate of consumption of natural resources and energy in man's activities has increased during the past few years as shortages have developed and the world's demands have been assessed. Because the operation of wastewater management facilities is to some extent energy- and resource-dependent, it is important to have a realistic appraisal of their requirements. The energy and resource requirements involved in the construction of these facilities are also important. Finally, alternative facilities and concepts derived from a careful analysis of energy and resource consumption must be implemented.

The purpose of this chapter is to explore these topics. To do this, the material to be presented has been organized into sections dealing with 1) the need for information, 2) the nature of the information that is needed, 3) the application of energy and resource information, and 4) the implications to be derived from energy considerations in wastewater management.

NEED FOR INFORMATION

Energy and resource information must be available if intelligent decisions concerning wastewater management are to be made with respect to existing federal laws and regulations and the need to design and construct systems that are cost-effective. The primary need for energy and resource data occurs at the planning stage when alternative treatment systems must be compared with respect to economic and resources impact. At this stage, it is important to be able to estimate energy consumption without making a detailed

---
*Assistant Engineering Specialist (Sanitary), State Water Resources Control Board, Sacramento, California.
**Associate Professor, University of California at Davis, Davis, California.

design of each alternative. The nature and form of such information is discussed in the following section.

BASIC DATA AND INFORMATION

The operation of facilities is the major component of energy consumption of treatment facilities. Because unit processes and operations have greatly varying energy consumptions and because there are innumerable combinations of process flow sheets, there is a need for planners to have energy data for each prospective treatment process or operation. Data sheets for determining operations energy consumption may be found in Appendix A. Based primarily on the work of Smith (1), the data can be supplied after making very preliminary estimates of the sizes of the selected unit processes and operations.

Other significant quantities of energy consumption are involved in the construction and replacement of the treatment facilities and equipment and the maintenance supplies required for the plant. An approximation of this energy consumption is made based on the costs of these activities. The energy consumption for various construction activities is given in Appendix B. Energy required for the maintenance of facilities may be estimated from 1) the combined estimated costs of maintenance labor, replacement parts, and supplies (excluding major chemicals) and 2) the energy intensity factor given in Appendix B for "Maintenance and repair construction." Energy consumption related to operation labor may be neglected.

When considering energy consumption, the total impact on energy resources must be considered. Direct energy consumption at a treatment plant is only a fraction of the total resources required. Energy is consumed in the manufacture and transportation of electricity and fuels. The mining and manufacture of chemicals are also significant factors in the total energy picture. The total quantity of energy resources consumed to produce a product or undertake an activity is called primary or source energy. For some items in Appendixes A and B primary energy consumption values have been given. When quantities of secondary energy sources, such as electricity or fuels, are given, primary energy values may be computed using conversion factors given in Appendix C. For reference, heating values for various fuels are also given in Appendix C.

The general format used in the preparation of the appendixes was to divide the data and information for each item into four sections: 1) assumptions, 2) energy consumption, 3) metric conversion, and 4) sources. Energy consumption data are usually expressed in graphical or equation form and are based on the assumptions given. While it is felt that the assumptions are generally appropriate for planning purposes, energy consumption data must be adjusted if it is known that conditions differ from the assumptions. For reference, metric conversion factors and sources of the data and information are also given for each item.

It is realized that data are not included in this report for many treatment operations and processes, chemicals, fuels, and materials. As more data become available, it is expected that users of this report can add it, utilizing the format presented in the Appendix for ease of future use.

ENERGY CONSUMPTION ANALYSIS FOR TREATMENT PLANTS

The application of the data in the various appendixes is demonstrated by comparing two hypothetical wastewater treatment plant alternatives. Basic waste load and flow assumptions are as follows. A list of abbreviations may be found following the reference list.

1. Average dry weather flow = 0.75 MGD
2. Average annual flow = 1.125 MGD
3. Peak hourly wet weather flow = 4.54 MGD
4. Average dry weather BOD_5 = 225 mg/1
5. Average dry weather suspended solids = 250 mg/1

The alternative treatment process flowsheets are shown in Figures 1 and 2. In Alternative 1, primary settling is omitted and an activated sludge process utilizing a lagoon with surface aerators and recycled sludge is employed. In Alternative 2, a high rate biofilter treatment process with rock media filters and primary sedimentation is employed. Preliminary estimates of component sizes have been determined using simplified design criteria.

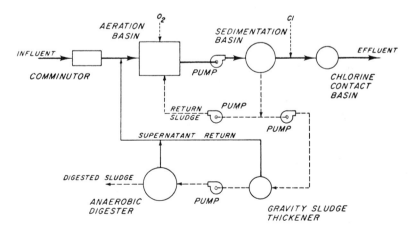

Figure 1. Flowsheet for Alternative 1, activated sludge treatment.

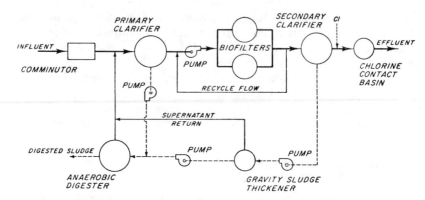

Figure 2. Flowsheet for Alternative 2, biofilter treatment.

Energy and Fuel Consumption

Electrical and fuel energy consumption and the controlling size parameters for the two alternatives are shown in Tables 1 and 2 and in Tables 3 and 4 respectively.

Table 1. Electrical energy consumption for operation in Alternative 1.

Item and Controlling Parameters	Electrical Energy Consumption, kWh/day
Comminutor, 1.125 MGD average flow	17
Aerated lagoon, mixing controls, 150,000 ft³ volume	2295
Intermediate pump, 1.69 MGD average flow, 5 ft TDH	41
Sedimentation basin, 3320 ft² surface area	48
Chlorination, 2000 lb/day size, 94 lb/day average flow	4
Sludge recycle pump, 0.56 MGD average flow, 5 ft TDH	14
Waste sludge pump to thickener, 15 gpm average flow, 7 ft TDH	1
Gravity sludge thickener, 314 ft² surface area	12
Thickened sludge pump to digester, 4 gpm average flow, 30 ft TDH	1
Anaerobic digester, 264,000 Btu/hr heat exchanger capacity, 50 ft diameter	183
Lighting and miscellaneous electricity uses, 0.75 MGD average dry weather flow	55
Total	2671

Preliminary estimates of chlorine use, construction costs, and maintenance labor, parts, and supplies costs have also been made.

Based on these estimates and the data in the appendixes, primary energy consumption for these items, as well as for electricity and fuels consumption, has been estimated as shown in Tables 5 and 6.

Table 2. Fuel energy consumption for operation in Alternative 1.

Item and Controlling Parameters	Fuel	Fuel Energy Consumption, Btu/day
Anaerobic digester, 1,440 lb dry sludge/day, 3% solids concentration, 60°F sludge, 6000 Btu/hr/1000 ft³, 0.75 thermal correction factor, 43,115 ft³ volume	Propane	8,450,000
Total	Propane	8,450,000

Table 3. Electrical energy consumption for operation in Alternative 2.

Item and Controlling Parameters	Electrical Energy Consumption, kWh/day
Comminutor, 1.125 MGD average flow	17
Primary clarifier, 3,320 ft² surface area	48
Intermediate pumping, 5.63 MGD average flow, 13 ft TDH	354
Trickling filters	0
Secondary clarifier, 3,850 ft² surface area	51
Chlorination, 2000 lb/day size, 94 lb/day average flow	4
Waste sludge pump to thickener, 3.00 gpm average flow, 7 ft TDH	0
Gravity sludge thickener, 314 ft² surface area	12
Thickened sludge pump to digester, 1.50 gpm average flow, 30 ft TDH	0
Anaerobic digester, 220,000 Btu/heat exchanger capacity, 50 ft diameter	182
Lighting and miscellaneous electricity uses, 0.75 MGD average dry weather flow	55
Total	723

Table 4. Fuel energy consumption for operation in Alternative 2.

Item and Controlling Parameters	Fuel	Fuel Energy Consumption, kWh/day
Anaerobic digester, 1,440 lb dry sludge/day, 8% solids concentration, 60°F sludge, 6000 Btu/hr/1000 ft³, 0.75 thermal correction factor, 43,115 ft³ volume	Propane	7,050,000
Total	Propane	7,050,000

Table 5. Primary energy consumption for Alternative 1.

Item	Energy Parameter	Conversion Factor	Primary Energy Consumption	Average Annual Energy Consumption, 10^9 Btu/yr
Construction	$1,050,000[a]	$\dfrac{900.96}{2100}$ (75,534 Btu/$)	34.00×10^9 Btu	1.134[b]
Operations electricity	2671 kwh/day	13,208 Btu/kwh	35.30×10^6 Btu/day	12.877
Operations propane	8,450,000 Btu/day	1.208	10.21×10^6 Btu/day	3.726
Operations chlorine	94 lb/day	70×10^6 Btu/ton	3.29×10^6 Btu/day	1.201
Maintenance labor parts and supplies[c]	$27,100/yr[a]	$\dfrac{900.96}{2100}$ (67,117 Btu/$)	780.00×10^6 Btu/yr	0.780
Total				19.718

a Estimated using an Engineering News-Record Construction Cost Index of 2100.

b Based on a service life of 30 years.

c It is assumed that the category "Maintenance and repair construction" listed in Appendix B for Construction Energy is a close approximation of the maintenance energy for a treatment plant. Labor costs are included because they are apparently included in the conversion factor. It is assumed that negligible energy consumption is associated with the labor required for operation.

Table 6. Primary energy consumption for Alternative 2.

Item	Energy Parameter	Conversion Factor	Primary Energy Consumption	Average Annual Energy Consumption, 10^9 Btu/yr[b]
Construction	$1,244,000[a]	$\dfrac{900.96}{2100}$ (75,534 Btu/$)	40.30×10^9 Btu	1.344[b]
Operations electricity	723 kwh/day	13,208 Btu/kwh	9.55×10^6 Btu/day	3.486
Operations propane	7,050,000 Btu/day	1.208	8.51×10^6 Btu/day	3.108
Operations chlorine	94 lb/day	70×10^6 Btu/ton	3.29×10^6 Btu/day	1.201
Maintenance labor parts and supplies[c]	$35,900/yr[a]	$\dfrac{900.96}{2100}$ (67,117 Btu/$)	1.03×10^9 Btu/yr	1.033
Total				10.172

a Estimated using an Engineering News-Record Construction Cost Index of 2100.

b Based on a service life of 30 years.

c It is assumed that the category "Maintenance and repair construction" listed in Appendix B for Construction Energy is a close approximation of the maintenance energy for a treatment plant. Labor costs are included because they are apparently included in the conversion factor. It is assumed that negligible energy consumption is associated with the labor required for operation.

Comparison of Alternatives

The basis of comparison of the alternatives is the consumption of primary energy per unit of time. Another possible basis is the consumption of primary energy per unit of plant capacity or unit of wastewater treated. The latter was not selected for two reasons. Several alternatives being compared may involve different flow-rates as in the case where several sewer rehabilitation alternatives are being considered. The impact on resources can also be assessed more effectively in terms of rate of consumption with respect to time.

Process Selection

If process selection were based only on energy consumption, it is clear that the biofilter alternative would be selected because, as shown in Tables 5 and 6, it requires about half the energy of the activated sludge alternative. However, energy consumption is but one of the many factors that must be considered before a decision can be made concerning process selection. Included among the factors that must be considered are 1) environmental effects, 2) least monetary costs, 3) implementation capability, 4) contributions to objectives, 5) energy and resource consumption, 6) reliability, and 7) public acceptance. The relative importance of these factors will vary with each situation.

ENERGY AND RESOURCE CONSUMPTION IMPLICATIONS

The steps involved in completing an assessment of energy and resource consumption for alternative wastewater treatment concepts has been demonstrated in the previous section of this report. The implications of such an analysis are related to 1) the actual use of energy and resource, 2) the use of alternative treatment facilities and concepts for reducing energy and resource use.

Energy and Resource Use

Because both energy and resources are consumed in the collection and treatment of wastewater it could be argued that savings could be achieved if such a course of action were followed, it would be inconsistent with our stated national goals. On an individual basis, it has been estimated that of the 11.6 equivalent barrels of fuel oil used by each individual in the United States in 1969, only 0.033 barrels were used for wastewater treatment (2). Further, the amount of fuel oil used for wastewater treatment is dwarfed by the amount devoted to transportation-using vehicles that are not designed to minimize fuel oil consumption. This is not to say that any savings that can be made should not be made (see following discussion), but rather, that energy and resource consumption should be based on national priorities. Because the consumption of energy and resources for wastewater management is relatively insignificant when compared to the total consumption, suggestions to curtail treatment effectiveness to conserve energy do not appear to be well-founded in fact or consistent with national goals.

Alternative Facilities and Concepts

One of the beneficial aspects of the concern over the use of energy and resources for wastewater treatment is that it has led to the development and examination of alternative facilities and concepts for achieving the same objectives. The purpose of the following discussion is to present some alternative facilities and concepts for saving energy. The discussion is not meant to be exhaustive, but rather, to illustrate some of the many opportunities that exist for achieving savings in the consumption of energy and resources. Because the actual quantities of energy and resources that can be saved will vary in each individual situation, detailed computations are not presented. However, the approach to be used in analyzing the savings would be as illustrated in the previous section.

Construction of Facilities. The use of alternative facilities and equipment often makes possible savings in both construction energy requirements and resource use. To illustrate these points the recently constructed treatment plant at Hickmott Foods, Inc., a small tomato cannery located in Antioch, California, will be used (3). The plan view of the treatment plant is shown in Figure 3. Photographs of the plant are shown in Figure 4. The plant is essentially an extended aeration activated sludge process. Waste activated sludge is thickened before being hauled for land disposal. Additional details may be found in reference (3). The most interesting aspect of this plant is that a Hypalon-lined earthen basin is used as a sedimentation tank. The estimated cost for this facility is $300,000 (ENRCC Index for San Francisco Bay Region = 2240). To achieve the same overflow rate in a conventional sedimentation tank constructed of steel or concrete, the equivalent cost would have been about $1,200,000. If it is assumed that construction energy and resource consumption are related to construction cost, the savings associated with the use of a lined earthen basin are significant. The same is true for the aeration basin. Although this plant is used for the treatment of cannery wastes, the concepts are directly applicable to the design and construction of facilities for small municipalities.

Piping. Another aspect of treatment plant design where energy savings are potentially possible is in the use of above-ground piping (see Figures 3 and 4). While this type of construction may not be

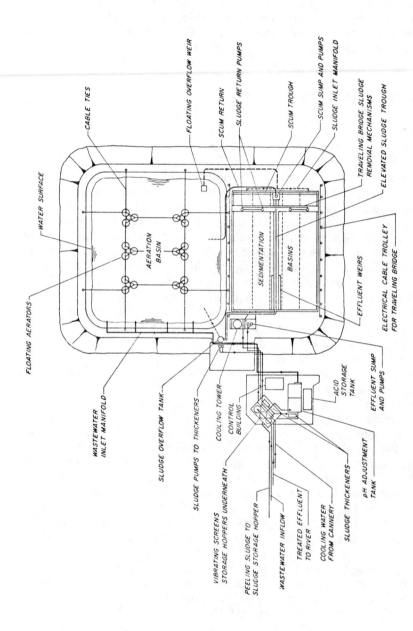

Figure 3. Plan view of wastewater treatment facilities at Hickmott Foods, Inc. (3).

Hypalon-lined Aeration Basin with Surface Aerators in Operation   Hypalon-lined Sedimentation Basin—Cooling Tower in Foreground

Central Sludge Trough and Traveling Bridge Sludge Removal Mechanisms   Traveling Bridge, Sludge Pickup Assembly, and Effluent Weirs

Figure 4. Photographs of wastewater treatment facilities at Hickmott Foods, Inc. (3).

possible in some of the northern areas because of freezing, it is applicable to a large portion of the United States. In addition to savings in construction energy, corrosion losses are also minimized. The use of above-ground piping also facilitates the making of process modifications with a minimum expenditure of energy.

A final point in support of the use of above-ground piping is that there is no good reason for burying piping other than tradition. Also, oil refineries and chemical plants have been very successful using exposed piping.

Alternative Facilities and Equipment. The use of alternative facilities and equipment offers significant potential for savings. The use of earthen lined basins for aeration and sedimentation operations has been discussed previously.

Alternative Equipment and Flowsheets. Energy and resource savings through the use of alternative equipment and flowsheets are also possible. An alternative approach for heating small digesters using commonly available supplies and equipment is detailed in Figure 5. Based on actual operating records, significant energy

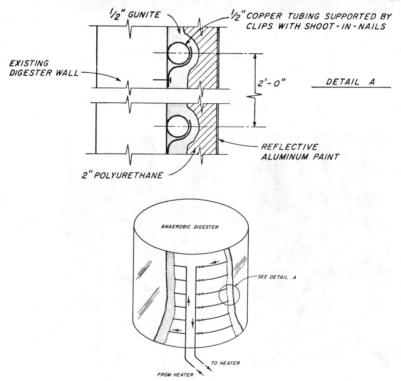

Figure 5. Digester heating system (4).

savings have been achieved with this system as compared to more conventional commercial digester heating equipment (4).

Another means by which energy and resource savings can be achieved in existing and new treatment plants is through the use of screening devices. Two flowsheets involving the use of screens are shown in Figures 6a and 6b. In Figure 6a a screening device is used to improve the performance of an overloaded primary settling tank, whereas in Figure 6b the screening device is used to improve the performance of an overloaded secondary settling tank. Details concerning the latter application may be found in reference (5). Here again the approach has been to use alternative equipment in modified flowsheets.

Chemical Usage. Because of the many unanswered questions re-

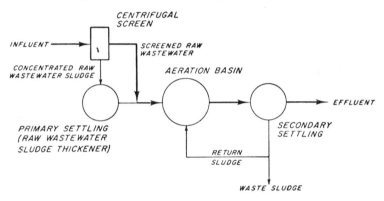

a. OVERLOADED PRIMARY

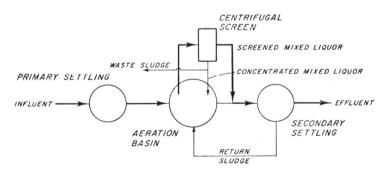

b. OVERLOADED SECONDARY

Figure 6. Flowsheets incorporating centrifugal screens to relieve overloaded treatment facilities (5).

lated to the future availability of chemicals, careful attention must be given to the use of chemicals in treatment plant operations. In many cases, it may be more cost and energy and resource effective to consider the adoption of an alternative process flowsheet even though it may be more expensive in the short run.

Treatment Plant Operation. The operation of treatment plants offers another potential area where energy and resource savings may be possible. Although the recent trend has been to incorporate more automation in treatment plants, it is suggested here that the use of more person-power and less automation may be more cost-effective and may lead to savings in energy and resource consumption. If more person-power is used, cost-effective computations must be based on a larger framework in which the incremental contribution of each person is considered.

Discussion

The purpose of this section has been to illustrate some of the many techniques that can be used to reduce the consumption of both energy and resources in the design, construction, and operation of treatment plants without altering quality objectives. If the reader's imagination has been stirred then the objective of this section has been met.

SUMMARY

Reliable data and information on energy and resource consumption in a usable format are necessary if intelligent decisions are to be made in the selection of treatment alternatives. While the consumption of energy for wastewater treatment is not significant when compared to other uses, every effort should be made to achieve savings which are consistent with national water quality objectives. To this end, consideration should be given to the use of alternative flowsheets, equipment, facilities, and methods of construction. In the long run, what is required is a commitment to achieve our water quality objectives with a minimum expenditure of energy and resources.

APPENDIX A

OPERATIONAL ENERGY REQUIREMENTS
FOR UNIT OPERATIONS AND PROCESSES

A-1 PUMPING

Assumptions

1. Efficiency factors for pumps

Flowrate		Pump Hydraulic
gpm	MGD	Efficiency, e_h
0 - 999	0 - 1.43	0.70
1000 - 7000	1.44 - 10	0.74
Over 7000	Over 10	0.83

2. The motor electrical efficiency is 0.877.

Energy Consumption

1. Electrical power =

$$= \frac{Q\gamma H}{k e_h e_m} = \frac{Q(1.55 \text{ ft}^3/\text{sec}/\text{MGD})\,(62.4 \text{ lb/ft}^3)H}{(550 \text{ ft-lb/sec-hp})e_h(0.877)}$$

$$= \frac{(0.2 \text{ hp/MGD-ft})QH}{e_h}$$

$$= \frac{(0.000288 \text{ hp/gpm-ft})QH}{e_h}$$

where
Q = volume flow rate
γ = weight density = 62.4 lb/ft^3
H = total dynamic pumping head
e_h = pump hydraulic efficiency
e_m = motor electrical efficiency
k = conversion factor = 550 ft-lb/sec-hp

2. Electrical energy consumption =

$$= \frac{(0.149 \text{ kw/MGD-ft})QH(\text{hr pumping/day})}{e_h}$$

$$= \frac{(0.000215 \text{ kw/gpm-ft})QH(\text{hr pumping/day})}{e_h}$$

Metric Conversion

1 gpm = 0.061 1/sec
1 MGD = 43.8 1/sec = 3790 m^3/day
1 ft = 0.305 m
1 hp = 0.746 kw

Sources

Derived from previously published information (1, 6).

A-2   MECHANICALLY CLEANED BAR SCREENS

Assumptions
1. A motor size of 0.75 hp is used for mechanically cleaned bar screens of 0 to 15 MGD capacity.
2. The motor electrical efficiency is 0.877.
3. Maximum operation is 6 min/hr.
4. A bar screen will be required for each 15 MGD increment of plant capacity.

Energy Consumption
1. For each 15 MGD increment of plant capacity, electrical energy consumption =
        = 1.53 kwh/day.

Metric Conversion
1 MGD = 43.8 1/sec = 3790 m³/day
1 hp   = 0.748 kw
Sources
Obtained from previously published information (1, 6).

A-3   COMMINUTORS

Assumptions
1. Motor sizes (See following page)

Average Flow Range, MGD	Motor Size, hp
0.25 - 1.82	0.75
0.97 - 5.10	1.50
1.0 - 9.4	1.50
1.3 - 20	2.00

2. The motor electrical efficiency is 0.877.
3. Operation is continuous.

Energy Consumption
1. Graph

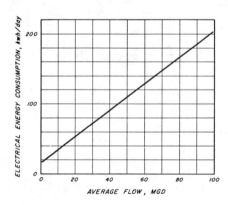

Metric Conversion
1 MGD = 43.8 1/sec = 3790 m³day
1 hp   = 0.746 kw
Sources
Derived from previously published information (1).

A-4 GRIT REMOVAL
Assumptions
1. A grit removal unit is required for each 5 MGD increment of plant capacity.
2. The motor size for each unit is 0.5 hp.
3. The motor electrical efficiency is 0.877.
4. The equipment is operated during the high flow period between 8 to 12 a.m.

Energy Consumption
1. For each 5 MGD increment of plant capacity, electrical energy consumption =
$$= 1.70 \text{ kwh/day.}$$

Metric Conversions
1 MGD = 43.8 1/sec = 3790 m³/day
1 hp   = 0.746 kw
Sources
Obtained from previously published information (1, 6).

A-5 MICROSCREENS
Assumptions
1. Microscreen has high pressure spray backwash system.
2. Operation is continuous.

Energy Consumption
1. Electrical energy consumption =
$$= [(0.4 \text{ ft}^{-2}) \text{ (Submerged screen area)} + 1.25] \times$$
$$\left(\frac{\text{Rotation speed}}{5 \text{ rpm}} \text{ kwh/day} \right)$$

2. Graph

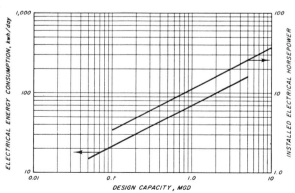

Metric Conversion
1 hp   = 0.746 kw
1 ft²   = 0.0929 m²
1 MGD = 43.8 1/sec = 3790 m³/day

Sources
Obtained from previously published information (1, 7).

A-6 SEDIMENTATION TANKS

Assumptions
1. Sedimentation tank shape is rectangular.
2. The motor size of the sludge collectors is dependent on the clarifier length.

Length, ft	Motor Size, hp
0 - 75	0.5
75 - 150	0.75
150 - 250	1.00
250 - 325	1.50

3. The motor electrical efficiency is 0.877.
4. Operation of collectors is continuous.

Energy Consumption
1. Graph

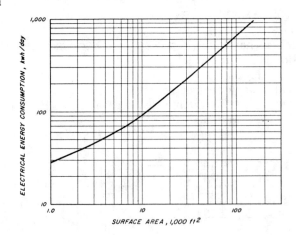

Metric Conversion
1 ft   = 0.305 m
1 ft² = 0.0929 m²
1 hp = 0.746 kw

Sources
Obtained from previously published information (1).

A-7 TRICKLING FILTERS

Assumptions

1. Electrical power is required for trickling filters to pump plant flow and recycle flow to the top of the filters and through the distributor arms. The distributor arms are rotated by hydraulic forces.
2. Depth of rock media filters typically ranges from 4 to 10 ft. Depth of plastic or redwood slat media filters typically ranges from 20 to 40 ft.
3. The distributor is about 1 ft above the top of the media.
4. The hydraulic head loss through the distributor is about 3 ft.
5. The hydraulic head loss through the underdrains is about 2 ft.
6. The ratio of recycle flow, R, to plant flow, Q, ranges from 0 to 4:1.

Energy Consumption

1. Electrical energy consumption =

$$= \frac{(0.149 \text{ kw/MGD-ft})(R + 1)QH(\text{hr pumping/day})}{e_h}$$

where e_h = pump hydraulic efficiency

Metric Conversion

1 ft = 0.305 m
1 hp = 0.746 kw

Sources

Derived from previously published information (1, 8).

A-8 ROTATING BIOLOGICAL DISKS

Assumptions

1. The rotating biological disk process consists of multiple polyethylene disks mounted on a horizontal shaft. The disks are half submerged in a primary settled waste and are rotated at 1.6 rpm.
2. Wastewater temperature is greater than 55°F.
3. Operation is continuous.

Energy Consumption

1. Graph

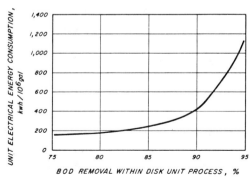

2. Electrical energy consumption =
$$\left(\begin{array}{l}\text{Unit electrical energy} \\ \text{consumption, kwh/}10^6 \text{ gal}\end{array}\right) \text{(Average plant flow, MGD)}$$

Metric Conversion
1 gal   = 3.785 1
1 MGD = 43.8 1/sec = 3790 m^3/day
°F   = 32 + 1.8(°C)

Sources
Derived from previously published source (9).

A-9 DIFFUSED-AIR AERATION FACILITIES

Theory
1. The blower power required for adiabatic compression is given by the following relationship:

$$\text{Power} = \frac{wRT_1 k}{e_1 e_2 e_3 (k-1)}\left[\left(\frac{p_2}{p_1}\right)^{\frac{k-1}{k}} - 1\right]$$

$$= \frac{wC_p T_1}{e_1 e_2 e_3}\left[\left(\frac{p_2}{p_1}\right)^{\frac{k-1}{k}} - 1\right]$$

where w = theoretical mass flow of air, based on oxygen requirements of organisms in aeration tank

   R = gas constant of air = 53.5 ft-lb$_f$/lb$_m$-°R
   T_1 = absolute inlet temperature
   k = $C_p C_v$ = 1.395 for air
   C_p = specific heat of air at constant pressure
        = 0.240 Btu/lb$_m$-°R = 0.240 cal/g-°k
   C_v = specific heat of air at constant volume
   p_1 = absolute inlet pressure
   p_2 = absolute outlet pressure
   e_1 = blower efficiency
   e_2 = oxygen transfer efficiency
        = oxygen dissolved/oxygen supplied
   e_3 = motor efficiency

2. The outlet pressure is the sum of the head losses in the air filter, silencers, diffusers, piping, and pipe fittings and the aeration tank pressure at the diffuser depth.

Assumptions
1. The blower efficiency ranges from 0.7 to 0.8; assume 0.75.
2. The oxygen transfer efficiency ranges from 5 to 15 percent, with 8 percent probable from porous tube diffusers and 6 percent from coarse-bubble diffusers; assume 6 percent.

3. The motor electrical efficiency is 0.877.
4. For preliminary planning estimates, an assumed outlet pressure of 8 psig is appropriate.
5. Absolute inlet pressure is generally atmospheric pressure.

Elevation, ft	Atmospheric Pressure, psi
0	14.7
1,000	14.2
2,000	13.7
4,000	12.7

6. For energy computations, the absolute inlet temperature should be the annual mean ambient temperature.
7. The air requirement for a completely mixed flow regime is from 20 to 30 scfm/1000 ft^3 of tank volume. (Standard air has a density of 0.0750 lb$_m$/ft^3.)
8. Operation is continuous.

Energy Consumption
1. Electrical energy consumption at sea level =

$$= \frac{w(0.240 \text{ Btu/lb}_m\text{-}°R)T_1}{(0.75)(0.06)(0.877)} \left[\left(\frac{22.7 \text{ psia}}{14.7 \text{ psia}}\right)^{\frac{1.395-1}{1.395}} - 1 \right] \times$$

$$(24 \text{ hr/day})(2.93 \times 10^{-4} \text{ kwh/Btu})$$
$$= (0.00559 \text{ kwh-hr/day-lb}_m\text{-}°R)wT_1$$

where w = mass flow of air based on organism requirements or based on mixing requirements, whichever is greater, lb$_m$/hr
T_1 = °R

Metric Conversions
1 lb$_m$ = 4.454 kg
1 °R = 1.8 °K
1 ft = 0.305 m
1 ft^3 = 28.3 1

Sources
Derived from previously published information (1, 6).

A-10   MECHANICAL AERATORS
Theory
1. The oxygen transfer rate for a mechanical aerator under field conditions can be determined using the following equation.

$$N = N_0 \left[\frac{\beta C_{DS} - C_0}{C_S} (1.024)^{T-20} \right] a$$

where N   = oxygen transfer rate under field conditions

N_0 = oxygen transfer rate in water at 20°C and zero dissolved oxygen concentration, generally expressed as lb O_2/hp-hr and supplied by manufacturer

β = $\dfrac{\text{saturation concentration of oxygen in waste}}{\text{saturation concentration of oxygen in clean water}}$

C_{DS} = saturation concentration of oxygen in clean water at design temperature and altitude

C_0 = operating oxygen concentration 20°C and sea level = 9.17 mg/1

C_S = saturation concentration of oxygen in clean water at 20°C and sea level = 9.17 mg/1

a = oxygen transfer rate correction factor for waste

2. The power required

$$= \frac{w}{N(e)}$$

where w = theoretical weight flow rate of oxygen, based on oxygen requirements of organisms in aeration tank

e   = motor electrical efficiency

Assumptions

1. Values for N_0 are provided by manufacturers based on field tests. If unknown, N_0 can be assumed to be 3 lb O_2/hp-hr for surface aerators and 2 lb O_2/hp-hr for turbine aerators.
2. Unless otherwise known, for municipal wastewater assume $\beta = 1$, $a = 0.8$, T = 15°C, $C_{DS} = 9.6$ mg/1.
3. The desired oxygen concentrator in an aeration basin or aerated lagoon is $C_0 = 1.9$ mg/1.
4. Power requirements to maintain a completely mixed flow regime with solids suspension is 0.75 hp/1000 ft^3, without solids suspension is 0.07 hp/1000 ft^3.
5. The motor electrical efficiency is 0.877.
6. Operation is continuous.

Energy Consumption

1. Electrical energy consumption is the greater of the following two values:
   a. To satisfy oxygen requirements, electrical energy consumption

$$= \frac{w(0.746 \text{ kw/hp})}{N(0.877)}$$

$$= \frac{(0.851 \text{ kw/hp})w}{N}$$

= (0.425 kwh/lb 0_2)w for surface aerators

= (0.638 kwh/lb 0_2)w for turbine aerators

where w = weight flow rate of oxygen required, lb/day

b. To satisfy complete mixing with solids suspension requirements, electrical energy consumption

= (0.0153 kwh/day-ft³)(Liquid volume in ft³)

Metric Conversion

1 lb = 0.454 kg

1 hp = 0.746 kw

1 ft³ = 28.3 1

Sources

Derived from previously published information (1, 6, 10).

A-11 GRAVITY SLUDGE THICKENERS

Assumptions

1. A 0.5 hp motor is required for each thickening tank.
2. The motor electrical efficiency is 0.877.
3. Operation is continuous.

Energy Consumption

1. Graph

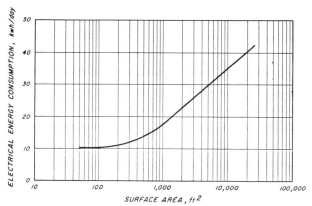

2. If the number of tanks is known, electrical energy consumption

= (10.2 kwh/day/tank)(Number of tanks).

Metric Conversion

1 hp = 0.746 kw

1 ft² = 0.0929 m²

Source

Derived from previously published information (1).

A-12   VACUUM FILTRATION
Assumptions
1. Vacuum filter horsepower requirements

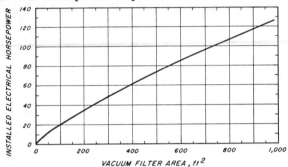

2. The motor electrical efficiency is 0.877.
3. The period of operation varies with plant size.

Energy Consumption
1. Electrical energy consumption

$$= \frac{(\text{Installed hp})(0.746 \text{ kw/hp})}{0.877} \text{ (hr/day of operation)}$$

Metric Conversion
1 hp = 0.746 kw
1 ft² = 0.0929 m²

Sources
Obtained from previously published information (1, 6).

A-13   DISSOLVED-AIR FLOTATION THICKENERS
Assumptions
1. Dissolved-air flotation thickener horsepower requirements

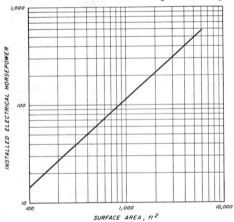

2. The motor electrical efficiency is 0.877.
3. For small plants operation is intermittent, and therefore surface area and period of operation are interrelated variables of design.

Energy Consumption
1. Electrical energy consumption
= (0.851 kwh/hp-hr)(Installed electrical hp)×
(Hours of operation/day)

Metric Conversion
1 hp = 0.746 kw
1 ft² = 0.0929 m²

Sources
Obtained from previously published information (1).

A-14 ANAEROBIC DIGESTERS

Assumptions
1. The following energy data apply to the first stage tanks of a two-stage digestion process. The second stage consists primarily of sludge concentration and supernatant clarification and has no energy consumptive processes, with the possible exception of pumping into second-stage tanks.
2. The addition of heat to the contents of an anaerobic digester is required to raise the incoming sludge to digestion-tank temperature and to compensate for heat losses through digester walls, floor, and roof. Various fuels, including digester gas, are the source of heat. Electrical energy is required during heat addition to operate a water pump, a sludge pump, a blower fan, and perhaps a fuel oil pump. Electrical energy is also required for mixing the digester contents, either by gas recirculation or by sludge mixing.
3. Heat required to raise incoming sludge to digestion-tank temperature is

$$q = \frac{S(T_2 - T_1)}{P} \quad (1 \text{ Btu/lb}^\circ\text{F})$$

where q = energy required per unit of time
S = undigested dry sludge solids weight flow rate
T_1 = undigested incoming sludge temperature
T_2 = digester temperature, normally 95°F
P = percent solids in incoming sludge
When basic design data are unavailable, assume
q = (90,000 Btu/hr-MGD)(Average plant flow) for primary plants
q = (150,000 Btu/hr-MGD)(Average plant flow) for secondary plants

4. Heat required to offset digester heat losses to the environment
for the coldest month of the year can be estimated by multiplying
the appropriate value in the following table by a thermal factor
dependent on location.

	Heat Required, Btu/hr per 1000 ft³ of digester capacity			
	Well Insulated Side Walls		Exposed Side Walls	
	Without Gas Recirculation	With Gas Recirculation	Without Gas Recirculation	With Gas Recirculation
Well Insulated Floating Cover	2600	3400	4700	6000
Uninsulated Floating Cover	3300	5500	5400	8100

Taken from (11).

Location	Thermal Correction Factor
Havre, MT; Minneapolis; Sault Ste. Marie, Mi; Montreal	1.00
Billings, MT; Omaha; Milwaukee; Albany; Portland, ME	0.90
Reno, NV; Albuquerque; St. Louis; Cincinnati; Philadelphia	0.80
Portland, OR; Las Vegas; Roswell, NM; Nashville; Richmond, VA	0.73
Sacramento; El Paso; Birmingham; Charlotte, NC; Norfolk, VA	0.67
Los Angeles; San Antonio; New Orleans; Jacksonville	0.54

Taken from (11).

5. Installed horsepower for heat exchanger

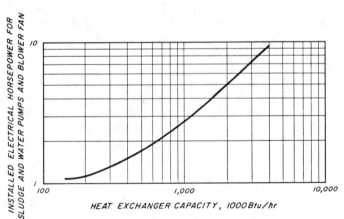

6. Installed horsepower for mixing

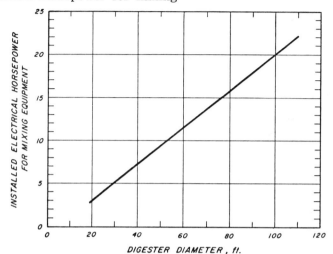

DIGESTER DIAMETER, ft.

7. The motor electrical efficiencies are 0.877.
8. The heat exchanger motors operate about 75% of the day. The mixing equipment operates continuously.

Energy Consumption
1. Fuel energy consumption to heat incoming sludge
   = q (See Assumption 3)
2. Fuel energy consumption to replace digester losses
   = (Heat required, But/hr/1000 ft³)×
   (Digester capacity, ft³)(Thermal correction factor)
3. Electrical energy consumption
   = [(Heat exchanger hp)(0.75) + (Mixing hp)]×
   (0.746 kw/hp)(24 hr/day)
   = (13.4 kwh/hp-day)(Heat Exchanger hp) +
   (17.9 kwh/hp-day)(Mixing hp)

Metric Conversion
1 Btu	=	1050 joules = 252 cal
1 lb	=	0.454 kg
1 °F	=	1.8°C
°F	=	32 + 1.8(°C)
1 MGD	=	43.8 1/sec = 3790 m³/day
1 ft	=	0.305 m
1 ft³	=	28.3 1
1 hp	=	0.746 kw

Sources
Derived from previously published information (1, 6, 11-13).

A-15 MULTIPLE-HEARTH SLUDGE INCINERATORS

Assumptions

1. Electrical power is required for operation of shaft for rotating rabble arms; fan and blowers for introducing cooling, combustion and auxiliary air; and water pump and induced draft fan of pre-cooler and wet scrubber for treatment of exit gases.
2. Auxiliary fuel in the form of oil, natural gas, or digester gas is required for warming up the incinerator to the required combustion temperature, for maintaining the desired temperature during combustion, and for maintaining a temperature during short standby periods, such as overnight, when the incinerator is not operating. Auxiliary fuel requirements are highly variable, dependent upon the heat content of the sludge and the length of periods of standby. Detailed procedures for computing fuel requirements are found in Reference (14). For preliminary planning purposes, the auxiliary fuel requirement reported below is an average of amounts used at six installations The sludges averaged about 26% solids by weight, and the dry solids averaged about 56% volatiles by weight and about 5,700 Btu heat content per pound.

Energy Consumption

1. Graph

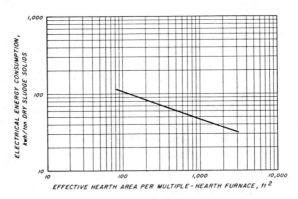

2. Auxiliary fuel energy consumption
$$= (4.03)(10^6) \text{Btu/ton dry sludge solids}$$

Metric Conversion
1 lb   = 0.454 kg
1 ton = 907 kg
1 Btu = 1050 joules = 252 cal
1 ft² = 0.0929 m²

Sources
Derived from previously published information (1, 4).

A-16   CHLORINATION EQUIPMENT

Assumptions

Chlorinator Size, lb/day	Power Requirement, kw
400	0.030
2000	0.030
8000	0.075

1. Chlorinator power requirements
2. Chlorinator operation is continuous.
3. For plants requiring more than 2,000 lb/day of chlorine, the chlorine is often acquired in liquid form and an evaporator is required. The energy required to bring chlorine to ambient temperature is about 135 Btu/lb.
4. Allowances must be made for local conditions such as for the pumping of injection water and chlorine residual analyzers.

Energy Consumption
1. Chlorinators
   Electrical energy consumption
   = (Power requirement/chlorinator)(No. of chlorinators)×(24/hr/day)
2. Evaporators
   Electrical energy consumption
   = (0.0396 kwh/lb)(Average weight flow rate of chlorine)

Metric Conversion
1 lb = 0.454 kg
1 Btu = (2.93)(10⁻⁴) kwh
Sources
Obtained from previously published information (1).

A-17   CHEMICALS

Primary Energy Consumption
1. Chlorine: About 70,000,000 Btu/ton
2. Lime: In excess of 5,500,000 Btu/ton
3. Ferric Chloride: Minimal energy consumption because it is a waste product of manufacturing.

Metric Conversion
1 Btu = 1050 joules = 252 cal
1 ton = 907 kg

Source
Obtained from previously published information (15).

A-18   LIGHTING AND MISCELLANEOUS ELECTRICITY USES

Assumptions

1. Items included in this category are indoor and outdoor lighting, operation of tools and office equipment, and other miscellaneous uses.

Energy Consumption

1. Graph

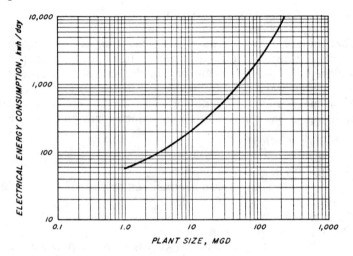

Metric Conversion

1 MGD = 43.8 1/sec = 3790 m³/day

Sources

Obtained from previously published information (1).

APPENDIX B

CONSTRUCTION ENERGY REQUIREMENTS
FOR VARIOUS ACTIVITIES

Assumptions

1. Energy requirements for construction activities

Sector of the U.S. Economy	Primary Energy Intensity, Btu/$ in 1963
Water supply and sewerage construction	75,534
Multi-purpose water resource projects	107,500
Maintenance and repair construction	67,117
Agricultural, forestry and fishery services	49,476
Stone and clay mining and quarrying	94,414
New construction, nonresidential buildings	66,304
Public utilities construction	75,507
Highway construction	98,216
Office and public building furniture	65,444
On-site, ready mixed concrete	66,200
Fabricated structural steel	123,040
Fabricated plate work	115,600
Architectural metal work	142,060
Miscellaneous metal work	133,730
Nonferrous wire drawing and insulating	96,340
Pipe, valves, and pipe fittings	73,742
Steam engines and turbines	84,232
Construction machinery	73,089
Industrial trucks and tractors	64,279
Pumps and compressors	58,254
General industrial machinery	62,891
Transformers	81,030
Motors and generators	65,378
Industrial controls	38,808
Wholesale trade	33,250

2. Indexes for adjusting costs

Index Source	1963 Cost Index	Base Year (Index = 100)
U.S. Environmental Protection Agency		
Sewage Treatment Plant Construction	108.07	1957 - 59
Sewer Construction	113.07	1957 - 59
Engineering News-Record		
Building	594.12	1913
Construction	900.96	1913
Marshall & Swift Equipment	239.2	1926
Chemical Engineering		
Plant	102.4	1957 - 59
Equipment, machinery, and supports	100.5	1957 - 59
Obtained from References (16-18).		

Energy Consumption
1. Primary energy consumption

$$= \frac{(\text{Construction or materials cost}) \times (1963 \text{ Index})}{(\text{Index at time of construction})}$$

Metric Conversion
1 Btu = 1050 joules = 252 cal

Sources
Obtained from previously published information (19).

APPENDIX C

PRIMARY ENERGY EQUIVALENT
OF ELECTRICITY AND FUELS

C-1 PRIMARY ENERGY EQUIVALENT OF ELECTRICITY

Energy Consumption

1. Primary energy consumption to produce and transmit electricity
   $= 3.87$(Electrical energy consumption)
   $= 3.87$(3713 Btu/kwh)(Electrical energy consumption)
   $= (13,208$ Btu/kwh)(Electrical energy consumption)

Source

Obtained from previously published information (19).

C-2 HEATING VALUES AND PRIMARY ENERGY
EQUIVALENTS OF FUELS

Fuel	Heating Value	Primary Energy / Delivered Fuel Energy
Anthracitic and		
Bituminous coal	13,500 Btu/lb	
Fuel oil No. 1	19,700 Btu/lb	
	137,000 Btu/gal	
Fuel oil No. 2	19,500 Btu/lb	
	142,000 Btu/gal	
Fuel oil No. 4	19,200 Btu/lb	
	145,000 Btu/gal	
Fuel oil No. 5	19,000 Btu/lb	
	148,000 Btu/gal	
Fuel oil No. 6	18,700 Btu/lb	
	151,000 Btu/gal	
Gasoline	20,000 Btu/lb	1.208
	124,000 Btu/gal	
Diesel fuel	19,800 Btu/lb	1.208
	140,000 Btu/gal	
Natural gas	1,050 Btu/ft^3	1.169
Propane		1.208
Liquid	21,600 Btu/lb	
	91,500 Btu/gal	
Gas	2,500 Btu/ft^3	
Butane		1.203
Liquid	21,200 Btu/lb	
	102,600 Btu/gal	
Gas	3,300 Btu/gal	
Kerosene	20,500 Btu/gal	

Adapted from References (19-21).

ABBREVIATIONS AND SYMBOLS

BOD_5	—	biochemical oxygen demand, 5 day, 20°C
Btu	—	British thermal unit
cal	—	calorie
ft	—	foot
g	—	gram
gal	—	gallon
gpm	—	gallon per minute
hp	—	horsepower
hr	—	hour
kg	—	kilogram
kw	—	kilowatt
kwh	—	kilowatt-hour
l	—	liter
lb	—	pound
lb_f	—	pound force
lb_m	—	pound mass
m	—	meter
mg	—	milligram
MGD	—	million gallons per day
psi	—	pounds per square inch
psig	—	pounds per square inch gage
rpm	—	revolutions per minute
sec	—	second
TDH	—	total dynamic head
yr	—	year
°C	—	degree Centigrade
°F	—	degree Farenheit
°K	—	degree Kelvin
°R	—	degree Rankine
%	—	percent

REFERENCES

1. Smith, R. *Electrical Power Consumption for Municipal Wastewater Treatment,* National Environmental Research Center, Office of Research and Monitoring, U.S. Environmental Protection Agency, Environmental Protection Technology Series EPA-R2-73-281, U.S. Government Printing Office, Washington, D.C., July 1973.
2. Maughan, W. D., *et al. Effects of Energy Shortage on the Treatment of Wastewater in California,* State Water Resources Control Board, State of California, April 1974.
3. Tchobanoglous, G., B. Ostertag, and E. Fernbach. *Wastewater Management at Hickmott Foods, Inc.,* presented at the Sixth National Symposium on Food Processing Wastes, Madison, Wisconsin, April 1975.
4. Harrison, G. L. Personal communication, March 1975.

5. Fernbach, E., and G. Tchobanoglous. "Centrifugal Screen Concentrator for Activated Sludge Process," *Water & Sewage Works*, Part I, Vol. 122, No. 1, January 1975, Part II, Vol. 122, No. 2, February 1975.

6. Metcalf & Eddy, Inc. *Wastewater Engineering: Collection, Treatment, Disposal*, McGraw-Hill Book Company, New York, 1972.

7. Engineering-Science, Inc.: *Investigation of Response Surfaces of the Microscreen Process*, prepared for the Office of Research and Monitoring, U.S. Environmental Protection Agency, Water Pollution Control Research Series 17090 EEM 12/71, U.S. Government Printing Office, Washington, D.C., December 1971.

8. Tchobanoglous, G. 'Wastewater Treatment for Small Communities," *Public Works*, Part I, Vol. 105, No. 7, July 1974, Part II, Vol. 105, No. 8, August 1974.

9. *Bio-Surf Design Manual*, Autotrol Corporation, Bio-Systems Division, Milwaukee, Wisconsin, 1972.

10. "Pollution Control with Aqua-Jet Aerators," Aqua-Aerobic Systems, Rockford, Illinois, February, 1974.

11. *PFT Water Quality Control Equipment*, Binder No. 340, Envirex, Inc., Water Quality Control Division, Waukesha, Wisconsin, August 1973.

12. "Dorr-Oliver Draft Tube Digester for Municipal Waste Treatment Plants," Bulletin Dig. 67-1, Dorr-Oliver Incorporated, Stamford, Connecticut, 1967.

13. "Eimco Digester Covers, Sludge Mixers, Digester Heaters," Eimco Processing Machinery Division, Salt Lake City.

14. Unterberg, W., R. J. Sherwood, and G. R. Schneider. *Computerized Design and Cost Estimation for Multiple-Hearth Sludge Incinerators*, prepared for the Office of Research and Monitoring, U.S. Environmental Protection Agency, Water Pollution Control Research Series 17070 EBP 07/71, U.S. Government Printing Office, Washington, D.C., July 1971.

15. Voegtle, J. A. "Be Conservative About Energy," *Deeds & Data*, published by Water Pollution Control Federation, February 1975.

16. "Better Control to Cut Cost Inflation in Half," *Engineering News-Record*, Vol. 188, No. 12, March 1972.

17. *Chemical Engineering*, Vol. 79, No. 8, November 1972.

18. "Sewage Treatment Plant and Sewer Construction Cost Indexes," Program Assessment Branch, Municipal Waste Water Systems Division, Office of Water Program Operations, U.S. Environmental Protection Agency.

19. Roberts, E. B., and R. M. Hagan. *Energy Requirements of Alternatives in Water Supply, Use, and Conservation*, Department of Water Science and Engineering, University of California, Davis, California, November 1974.

20. Perry, J. H., C. H. Chilton, and S. D. Kirkpatrick, editors, *Chemical Engineers' Handbook*, 4th Edition, McGraw-Hill Book Company, New York, 1963.

21. *Sewage Treatment Plant Design*, ASCE Manual of Engineering Practice No. 36, American Society of Civil Engineers and Water Pollution Control Federation, New York, 1959.

13.

Total Wastewater Recycling and Zero Discharge in St. Petersburg, Florida

Lloyd A. Dove*

The massive municipal wastewater treatment construction program administered by the U.S. Environmental Protection Agency will have a direct and vital effect on agricultural waste management.

The $18 billion Federal Water Pollution Control Act Amendments of 1972, hereafter referred to as Public Law 92-500, are designed to restore and maintain the integrity of the nation's waters. The wastewater technology utilized in this program will determine the amount of energy and natural resources consumed in the operation and maintenance of the new or improved wastewater facilities for many years to come.

The unique technology of the St. Petersburg zero discharge and total wasterwater recycling system will minimize these requirements and will return the treated wastewater and digested sludge to the land in the form of critically needed water and fertilizer.

Agriculture should be concerned with the technology utilized in the implementation of Public Law 92-500 as the energy, water, chemicals and fertilizer conserved or wasted in the program will have considerable impact on the limited natural resources that our urban and rural communities need to sustain our quality of life.

1. It is estimated that the national water consumption of 16 billion gallons per day for domestic use in 1970 will double by the year 2000. Water quality and quantity problems can now be cited in

---

* P.E., Black, Crow & Eidsness, Inc., Clearwater, Florida

every state. In Florida, with a continued weekly increase of 6,000 new residents, all readily available water could be used up by 1985, if strict conservation and reuse is not practiced.

2. The consumption of oil already exceeds our national production and we must import millions of barrels annually. The proposed wastewater systems will require awesome quantities of oil for power. This requirement can be minimized with cost-effective approaches.

3. It has been estimated that it would take 2¼ billion gallons of crude oil per year to produce the nitrogen fertilizer discharged to the nation's waterways each year from wastewater systems.

Further examples could be cited, but the point is that technology is available to conserve nonrenewable natural resources in the water pollution control program. Every wastewater facility to be constructed or improved should ultimately incorporate the reuse of the treated wastewater and digested sludge it produces and should minimize the energy and chemicals required in the collection and treatment processes.

Agriculture must take a more active role in the water pollution control program. It should demand the conservation of our limited resources and contribute its technology relative to the utilization of nutrient-rich treated wastewater and digested sludge for land applications. Only minimal progress has been made toward this objective to date.

The Sierra Club suggests that we make nature a partner. I would suggest that the urban and agricultural interests of the nation become partners in the continued implementation of Public Law 92-500. If they do not, the current scattered legal skirmishes for water, energy, and fertilizer between the two sectors will become critical and widespread in the decade ahead. The St. Petersburg project described hereafter is an example of what can be accomplished with cooperation.

Wastewater recycling can provide a safe, dependable supply of water for irrigation, heating and air conditioning, power plants, industries, and other nonpotable urban purposes. It is the next natural step toward recycling wastewater for potable purposes.

The City of St. Petersburg, Florida, committed itself to total recycling of its wastewater and zero discharge to its surrounding bays in 1971. Some of the incentives for this action included:

1. Advanced waste treatment, or an approved alternate, is required by Florida's Wilson-Grizzle Act of 1972, for all treatment facilities in the Tampa Bay Region.

2. Typical advanced treatment facilities require massive use of scarce chemicals and electric power.

3. Wastewater nutrients can reduce the need for commercial fer-

tilizers in urban areas if the wastewater is disinfected and applied as an irrigant at appropriate spraying rates.

4. Digested sludge can also be used as a soil conditioner and fertilizer if processed and applied in an environmentally acceptable manner.

5. Utilizing wastewater will reduce the damaging drawdown on the Floridan Aquifer, from which the State withdraws about 90% of its potable water.

RESEARCH AND DEVELOPMENT

The reuse of treated wastewater by landspreading has been practiced in the United States since late in the 19th century. Regarding past experience in landspreading, it is noted that most existing systems have not employed acceptable technology, monitoring and management, and have been potentially harmful to man and his environment. The St. Petersburg research program has provided hard data for improvement of existing systems and design of acceptable new systems.

The protection of both man and his environment has been the guiding principle in the St. Petersburg wastewater recycling program. Every feasible safeguard is being employed in an attempt to assure a "fail-safe" technology.

Fortunately for its research and development program, St. Petersburg had no public water supply drawing from the aquifer beneath it. This provided an excellent setting for landspreading. Although most of the cost for the three years of research was borne by the City of St. Petersburg, many regional, state and federal agencies participated, as shown in Figure 1.

The State Department of Health and Rehabilitative Services shared the cost of the viral research program. A dependable process had to be developed to inactivate the human virus in the wastewater, due to public exposure in lawn sprinkling and other potential urban reuses.

Dr. Flora Mae Wellings, Administrator of the State Epidemiology Research Center in Tampa, concluded from her studies that, "One would anticipate a virus-free effluent as a finished product if turbidity can be held to 0.5 JTU by a breakpoint chlorination of 0.5 ppm for at least 60 minutes." Her studies have been critical of normal spray irrigation procedures because human viruses have been detected in groundwater to a 20-ft (6-m) depth. She fully approves the St. Petersburg system since it includes virus inactivation prior to reuse of the effluent. The results of some of her many tests conducted at the City's Northwest Wastewater Treatment Plant and effluent irrigation site are summarized in Figure 2.

CITY: PARKS AND RECREATION ADMINISTRATION
             ECKERD COLLEGE

COUNTY: PINELLAS COUNTY HEALTH DEPARTMENT

REGIONAL: TAMPA BAY REGIONAL PLANNING COUNCIL
             SOUTHWEST FLORIDA WATER MANAGEMENT
             DISTRICT

STATE: DIVISION OF HEALTH, DEPT. OF HEALTH
             AND REHABILITATIVE SERVICES
             DEPARTMENT OF POLLUTION CONTROL
             DEPARTMENT OF TRANSPORTATION

FEDERAL: U.S. GEOLOGICAL SURVEY
             SOIL CONSERVATION SERVICE, U.S. DEPT.
             OF AGRICULTURE

CONSULTANTS:
             BLACK, CROW & EIDSNESS, INC.

Figure 1. Participants in effluent irrigation research program, City of St. Petersburg, Florida.

SOURCE	TOTAL ISOLATES/ LITERS TESTED	AVERAGE No. PFU/LITER (I)
RAW WASTEWATER	19/2.5	7.600
AERATOR EFFLUENT	15/1	15.000
GREEN SLUDGE	33/6.78	4.867
FINISHED EFFLUENT	165/9	18.333
CHLORINATED EFFLUENT	12/551.88	0.022
SPRAY	3/378.0	0.007
WEIR FOLLOWING SUBDRAINS	0/869.0	0
LAKE	0/283.5	0
(I) PFU — PLAQUE FORMING UNITS (VIRUSES)		

Figure 2. Example of virus analyses by F. M. Wellings.

The County Agricultural Officer provided agronomy and soil studies and recommendations on the effluent irrigation site plan, shown in Figure 3. The U.S. Geological Survey shared the cost and conducted the spray irrigation tests. Acceptable lawn sprinkling rates of 2 to 4 in. (5.08 to 10.16 cm) per week were established for the typical native grasses. The rates were based on the reduction of nutrients and possible toxic elements to an acceptable level prior to reaching the groundwater table. The results of one of the spray irrigation test series are given in Figure 4.

The State Department of Natural Resources shared the cost and participated in the deep well injection research program. To assure zero discharge, it proved necessary to provide a 100% backup for the wastewater recycling system for periods when treated wastewater supply exceeded demand, or when effluent quality standards were not met. Injection tests have shown that a suitable zone for injection is present at a depth of 1180 to 1270 ft (360-390 m). This

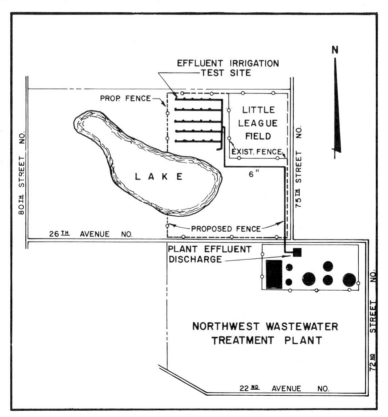

Figure 3. Effluent irrigation research site, City of St. Petersburg, Florida.

SAMPLING POINT:	NITROGEN AND PHOSPHORUS CONTENTS OF GROUNDWATER, MG./l, AFTER:	
	2 WEEKS SPRAYING AT 2 IN./WEEK	
	TOTAL N - N	TOTAL PO_4-P
PLANT EFFLUENT	17.0	2.9
5 FT. WELL	----	----
IO FT. WELL	4.8	0.2
20 FT. WELL	1.1	0.5
WEIR FROM DRAIN TILES	5.2	0.7
LAKE	15.0 (1)	0.5
	6 WEEKS SPRAYING AT 4 IN./WEEK	
	TOTAL N - N	TOTAL PO_4-P
PLANT EFFLUENT	19.0	2.6
5 FT. WELL	9.6	3.0
IO FT. WELL	5.4	0.2
20 FT. WELL	0.8	0.5
WEIR FROM DRAIN TILES	3.3	2.4
LAKE	1.4	0.2
	6 WEEKS SPRAYING AT 11 IN./WEEK	
	TOTAL N - N	TOTAL PO_4-P
PLANT EFFLUENT	20.0	4.2
5 FT. WELL	13.0	5.4
IO FT. WELL	12.0	3.0
20 FT. WELL	3.3	0.6
WEIR FROM DRAIN TILES	6.2	3.0
LAKE	1.0	0.1
(1) HIGH ORGANIC - NITROGEN VALUE RESULT OF ALGAL BLOOM.		

Figure 4. Example of spray irrigation analyses of U.S. Geological Survey.

zone has a high transmissivity of 800,000 gpd/ft (3000 m³/day per m) and contains water that is similar in quality to sea-water. Analysis of data from the test drilling and related operations indicates that injection of the highly treated wastewater at a rate of 6 mgd (23,000 m³/day) per well is technically feasible. Further testing and monitoring is scheduled for the year ahead.

Black, Crow & Eidsness, Inc., served as consulting engineers in all facets of the research and development program.

THE WASTEWATER RECYCLING PLAN

St. Petersburg has four regional wastewater treatment plants which presently provide secondary treatment. They are being designed to accommodate a projected population of 700,000 for the year 2000.

The first improvement under the regional facilities plan will be the expansion of the existing secondary Southwest Wastewater Treatment plant from 9 mgd (34,000 m³/day) to 20 mgd (76,000 m⁹/day. A 20-mgd (76,000-m³/day) multimedia filter, flash mixing of chlorine and extended contact time, a 14-million-gallon (53,000-m³) chlorinated retention lake, a 14-mile (23-km) wastewater distribution system, and a 20-mgd (75,000-m³) deep well injection system are added design elements. Similar improvements should be underway within the year to serve the other three regional plants.

The letting for the Southwest Plant improvements was held on November 15, 1974. Lowest and best bids were as follows:

a) Expansion and Improvement of Southwest Wastewater Treatment Plant, Gulf Contracting, Inc. $11,689,000.00

b) Construction of Deep Monitoring and Disposal Wells, Singer-Lane Atlantic Co., Inc. $ 1,639,512.50

c) Treated Wastewater Distribution System for Southwest District, Boyce Company $ 1,957,727.00

     Total Construction Cost $15,286,239.50

It is estimated that a conventional advanced wastewater treatment plant would cost $4 million more to construct and $0.5 million more annually to operate than the unique process utilized. Conventional AWT would also require the massive quantities of critical chemicals and would remove rather than preserve the valuable nutrients.

Cost-effectiveness includes the consideration of the environment as well as dollars. Figure 5 provides a list of the green areas to be irrigated in the first 6 to 12 months of operation of the St. Petersburg Southwest Treatment Plant. The irrigation will be conducted under strict R&D conditions at no cost to the users. Thereafter, the treated wastewater will be sold to all users in the Southwest district at a price not to exceed the cost of treatment. A city ordinance has been passed to this effect without objections.

Studies have also proven the feasibility of using this same treated wastewater as condensing water for low-energy air conditioning systems along the route of the distribution lines. Considerable interest has been generated in the potential savings in electrical power

LOCATION

POINT BRITTANNY 18-HOLE GOLF COURSE	54 TH AVENUE, SOUTH PARKWAY
	LAKEWOOD COUNTRY CLUB
ECKERD COLLEGE	
	BOYD HILL NATURE TRAIL ——
SOUTHWEST PLANT	LAKEWOOD PARK
MAXIMO PARK	I-75 RIGHT - OF - WAY
	FLORIDA POWER CORPORATION
MAXIMO MOORINGS RESIDENTIAL SUBDIVISIONS	
	ALLSTATE INSURANCE
POINT BRITTANNY CONDOMINIUMS AND BAYWAY ISLES RESIDENTIAL SUBDIVISIONS	LAKE VISTA PARK
	LAKEWOOD HIGH SCHOOL

Figure 5. First phase spray irrigation areas Southwest Plant Service District.

for the water-type air conditioning. For those of you in the colder climes, the potential power savings of using treated wastewater in heat pumps is even greater.

THE DIGESTED SLUDGE RECYCLING PLAN

The usually superb television show "60 Minutes" recently did a rather humorous but technically weak feature entitled "Here Comes The Sludge." Unfortunately, even some of our engineers take sludge treatment and handling too lightly. About 40 to 50% of the cost of a modern wastewater treatment system is involved in the processing of sludge.

For a comprehensive look at this subject I would recommend the U.S. Environmental Protection Agency Process Design Manual For Sludge Treatment and Disposal, which was prepared by our firm. I will limit my comments to our St. Petersburg experiences.

Outstanding success has been achieved in the truck-spraying of digested sludge on green areas of the city. It is seen in Table 1 that 28.5% of the production from the four city treatment plants has been recycled in this manner.

Good progress has also been made in the establishment of an experimental sod farm and nursery. The first phase of operations at this site contains 23 tillable acres. Sludge from the city and much of the remainder of Pinellas County is trucked to the sod farm and sprayed on the surface by an underground sprinkler system. The sludge is then disced or rototilled into the soil.

The preceding methods of sludge handling are performed with

Table 1. Summary of digested sludge production in St. Petersburg for the year 1974.

W.W.T.P. No. & Description	Digested Sludge Production Parameters	January	February	March	April	May	June	July	August	September	October	November	December	TOTALS
No. 1—Albert Whitted	Total Solids, %	1.2	1.4	1.7	1.12	1.15	1.3	1.8	1.54	1.28	1.5	1.4	1.88	1.44 (Avg)
	Sludge, to Landfill, gal	1,512,800	1,262,100	1,413,900	1,742,600	2,057,900	1,363,000	1,530,300	841,400	846,800	1,455,100	752,400	1,367,300	16,145,600
	Sludge, Truck Sprayed, gal	479,500	501,700	582,600	1,091,800	454,000	290,900	354,000	247,400	329,100	507,500	524,100	590,000	5,852,600
	Sludge, to Landfill, Dry Tons	76	74	100	81	99	74	115	54	45	91	44	107	960
	Sludge, Truck Sprayed, Dry Tons	24	29	41	51	22	16	27	16	18	32	31	46	353
No. 2—Northeast	Total Solids, %	1.4	2.1	1.5	1.5	1.6	1.45	1.7	1.65	1.6	1.55	1.6	1.55	1.6 (Avg)
	Sludge, to Landfill, gal	912,000	664,800	879,400	879,400	861,700	926,200	947,800	1,032,000	831,200	790,200	853,000	766,700	10,344,400
	Sludge, Truck Sprayed, gal	—	—	—	—	—	—	—	—	—	—	—	—	—
	Sludge, to Landfill, Dry Tons	53	58	55	55	58	56	67	71	56	51	57	77	714
	Sludge, Truck Sprayed, Dry Tons	—	—	—	—	—	—	—	—	—	—	—	—	—
No. 3—Northwest	Total Solids, %	1.7	1.7	1.8	2.0	1.8	1.95	1.95	2.0	2.3	2.6	2.0	1.9	1.98 (Avg)
	Sludge, to Landfill, gal	919,200	995,600	948,000	672,000	964,000	761,800	761,800	700,000	711,400	718,900	987,500	1,035,200	10,175,400
	Sludge, Truck Sprayed, gal	42,000	—	—	40,600	—	—	—	—	—	—	—	—	82,600
	Sludge, to Landfill, Dry Tons	65	71	71	56	72	62	62	58	68	78	82	82	827
	Sludge, Truck Sprayed, Dry Tons	3	—	—	3	—	—	—	—	—	—	—	—	6
No. 4—Southwest	Total Solids, %	1.47	3.0	1.86	1.64	1.73	1.66	1.48	1.25	1.19	2.68	1.38	1.63	1.75 (Avg)
	Sludge, to Landfill, gal	—	—	—	—	82,700	598,200	135,600	—	—	—	—	—	816,500
	Sludge, Truck Sprayed, gal	490,100	588,700	330,600	147,900	865,100	668,700	708,400	1,264,100	1,264,100	781,700	925,200	849,700	8,884,300
	Sludge, to Landfill, Dry Tons	—	—	—	—	6	41	8	—	—	—	—	—	55
	Sludge, Truck Sprayed, Dry Tons	30	73	26	10	62	46	44	66	63	87	53	58	618
TOTALS FOR EACH MONTH	Total Solids, Avg	1.44	2.05	1.72	1.57	1.57	1.59	1.73	1.61	1.59	2.08	1.6	1.74	1.69
	Sludge, to Landfill, gal	3,344,000	2,922,500	3,241,300	3,294,000	3,966,300	3,649,200	3,375,500	2,573,400	2,389,400	2,964,200	2,592,900	3,169,200	37,481,900
	Sludge, Truck Sprayed, gal	1,011,600	1,090,400	913,200	1,280,300	1,319,100	959,600	1,062,400	1,511,500	1,593,200	1,289,200	1,449,300	1,439,700	14,919,500
	Sludge, to Landfill, Dry Tons	194	203	226	192	235	233	252	183	169	220	183	266	2,556
	Sludge, Truck Sprayed, Dry Tons	57	102	67	64	84	62	71	82	81	119	84	104	977

careful consideration of the environment under permits from the State Department of Pollution Control. Nevertheless, we must consider the long-term potential of these methods in this dense urban setting. The production of digested sludge from the city plants will increase from 3533 dry tons in 1974 to an estimated 8000 dry tons in the year 2000, as shown in Table 2.

Table 2. Projected digested sludge production in St. Petersburg for the period 1974-2000. (Based on population projections of the Tampa Bay Regional Water Quality Management Plan.)

WWTP No. & Description	Act. Prod 1974 Million Gallons	Dry Tons	Est. Prod. 1980 Million Gallons	Dry Tons	Est. Prod. 1990 Million Gallons	Dry Tons	Est. Prod. 2000 Million Gallons	Dry Tons
No. 1—Albert Whitted	22.1	1313	26.2	1562	33.2	1970	39.9	2371
No. 2— Northeast	10.3	714	15.2	1057	23.4	1621	31.5	2182
No. 3— Northwest	10.3	833	12.4	1000	15.9	1283	19.4	1566
No. 4— Southwest	9.7	673	13.6	942	20.0	1390	26.5	1837
TOTALS, All Plants	52.4	3533	67.4	4561	92.5	6264	117.3	7956

The current recommended practice for truck-spraying of digested sludge in the city is for the user to follow the sludge application with sprinkling by potable water to wash the organics into the soil and minimize the period in which the musty odor is emitted by the digested sludge. Nevertheless, the digested sludge does contain viruses, pathogens and other elements which could be hazardous to health. The EPA has reported no clinical evidence of disease caused by the landspreading of sludge in the United States, but is studying this practice closely. The Florida Department of Health prohibits the landspreading of sludge on green areas where contact sports may be played. We must consider the possibility that disinfection of digested sludge prior to truck-spraying might be required in urban areas in the future, and the cost would be prohibitive.

The experimental sod farm and nursery has good potential for utilizing large quantities of digested sludge in an environmentally acceptable manner. The estimated future requirement of up to 300 acres of tillable land, plus added land for support facilities for this purpose in its current location, however, raises many questions. The sod farming operation may become objectionable to planned developments in this vicinity. Further, pressure to convert some of the city's landholdings in and around the sod farm to parks, schools, or other purposes may be forthcoming.

Thus, truck-spraying and the sod farm have excellent potential for satisfying the city's handling needs for the next 5 to 10 years. Presently, a study relative to long-term sludge handling plans is being conducted for the city.

At this point in time, the production and sale of dried sludge to commercial fertilizer manufacturers has great potential. The following facts are pertinent to this consideration:

1. There are 35 Florida commercial fertilizer firms which purchase from 80,000 to 100,000 tons of dried sludge annually, primarily from Chicago and Houston. Florida sludge is not purchased because the manufacturers require a low moisture content of about 5%. No Florida treatment plants dry their sludge to this low moisture level.

2. The 5% dried sludge is considered one of the most important of a dozen ingredients utilized in many commercial fertilizers as it is an excellent stabilizer plus it provides considerable nitrogen and phosphorus.

3. A review of potential air pollution problems from the sludge drying process leaves little concern that state and federal standards can be met. We are burning a clean fuel (natural gas, oil or methane gas) and drying a homogeneous substance (sludge).

4. The cost-effectiveness and market for the 5% dried sludge looks extremely good.

A combination of truck-spraying, sod farm and the sale of dried sludge to commercial manufacturers should assure the complete recycling of St. Petersburg's digested sludge for many years to come.

Table 3. Comparison of composition of dried sludge in Chicago, Houston and St. Petersburg.

Composition of Dried Sludge	St. Petersburg	Chicago	Houston
Moisture, %	4.6	5.0*	5.0
Nitrogen, as N	6.06	4.81-5.6	5.34
Phosphorus as P_2O_5	7.74	6.86-6.97	
Phosphorus as H_3PO_4			3.93

* % moisture reported when received in Florida. Moisture content from drying process in Chicago is reported at 8% to 10%, but further drying likely occurs in storage and shipping.

CONCLUSIONS

Wastes are resources that we are just beginning to utilize in the United States. The theme of Energy, Agriculture, and Waste Management is most appropriate for our time. The collective efforts of the agricultural and urban sectors of our nation can provide the technology and the resources to not only sustain, but improve on our quality of life.

REFERENCES

1. Dove, L. A. "Recycling Allows Zero Wastewater Discharge," *Civil Engineering Magazine* (February 1975).
2. U.S. Dept. of Interior. "The River of Life," Environmental Report, Conservation Yearbook Series, Volume No. 6.
3. Cherry, *et al.* "Hydro-Biochemical Effects of Spraying Waste-Treatment Effluent in St. Petersburg, Florida," United States Geological Survey, Tallahassee, Florida (1975).
4. Black, Crow & Eidsness, Inc. "Wastewater Quality Management Plan For St. Petersburg Wastewater District," Clearwater, Florida (June 1973).
5. Welling, F. M. Spray Irrigation Research Project. Epidemiology Research Center, Division of Health, Tampa, Florida. December, 1972.
6. Black, Crow & Eidsness, Inc. "Sludge Treatment and Disposal," EPA Technology Transfer Manual, Gainesville, Florida (October 1974).
7. *Compost Science,* Volume 16, No. 1, p. 22 (January-February 1975).
8. Florida Division of State Planning. "Study of Floridan Aquifer and Green Swamp Recharge Area" (1974).

14.

Automated Handling of Poultry Processing Wastes

W. K. Whitehead, R. E. Childs and E. J. Lloyd*

More than 3 billion head of broiler class poultry are slaughtered annually under federal inspection in the United States (1) using an average as reported by one study (2) of over 10 gal of water per bird. Water alone costs the industry about $15 million annually and in 1966 poultry processing industry expenditures for waste treatment cost about $6.6 million (3).

The poultry processing industry has rapidly progressed to a high degree of automation. Automated handling of the inedible wastes generated during processing (blood, heads, feet, viscera) is a vital part of the scheme since it not only reduces the labor and water requirements but can also reduce waste load of the water discharged.

The objective of this work was to develop improved waste handling systems to effectively collect and remove wastes from poultry processing plants. Requirements were that the systems be automated, reliable, reduce water requirements for processing and reduce pollution of plant effluent.

EXPERIMENTAL INSTALLATIONS

This paper reports on three experimental waste handling systems that were designed and installed in two commercial poultry processing plants. A vacuum system for handling collectable blood in

* Agricultural engineer, industrial engineer, and engineer technician respectively, Richard B. Russell Agricultural Research Center, Southern Region, Agricultural Research Service, U.S. Department of Agriculture, P.O. Box 5677, Athens, Georgia 30604.

the slaughter area and a negative pressure pneumatic system (4) for handling solid wastes (heads, feet, oil sacs, intestinal tracts and crops/windpipes) were installed in one plant. A third system, installed in a second processing plant, was similar in design to the vacuum blood system but was developed for handling both blood and lungs. These systems were designed with enough flexibility to permit developing basic design criteria for waste handling systems for processing plants having different production capacities. They were tested and modified as necessary to improve their reliability and effectiveness.

Vacuum System

The two vacuum systems were designed to 1) collect the blood and contain it in the slaughter area to prevent messiness and facilitate pick-up, 2) pick up and transport the blood and lungs to collection tanks at predetermined intervals, and 3) discharge the materials from the collection tank.

Basic components of the blood system (Figure 1) were two basins for collecting and holding blood in the slaughter area, a 500-gal collection-storage tank, a 400 cu ft/min vacuum pump with a 30-hp electric motor and necessary pipe, valves and fittings. Except for certain features of the collection tank, all components are standard parts or simple designs. System controls were designed to completely automate pickup and discharge of blood.

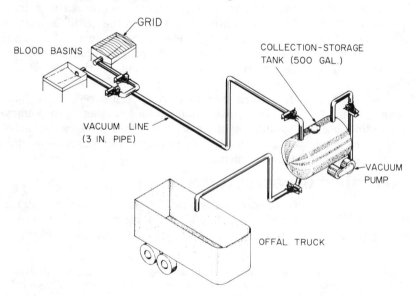

Figure 1. Vacuum blood collection system.

Because blood coagulates rapidly and contains loose feathers that fall from the slaughtered birds, a 3-in pipe was used to pick up the material. A vacuum pump with rated capacity of 400 cu ft/min at 24 in. of mercury was selected to move and lift the clotted blood. It was not necessary to determine the exact vacuum pump size for this experimental system since the vacuum normally would be supplied by a large pump used for other purposes in the processing plant.

The collection-storage tank was a modified 500-gal propane tank. Ports for air evacuation, material entrance and exit, visual inspection, gages, probes and other devices were installed. A baffle was installed inside the tank to prevent material from being pulled into the pump and an electrical probe (modified milk tank probe) was installed to extend 4 in. below the baffle to prevent overfilling the tank. A CIP (clean-in-place) assembly completed the installation.

Two catch basins (Figure 2), each placed under a defeathering line, were designed to hold about 80 gal of blood each. The two defeathering lines process a total of 9600 birds/hr which produces a total of about 125 gal/hr of collectable blood.

Blood is automatically transferred from the two blood basins to

Figure 2. Collection troughs and basins for blood system.

the collection-storage tank at predetermined intervals throughout the day; the volume of blood is too small for continuous pick-up. At the end of 1 hour operation the system is energized and the pump started to initiate the pick-up cycle, placing vacuum on the collection-storage tank. One of the quick-opening valves (Figure 1) to one basin opens and blood is evacuated and moved to the collection-storage tank. The valve to the first basin closes and the valve to the second opens. Blood in the second basin is moved to the collection-storage tank, the valve closes and the system then shuts down. After 1 hour the pick-up cycle is repeated and the system enters the discharge cycle. During the discharge cycle, the valve at the collection tank on the material inlet line and the valve between the tank and vacuum pump are closed and air pressure (15 psi) is applied to the tank. Before discharge, a warning whistle blows and a red light flashes in the offal room to insure that the offal truck is in place. At discharge time, the discharge valve opens and the blood discharges. Just before the tank is emptied, the CIP system is activated to spray the tank interior with cold water which mixes with and is discharged with the final portion of blood. This completes one overall 2-hour cycle.

The blood and lung system (Figure 3) has two defeathering lines (total 12,000 birds/hr) passing over one catch basin (Figure 4). It

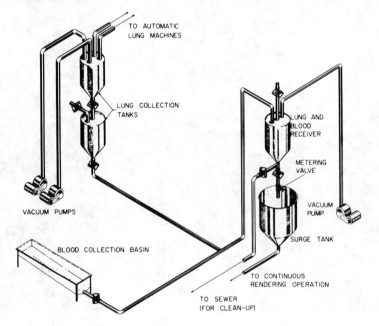

Figure 3. Vacuum blood and lung collection system.

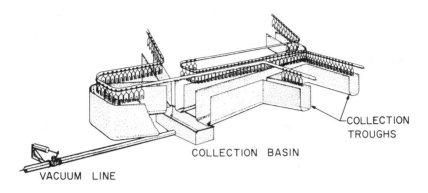

COLLECTION TROUGHS

COLLECTION BASIN

VACUUM LINE

Figure 4. Blood collection troughs and basin for blood and lung system.

picks up blood from the collection basin and lungs from the lung tank at 7.5 min intervals and dispenses blood and lungs from the receiver tank into a surge tank at the on-site continuous rendering operation.

Two tanks (Figure 3) are provided for lung collection so that lung pick-up by the waste system can be effected without interrupting continuous lung removal in the eviscerating room. Normally the valve between the lung collection tanks is open and lungs collected in the top tank drain into the lower tank. When the pick-up cycle begins, the valves on the receiver tank (Figure 3) are closed, the waste system vacuum pump started and vacuum is applied to the receiver. The valve on the blood collection basin opens, blood is removed from the basin and transported to the receiving tank and the valve closes. At the start of lung pick-up the valve between the two lung collection tanks closes to maintain the vacuum on the automatic lung guns and isolate the lower tank. The vacuum is then released on the lower lung tank, the valve on the tank discharge opens and lungs are pulled through the vacuum line into the receiver.

The receiver tank is mounted directly above the surge tank located in the rendering plant adjacent to the processing plant. The receiver is a cyclone-type tank with about 500-gal capacity. Vacuum is supplied by a vacuum pump with a rated capacity of 375 cu ft/min at 18 in. of mercury. Collected blood and lungs drain into the surge tank when the valve on the bottom of the receiver is opened. The metering valve is manually adjusted to allow the product to be continuously fed into the surge tank.

A warning system was developed for the mechanical and electrical controls so that malfunctions can be quickly located and corrected to avoid plant shutdown. The failsafe aspects of the system provide for normalizing the controls in the event of temporary interruption of air or electrical services.

Pneumatic System

The negative pressure pneumatic system (Figure 5) currently handles the heads and feet from all poultry processed (9600 birds/hr) and viscera (oil sacs, intestinal tracts, crops and windpipes) from one eviscerating trough (4800 birds/hr). At this production rate, the system as installed handles a waste load of: heads, 16 lb/min; feet, 22 lb/min; intestinal tracts, oil sacs and crops/windpipes, 25 lb/min.

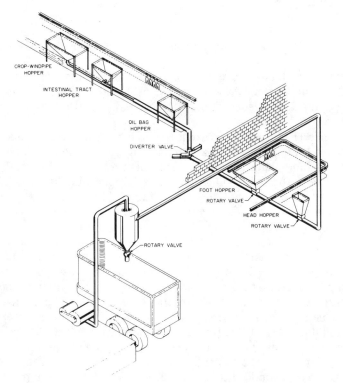

Figure 5. Layout of pneumatic waste handling system.

Vacuum is supplied with a 40-hp blower capable of 900 cu ft/min at 14 in. of mercury. The stainless steel separator or negative receiver is 30-in. OD by 65-in. high with a capacity of about 20 cu ft (Figure 6). It is located on the roof of the plant directly above the offal truck and discharges directly into the truck. A rotary airlock valve with a discharge capacity of 1.2 cu ft/rev operating at 20

Figure 6. Separator and discharge rotary valve for pneumatic system.

rev/min is mounted onto the bottom of the separator to discharge the collected waste. The pump and receiver are connected by a 5-in. OD aluminum tube (Figure 5).

A 5-in. OD aluminum tube extends horizontally about 152 ft from the separator to a 90° elbow, at a point above the head pick-up hopper, then descends vertically about 16 ft to the floor, goes through another 90° elbow and into the rotary valve of the head collection hopper. The head hopper is positioned below the head puller (Figure 7). The tube then extends along the floor into the rotary valve of the foot collection hopper (Figure 8) located directly beneath the foot discharge mechanism. The tube then passes through the wall separating the picking and eviscerating rooms and enters a diverter valve (Figure 5). The diverter valve will permit incorporating a second eviscerating line into the waste handling system.

The 5-in. OD tube extends from the diverter valve along the floor of the eviscerating room to the collection hopper for oil sacs at the beginning of one eviscerating line. The oil sac hopper (Figure 9) is connected directly to the 5-in. tube by a 2-in. OD plastic tube. The

Figure 7. Pneumatic system head collection hopper and rotary valve.

5-in. conveying tube then extends about 32 ft to a manifold. A 2.5-in. PVC pipe extends 12 ft from the manifold to the intestinal tract hopper. The bottom of this hopper has a built-in grid to allow water to drain when the sides are flushed. The air line extends down the side of the hopper and the intestinal tracts are picked up from the bottom through a 2.25-in. by 2.25-in. opening. A 2-in. OD aluminum tube extends from the manifold 23 ft to the far end of the eviscerating trough to a crop-windpipe hopper. Waste material is pulled through a 2-in. hole in the bottom of the hopper. The three hoppers in the eviscerating room do not require rotary valves to feed the material into the air line since all three are open all the time.

RESULTS AND DISCUSSION

The blood system has been operating satisfactorily for 2 years and the blood and lung system for 1 year. The pneumatic solid waste system has been operating successfully for 1 year but the total

Figure 8. Pneumatic system foot collection hopper and rotary valve.

system is not yet completely installed. Changes and refinements have been made to improve performance and reliability. It was not possible to investigate all conceivable variables but sufficient experience has been gained from operating these systems to establish needed information.

Vacuum Systems

Collection basins must be provided to collect and contain the blood in the slaughter area to prevent its getting into the plant effluent. When properly designed so that chickens hang below the sides of the troughs and where an electric stunner is used to minimize terminal struggle, very little blood splatters on the walls or floors. As much slope as possible should be built into the drain troughs and the bottom of the collection basin so that blood will drain into a centrally located sump. Depth of the blood influenced separation of the serum from the clot and a V-shaped bottom in the basin appeared to work better than a flat bottom.

The vacuum line must connect to the lowest position of the basin. The 3-in. pipe appeared to be the smallest practical size for nonclogging operation. A large number of wing feathers in the blood can bridge across a smaller pipe and clog the system. However, there are some hazards in using even the 3-in. pipe because whole

Figure 9. Pneumatic system oil sac hopper.

birds (especially small ones) can enter the line and jam valves and elbows. To avoid this possibility a grid can be placed over the collection basin to screen out large objects (Figure 1). Openings of 2.75 in. seemed to be satisfactory but increasing the opening to 3 in. permitted small chickens to pass through.

In the experimental blood system installation (Figure 1), the blood collected in the collection-storage tank was blown onto the feathers in the offal truck. In a commercial operation it might be desirable to hold the blood in the collection-storage tank for a longer time to expedite further processing such as batch cooking. Also the tank, if located directly over the offal truck, could be discharged by gravity as in the blood and lung system (Figure 3). The receiving tank on the blood and lung system is suitable for dispensing blood into a continuous rendering operation or into any type of transport truck.

When the final effluent of the plant with the blood system was sampled and analyzed (5) weekly for 10 weeks during a period when most of the blood was entering the effluent, the average BOD_5 was 1330 mg/liter. Average water use for processing during this period was about 9 gal/bird. Based on these figures the total plant waste discharge was about 100 lb $BOD_5/1000$ broilers processed. During typical operation the vacuum blood system collected about 13.5 gal of blood/1000 broilers. Based on data presented by Porges (6) showing poultry blood to have a BOD_5 of 92,000 mg/1, this represents about 10.4 lb $BOD_5/1000$ broilers. The blood system would reduce BOD_5 about 10% if all the collectable blood was kept out of the effluent.

Pneumatic System

Pollution abatement effects of the pneumatic waste handling system were measured in the eviscerating room. One eviscerating trough was operated in the conventional manner with all waste material being dropped into the water and the other with all waste being handled pneumatically without entering the water.

Water flow was measured in both eviscerating troughs with a 90° V-notch weir and by collecting it in a container for timed intervals. The trough equipped with the pneumatic waste handling system had a flow rate of 196 gal/min and the other 221 gal/min.

Table 1. Average eviscerating waste loads in lb/1000 broilers processed.

	Pneumatic System	Conventional Flow-Away System
BOD_5	12.5	50.5
Suspended Solids	4.9	21.4
Fat	5.2	43.3

Water samples from each eviscerating trough were collected weekly for 3 weeks and analyzed for BOD_5, suspended solids and fat (5). The total waste load for each system was obtained from the water flow and concentration of the waste in the stream (Table 1). Analyses of the plant's total effluents showed that the eviscerating room contributed 37% of the BOD_5, 32% of the suspended solids and 97% of the fat.

These data show that the pneumatic waste handling system reduced the eviscerating wastes by: BOD_5, 75%; suspended solids, 78%; and fat, 77%. Based on these figures, the pneumatic waste handling system reduced the total plant discharge by: BOD_5, 28%; suspended solids, 25%; and fat, 75%.

The pneumatic waste handling system satisfactorily removed wastes from the processing plant and notably reduced the amount of waste discharged to the sewer. Overall waste reduction cannot be fully assessed until the system is expanded to the other eviscerating line and to other areas where solid wastes accumulate. The amount of water used for processing, especially in the eviscerating area, has also been reduced.

REFERENCES

1. U.S. Department of Agriculture. Agricultural Statistics 1972. (1973).
2. Kerns, W. R. and F. J. Holemo. Cost of waste water pollution abatement in poultry processing and rendering plants in Georgia. Bulletin ERC-0673. University of Georgia. (1973).
3. Federal Water Pollution Control Administration. The Cost of Clean Water. Vol. III. Industrial Waste Profile No. 8: Meat Products. Part II. Poultry Processing. FWPCA Pub. No. I.W.P.-8. (1967).
4. Whitehead, W. K., R. E. Childs and E. J. Lloyd. Pneumatic waste handling system for poultry processing plants. Paper No. 74-6527 presented at the 1974 Annual Meeting American Society of Agricultural Engineers, Chicago, Ill. (1974).
5. American Public Health Association. Standard Methods for the Examination of Water and Wastewater. American Public Health Association, New York, N.Y. (1971).
6. Porges, R. Wastes from poultry dressing establishments. Sewage and Industrial Wastes 22:531-535 (1950).

15.

Energy Conservation at the Dairy Milking Center

D. R. Price*, R. R. Zall,** D. P. Brown,**
A. T. Sobel* and S. A. Weeks*

Energy resources are finally being recognized as finite. Temporary shortages have already surfaced. On-the-farm energy consumption is small (2-4%) compared to the total U.S. use. Nonetheless, energy conservation in all segments is important and the accumulation of savings can have a significant effect on energy use. While food production must be considered a priority use of energy resources, there is a need to operate as efficiently as possible.

The energy consumed on dairy farms for cleaning the milking system is a major component of the total energy use. New designs in parlors, clean-in-place systems, and demands from inspectors for better sanitation in general have resulted in increased consumption of hot water, detergents and sanitizing agents.

On dairy farms that use electric water heaters, approximately ¼ of the electrical power consumed is for heating water. Improved efficiency in cleaning milking equipment could, therefore, result in substantial savings in energy. Dairy farms are continually increasing in size and the economic benefits associated with salvaging water and cleaning supplies for reuse are becoming significant.

The objectives of this study are to investigate technologies to reduce hot water usage and to evaluate the feasibility of recycling cleaning and sanitizing chemicals. The use of soft water, various

---

* Department of Agricultural Engineering, New York State College of Agriculture and Life Sciences at Cornell University, Ithaca, N.Y.
** Department of Food Science, New York State College of Agriculture and Life Sciences at Cornell University, Ithaca, N.Y.

211

water temperatures, recycling detergents and the resultant effect on milk quality are all part of the study.

The project is a joint effort of the Food Science Department and the Department of Agricultural Engineering. The research facility being used in the experimental testing is the New York State College of Agriculture and Life Sciences Animal Science Teaching and Research Center, Harford, New York. The milking center includes a double-ten herringbone parlor with a separate CIP (clean-in-place) system for each half of the herringbone parlor. This arrangement made it possible to set up one side as the control and utilize the other side for experimentation with hot water usage and detergent recycling.

METHOD

The principal thrust of the study is to measure the effects of substituting soft water for hard water, lower water temperatures, and detergent reuse by recycling. Experimental equipment was necessary to permit the several combinations to be studied. All combinations of hard water, soft water, cold water, and hot water were made available at a manifold near the CIP cleaning system. A detergent recycle unit was designed and constructed to connect directly to one side of the CIP system for the double-ten herringbone parlor. (The recycle equipment is described in some detail later in this report.)

Twelve combinations of three water temperatures, (70°F, 100°F and 160°F) hard and soft water, and detergent recycle for six consecutive days were established as the experimental design. Twenty-four treatment combinations were necessary to complete one experiment. The combination of variables requires a one-year test period to complete two replicates of each test. The experiments were designed to obtain a 95% confidence level in the results.

Approximately 350 cows were milked in the double-ten herringbone parlor. Each half of the parlor has its own cleaning system with equipment from conventional "off-the-shelf" units available commercially. The control side cleaning procedure (as recommended by the equipment supplier) was to use a tepid (100°F) prerinse, hot water with detergent, and a tepid rinse after the detergent cycle. The system is sanitized prior to milking and acid washes were used on the control side once per week as recommended. One pound of CIP (standard brand) detergent was used at each wash cycle twice per day. The test side was cleaned with the same detergent (1½ lb initially and 1½ oz added each day) as the control side except that during recycle tests the 39 gallons of detergent solution were captured and recycled twice per day for six days. Recycling equipment was developed to automatically save the detergent.

The cleaning effectiveness was measured by bacteriological monitoring methods. Selected surface areas throughout the CIP system were swabbed and tested for bacterial numbers per unit of surface area. Standard plate counts and coliform counts were used as primary monitoring tools to detect bacterial populations. In addition, the recycled detergent was monitored to determine the dissipation per wash cycle and to measure its bacterial contamination. Visual inspection was made of the equipment at each sample time as a general observation of apparent cleanliness. Samples were collected from the final rinse water level leaving both sides of the system and tested for bacterial contamination. The quality and flavor of the milk was also measured.

DETERGENT RECYCLE EQUIPMENT

The basic concept of detergent recycle is illustrated in Figure 1. During the wash portion of the post-milk cycle the stored detergent is released into the CIP sink. This detergent is circulated through the CIP lines several times by returning and passing through the detergent storage. Thus the detergent solution does not leave the system during this portion of the cycle. Toward the end of the wash portion of the cycle the valve at the bottom of the storage tank is closed and the detergent captured. During the other portions of the pre- and post-milk cycles, the valve at the top of the storage diverts all returning water to waste.

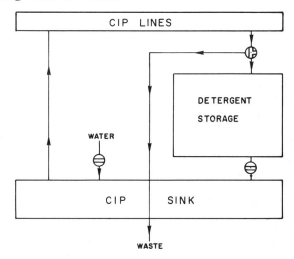

Figure 1. Basic concept of detergent recycle.

A proposed modification of the equipment would allow reuse of the sanitizing solution from the pre-milk cycle for the pre-rinse portion of the post-milk cycle. The basic concept for detergent recycle with sanitizer reuse is illustrated in Figure 2.

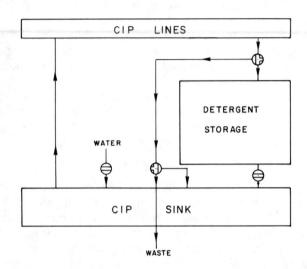

Figure 2. Basic concept of detergent recycle with sanitizer reuse.

The equipment configuration for the experimental unit utilized for the tests described in the study is shown in Figure 3. The detergent storage tank was insulated and a heater was provided to maintain the storage detergent at one of the test temperatures. Diverter valve V_1 allowed the solution returning from the receiving jar to either enter the storage or be diverted to waste. Valve V_5 emptied the storage into the CIP sink and introduced the stored detergent into the system. Water valve V_2 provided water for the initial changing of the detergent tank. Water valves V_3 and V_4 introduced water into the system for other portions of the cycles.

The timing schedule is presented in simplified form in Figure 4. The basic difference between this schedule and a standard CIP cycle occurs during the wash portion. Timing was accomplished electrically by a program timer which allowed experimental changes.

ENERGY CONSERVATION POTENTIAL

The detergent recycle unit would give a savings in detergent, water heating, and water usage as shown. These figures are based on a 7-day recycle period.

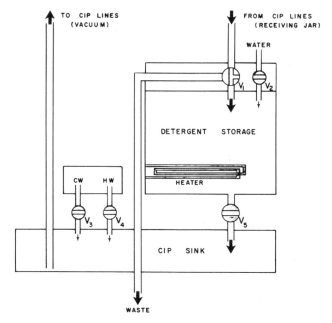

Figure 3. Detergent recycle equipment configuration.

	PRE-MILK	POST- MILK		
	SANITIZE	PRERINSE	WASH	POSTRINSE
VACUUM PUMP	▨▨▨▨▨	▨▨▨▨▨	▨▨▨▨▨▨▨▨	▨▨▨▨
MILK PUMP	▨	▨	▨	▨
STORE DETERGENT, V_1			▨▨▨▨▨▨▨▨	
WATER, V_3 V_4	▨	▨		▨
RECYCLE DET., V_5			▨▨▨▨▨▨	

Figure 4. Timing schedule for detergent cycle.

Detergent Usage

(Based on Detergent Price @ 42¢/lb)
CONTROL: 1 lb/milking → 14 lb/week
→ $5.88/week → $306/year
EXPERIMENTAL: (Based on Detergent Reuse for 1 week)
1½ lb initial + 1½ oz/milking → 2.8 lb/week
→ $1.18/week → $61/year
[Savings of $245/year]

Water Heating

(Based on Electricity Cost @ 2¢/kWh)
CONTROL: Tepid rinse-hot wash-tepid rinse
50 gal water/milking heated from 50°F to 160°F
➔ 43,815 Btu/milking ➔ 9775 kWh/year
➔ $195 + $33(standby loss) ➔ $228/year
EXPERIMENTAL: Cold Rinse-Cold Wash-Cold Rinse
➔ 0 cost
[Savings of $228/year]
      Cold Rinse-Hot Wash-Cold Rinse
39 gal — Initially heated from 50°F to 160°F, then reheated 13
times from 115°F to 160°F
➔ 3442 kWh year ➔ $69 + $20(standby loss)
➔ $89/year [Savings of $139/year]

Water Usage

CONTROL: Rinse-Wash-Rinse
75 gal water/wash = 150 gal water/day
➔ 1050 gal/week ➔ 54,600 gal water/year
EXPERIMENTAL: Rinse-Wash-Rinse
89 gal water/initial wash + 50 gal water/next 13 washes
➔ 739 gal/week ➔ 38,428 gal water/year
[Savings of 16,172 gal water/year]

Savings figures are for one side of the parlor only. They should
be doubled to find total savings for the double-ten herringbone parlor. The cold rinse-hot wash-cold rinse combination gave line wash
water temperatures approximately the same as for the tepid rinse-
hot wash-tepid rinse combination.

RESULTS

The experimental tests are currently about 50% completed. Large
quantities of data have been collected and has been punched for
digital computer analysis. Because the experimental testing is not
complete, the final recommendations as to the optimum technology
for proper sanitation with minimum energy and detergent consumption cannot be stated at this point.

Some preliminary observations and indications are that it may be
possible to save substantial amounts of energy (60%) and at least
80% of the detergents. In addition, it appears that water use for
cleaning and sanitizing may be reduced by about 30%. All of these
observations will, of course, have to be validated by the test results.
The use of soft water has the single greatest effect on the apparent
cleanliness of the system from a visual inspection standpoint.

The recommendations that eventually emerge from the experimental test results will be evaluated and released only as approved by the Milk and Food Sanitation Code Board. The potential economic benefit from energy and detergent conservation appear to be significant at this point in the project.

* * *

This investigation is partially supported from funds from the Dairy Chore Reduction Program, the New York Farm Electrification Council, Alfa-Laval, and Agway Inc.

16.

Integrated Systems for Power Plant Cooling and Wastewater Management

Douglas A. Haith*

The production of electrical energy has traditionally provided a rationale for many water resources development projects. Major U.S. multipurpose water resources projects initiated in the 1930s, including Hoover Dam and the work of the Tennessee Valley Authority, demonstrated that power production could be consistent with other water management goals such as flood control, navigation and provision of water supply. Implicit in these projects was the recognition that the development of energy resources, in the form of hydroelectric power, should be evaluated in the context of management of a second resource, namely water, and that multipurpose development of the water resource was not only possible but economically and socially desirable.

In most regions of the United States today power production relies much more on steam-generation plants than on hydroelectric facilities. The former are not often considered as components of an overall water resources management system. However, the water-related impacts of steam plants are far from insignificant. With current operating efficiencies in the order of 30% and 40% for nuclear and fossil-fueled units, respectively, it is necessary to dissipate enormous quantities of waste heat from these plants. Most of this heat is contained in cooling water which is used to condense

* Assistant Professor of Agricultural Engineering and Civil Engineering, Cornell University, Ithaca, NY 14853.

steam after it leaves the power plant turbines and subsequent to return to the boilers. In once-through systems the cooling water is returned to surface water bodies, where its elevated temperature constitutes a thermal pollution hazard. Closed-cycle cooling mechanisms such as towers and ponds avoid these direct heated discharges, but require large consumptive (make-up) water use to replace evaporative and blow-down losses. Thermal pollution and/or consumptive water use can impose serious limitations on a region's options for management of its water resource and of course are not incidental to public concern over siting of power plants.

It is reasonable to suggest that the concept of integrated management of energy and water resources which has been demonstrated in hydropower development may have applicability to steam-generated power, also. The mechanisms for such integration are not obvious, however. In the case of hydropower, water is used in a pure productive capacity as the source of energy. For steam plants water is the means of disposing of a waste product, which is unutilized energy in the form of heat. At least from the viewpoint of the power industry, there is of necessity less concern for management of a resource whose function is waste assimilation. Although the use of the water resource is much different for steam plants than for hydropower, at least one aspect of the rationale for considering integrated energy and water resource systems is the same for both. The use of a dam and reservoir for multiple purposes including hydropower increases the economic benefits above that which could be obtained from a single-purpose project. Moreover, the resulting benefits are distributed over a larger range of users, thus generating a higher level of political support. In a similar vein, an integrated management system which includes a steam-generation power plant may produce benefits in addition to those associated with energy production, and these supplemental benefits may in turn yield greater public support for an integrated energy/water resource management system than for a single-purpose power plant.

Although a variety of integrated systems for managing energy and water may be possible in a given region, any system must address the two key water-related issues associated with steam-generated power: thermal pollution from cooling water discharges and consumptive water use in evaporative cooling systems. One framework for the evolution of integrated systems is the consideration of possible technical linkages between power plant cooling and municipal wastewater management. Such linkages include the use of waste heat as a mechanism for enhancing wastewater treatment (1-4), the use of treated wastewater as make-up for evaporative cooling structures (5-7), and the use of a pond or reservoir for both cooling and waste stabilization (8). This chapter reports the results of a systematic evaluation of possible integrated systems for power plant cooling and wastewater management (9). The generalized system is shown in Figure 1. Alternatives were analyzed for each of the three components of the system, and several alternative configurations were evaluated for technical feasibility, monetary costs, environmental impacts and effects on water supply. These evaluations are in general limited to locations in the northeastern U.S. coastal region. Cost figures were based on an Engineering News Record Construction Cost Index of 2200. Finally, one of these systems was physically sized for the Long Island, New York, area.

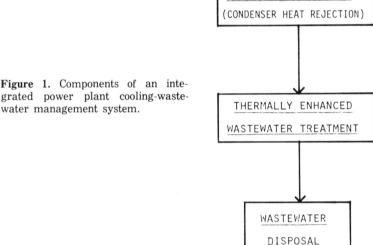

Figure 1. Components of an integrated power plant cooling-wastewater management system.

COMPONENTS OF THE INTEGRATED SYSTEM

The alternatives which were considered for the components shown in Figure 1 include four cooling options, three means of thermal enhancement of wastewater treatment and three wastewater disposal methods (Figure 2). These alternatives, while not all-inclusive, represent the most commonly used options for cooling (10-13), methods suggested in previous studies for waste heat transfer to wastewater treatment (1, 4, 8) and wastewater disposal options which are either in common use or of particular relevance to power plant cooling and water conservation. Technical details and bases for cost estimates are given in detail elsewhere (9) and

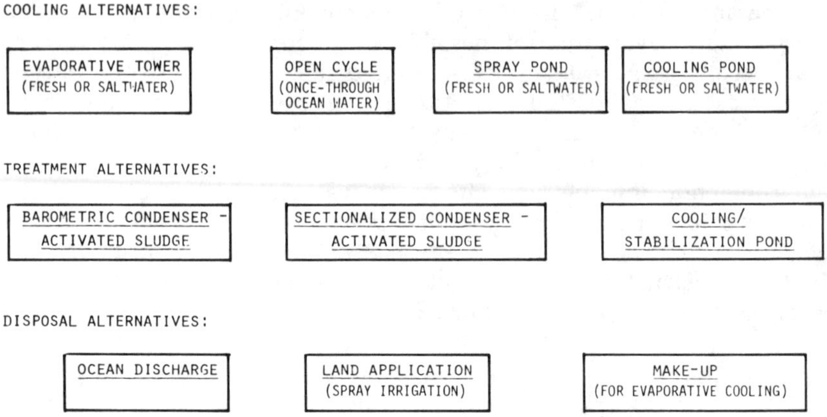

COOLING ALTERNATIVES:

EVAPORATIVE TOWER (FRESH OR SALTWATER)	OPEN CYCLE (ONCE-THROUGH OCEAN WATER)	SPRAY POND (FRESH OR SALTWATER)	COOLING POND (FRESH OR SALTWATER)

TREATMENT ALTERNATIVES:

BAROMETRIC CONDENSER – ACTIVATED SLUDGE	SECTIONALIZED CONDENSER – ACTIVATED SLUDGE	COOLING/ STABILIZATION POND

DISPOSAL ALTERNATIVES:

OCEAN DISCHARGE	LAND APPLICATION (SPRAY IRRIGATION)	MAKE-UP (FOR EVAPORATIVE COOLING)

Figure 2. Component alternatives for an integrated power plant cooling-waste-water management system.

are summarized in this section. To facilitate system comparisons, a "standard" 1100-megawatt (MW) nuclear power plant was selected. This plant is considered to have a nominal efficiency of 32%, indicating that 10,700 Btu of heat are required to produce one kilowatt-hour of electricity. A cooling water flow to 850 million gallons per day (mgd) is necessary to remove 7.5 billion Btu per hour of waste heat from the power plant condensers. The temperature differential between the cooling water entering and leaving the condensers is 25°F.

Power Plant Cooling

Once-through (open-cycle) cooling is currently the most common form of power plant heat dissipation, at least in the northeastern U.S. Its selection has been based on (relatively) low capital costs and minimal loss of power plant efficiency compared to other systems, as well as the availability of surface bodies of water. Environmental effects of once-through cooling can be quite severe. The physical movement of large quantities of water can result in entrainment and impingement of fish and other organisms on intake structures, and induced circulation currents may produce changes in the aquatic ecology near intakes and outfalls. In addition the large artificial heat input to the water body can result in fish kills due to high temperatures and lowering of dissolved oxygen concentrations.

Closed-cycle (recirculating) evaporative cooling structures (natural draft towers, cooling ponds and spray ponds) have features sharply contrasting with once-through cooling. They are generally more expensive, in terms of capital and operating costs as well as loss of power plant efficiency, and unlike open-cycle systems have significant consumptive water use. The principal advantage of closed-cycle evaporative cooling is the elimination of thermal discharges to receiving waters. Use of either fresh or saline waters is considered feasible, but current systems for large power plants are nearly exclusively freshwater based (14).

Natural draft towers are massive structures with base diameters and heights of up to 500 feet. An 1100-MW nuclear plant could require two of these towers. Cooling towers are not free of environmental effects. The visual impact of these structures is substantial, and in areas of low topographic relief, towers may be objectionable for aesthetic reasons. Blow-down water, which has been drained from the circulating flow when water is added to control dissolved solids, may be difficult to dispose of without environmental impact. Drift of entrained liquid water may result in deposition of salt particles within the immediate area of the tower. Finally, the increase in humidity or water vapor content of the air near a tower can produce fogging and ice formation at ground levels.

Cooling ponds are small artificial lakes which require substantial surface areas (1-2 acres per megawatt). Ideal design is a long narrow lake with cooling intake and outfall at opposite ends so that pond circulation approximates "plug flow" conditions. Freshwater cooling ponds require minimal blow-down. Apart from possible destruction or alteration of natural areas by pond construction, the principal environmental impact is possible ground-level fogging. Pond costs vary widely and are chiefly influenced by land values and soil permeabilities. The latter characteristic will determine whether or not artificial lining of the pond bottom is required.

Spray ponds or canals utilize sprays to enhance heat transfer to the atmosphere thus requiring only 1/10 to 1/20 of the surface area of a cooling pond (12). Depending on local wind conditions, drift and fogging may be comparable to natural draft towers. However, since water and water vapor are discharged closer to ground levels than with a tower, effects in the immediate vicinity of the power plant may be more severe.

Construction costs and consumptive water use of cooling alternatives for the 1100-MW nuclear plant are summarized in Table 1. Costs are based primarily on values reported by Woodson (11) and do not include make-up water costs.

Thermally-Enhanced Municipal Wastewater Treatment

Many of the treatment processes which are used to stabilize and renovate wastewaters are temperature-dependent. Within certain

Table 1. Construction costs and consumptive water use of cooling systems.

	Construction Cost ($10^6)	Consumptive Water Use (Freshwater) (mgd)
Once-Through	15.7	—
Natural Draft Tower		
Saline	23.6	—
Freshwater	20.4	18[1]
Spray Pond		
Saline	14.3	—
Freshwater	12.9	18[1]
Cooling Pond		
Saline	34.6[2]	—
Freshwater	33.0[2]	12[3]

1. Assumes negligible blow-down water. If make-up is high in dissolved solids, consumptive use could be doubled.
2. Costs for Suffolk County area of Long Island. Includes $17.6 million for land purchase and lining of pond bottom (1500-acre pond).
3. Based on Suffolk County climate. Annual precipitation of 5 mgd, evaporation of 17 mgd.

limits, solids settle more rapidly, chemical reactions proceed at faster rates and bacterial oxidation is enhanced at elevated temperatures. The limits of such improvements are not known with precision, since extensive field experience with warmed wastewaters is lacking. The sewage that is received by municipal treatment plants may typically be at 65°F, and there is seldom an economical heat source which could be used to elevate this temperature. The waste heat from a nuclear power plant is such a potential source, and although a large conventional treatment plant might be capable of utilizing only a fraction of the waste heat normally rejected to the environment, this heat should be sufficient to significantly enhance treatment efficiency. As a result, treatment process units could be reduced in size while maintaining a desirable effluent quality.

Design formulae for wastewater treatment processes generally contain temperature-dependent parameters. A recent study has used these formulae to determine the cost savings which could be realized in an activated sludge secondary treatment plant (1). With sufficient heat added to the wastewater to maintain a temperature of 86°F, the capital cost of a 50-mgd plant could be reduced from $25.3 million to $23.1 million (1, 9). Such results must be interpreted with care, since the applicability of standard design procedures to processes operated at elevated temperatures has not been established, and some evidence indicates that the procedures may not accurately predict the performance of biological treatment systems (15).

At any rate, in order to utilize waste heat in sewage treatment a technically feasible means of transferring the heat to the wastewater must be devised. The three possibilities considered in this study were a barometric condenser, a sectionalized condenser, and a combined cooling/stabilization pond. The first two options were considered as means of heating wastewater prior to its entry to an activated sludge treatment plant. An ambient wastewater temperature of 65°F was assumed, and sufficient heat was added to insure a temperature of 86°F in the activated sludge units, this being approximately the optimal temperature for the mesophilic bacteria which oxidize organic waste material (1).

A barometric condenser would condense spent steam from the power plant's steam cycle directly on the surface of the wastewater, thus transferring the steam's latent heat of evaporation to the water. Make-up water would be required for the plant's boilers to replace this steam. Required steam extraction, waste heat utilization and boiler make-up are indicated in Table 2. The capital cost of a condenser for a 50-mgd waste flow is estimated to be $300,000 (1, 9).

Table 2. Barometric condenser characteristics.

	Wastewater Flow (mgd, million gallons per day)				
	10	20	30	40	50
Required Steam Flow (10^6 lb/hr)	0.12	0.25	0.36	0.49	0.61
% of Waste Heat Utilized	1	3	4	5	7
Boiler Make-up (mgd)	0.32	0.72	1.04	1.41	1.75

The use of a barometric condenser to heat wastewater with spent steam from a nuclear power plant appears to be technically feasible. Economic feasibility would likely depend on the cost of supplying boiler make-up which must be relatively free of impurities. Problems associated with the alternative involve radiation safety and maintenance. Although the steam passing through the turbines in a pressurized water reactor nuclear plant is theoretically free of radiation, the possibility of radiation leaks to the steam and subsequent contamination of the wastewater cannot be ignored, and it is questionable that the Atomic Energy Commission licensing procedures would approve of the steam being removed for wastewater heating. Maintenance difficulties with the barometric condenser would be associated with the prevention of anaerobic conditions which could produce bacterial growth and odors.

A second option for heat transfer to wastewater is the sectionalizing of the power plant condensers to permit a portion of the steam to be condensed using wastewater, and the remainder with a conventional cooling water flow. The design of such a condenser is a direct function of the heat transfer coefficient of the wastewater. This coefficient depends on several factors, one of which is the fouling resistance of the cooling water. Since experience with wastewater use in condensers is minimal, fouling resistance can be estimated only from reported values for natural waters which are polluted (1, 16). Based on these values, heat transfer coefficients of 130 to 310 Btu/hr-ft^2-°F may be expected. Condenser costs are proportional to required surface areas which vary inversely with heat transfer coefficients. Since typical coefficients for saltwater coolant are 400-450 Btu/hr-ft^2-°F, costs may be substantially increased with wastewater coolant. Condenser areas and waste heat utilized by a sectionalized condenser are listed in Table 3, based on a heat transfer coefficient of 185 Btu/hr-ft^2-°F.

Table 3. Sectionalized condenser characteristics.

	Wastewater Flow (mgd)				
	10	20	30	40	50
Condenser Area (ft^2)	13,500	27,000	41,000	55,000	68,000
% of Waste Heat Utilized	1	3	4	5	7

The capital cost of a condenser for a 50-mgd wastewater flow is estimated at $1,300,000. A sectionalized condenser is a technically feasible alternative for wastewater heating. Principal technical uncertainties are condenser fouling and possible corrosion associated with the wastewaters. Stainless steel or copper nickle alloys used for condenser construction with saline cooling water may resist corrosion, but this is somewhat uncertain. Fouling resistance will have a major impact on financial feasibility, since lower resistance values may reduce surface areas to levels comparable to saltwater cooling. Maintenance problems associated with condenser fouling may occur, but their severity is again unknown.

A third alternative for the transfer of power plant waste heat to wastewater is the construction of a large pond or lake which would serve as both a cooling and secondary wastewater treatment system. Wastewater, which would have received primary treatment, and precipitation would be the only water inputs and cooling water would circulate through the pond. Two sets of inlets and outlets would be provided, one for wastewater influent and effluent (overflow) and the other for the circulating cooling water.

The cooling/stabilization pond considered in the study would have a surface area of 1500 acres and an average depth of 10 feet. The temperature characteristics of a pond were evaluated for Long

Island and are shown in Table 4. Evaporation and precipitation would average 17 and 5 mgd, respectively, for a net natural water loss (*i.e.*, exclusive of wastewater inputs) of 12 mgd.

Table 4. Temperature characteristics of a 1500-acre cooling pond for a 1100-MW nuclear power plant.

	January	July
Cooling Water Discharge Temperature (°F)	80	115
Cooling Water Intake Temperature (°F)	55	90
Average Pond Temperature (°F)	68	102

Wastewater treatment in a facultative stabilization pond is achieved in aerobic surface and anaerobic subsurface regions. Removal of organics is considered to be a function of pond detention time and a temperature-dependent reaction rate (17, 18). Typical BOD (biochemical oxygen demand) loading rates for unheated stabilization ponds which result in 80 to 90% BOD removal are 20-50 lb/day per acre in areas with winter ice cover and 50-150 lb/day per acre in regions with mild winters (17). Such loadings may not be appropriate for heated ponds, however, and since design and field experience is limited to unheated stabilization, the quantities of organics in a cooling/stabilization pond must be considered unknown. The concentrations and forms of phosphorus and nitrogen in the effluent are equally uncertain, since wide variations have been observed in existing (unheated) ponds (8). The treatment characteristics of the 1500-acre cooling/stabilization pond which can be reasonably estimated are given in Table 5. Concentrations of BOD, suspended solids, and dissolved solids are assumed to be 120 mg/1, 60 mg/1, and 200 mg/1, respectively, in the primary treated wastewater input to the pond.

Table 5. Waste treatment characteristics of 1500-acre cooling/stabilization pond.

	Wastewater Inflow (mgd)*			
	20	30	40	50
Detention Time (days)	610	270	180	130
Effluent Dissolved Solids (mg/1)	500	340	280	260
BOD Loading (lb/day-acre)	13	20	27	33
Sludge Build-Up (ft/yr)	0.05	0.08	0.11	0.13

* Wastewater is assumed to have received primary treatment.

The cooling-stabilization pond would essentially be a conventional cooling pond for which primary treated municipal wastewater was used as make-up water. If pond effluent is comparable to that achieved by secondary treatment there is little direct evidence to indicate severe power plant condenser difficulties. There would be a greater possibility of condenser fouling by algae and possible lower heat transfer coefficients than normally encountered in closed-cycle cooling systems. Possible adverse environmental impacts in addition to those related to a cooling pond could be associated with the disposal of the pond overflow. Also, since the pond would be highly eutrophic, problems of public acceptance may be anticipated.

The capital cost of a 1500-acre cooling/stabilization pond for the Long Island area (including land acquisition and pond lining) and a 30-mgd primary wastewater treatment plant is estimated to be $41,000,000. By comparison, the combination of saltwater natural draft evaporative cooling towers and a secondary treatment plant would provide comparable performance at an investment of $36,000,000. The latter cannot be assumed to be economically preferable to the cooling/stabilization pond, however. The cost of a make-up water system is not included in the cooling tower costs, and pond costs include high land values and artificial lining.

Wastewater Disposal

The three options considered for disposal of the heated wastewater were ocean disposal, use of the wastewater as make-up in closed-cycle cooling systems, and land application to agricultural crops. The alternatives were evaluated primarily for their impacts on water supplies and the environment. The costs of the disposal options are highly site-specific, and no generalizations can be made concerning relative costs.

The discharge of treated wastewater to the ocean or to bays and estuaries is frequently considered the most economical and feasible method of disposal in coastal areas. On Long Island, for example, present and proposed wastewater management systems rely almost exclusively on coastal waters for disposal (19). The discharge of sewage to saline waters essentially converts municipal water requirements in the sewered area from a conservative to a consumptive use. Water that previously may have passed through domestic use and into septic tanks and cesspools for leaching into the ground and subsequent reuse now is mixed with saline waters and lost to further supply use. Ocean disposal may result in a set of environmental impacts ranging from a temporary disturbance of beaches and wetlands during construction to the possible major disruption of a highly productive and valuable coastal ecology (see, for example, reference 19 for discussion of such impacts).

Cooling water make-up constitutes the largest direct use of treated wastewater in the U.S. today (7), although total use by steam-generation power plants is only 15 mgd (6). Make-up typically is effluent from a secondary treatment plant, and additional treatment is often required to control corrosion, fouling and slime growth, particularly in cooling towers. Since make-up water is a small fraction (2-3%) of a total cooling flow, the heat contained in warmed wastewater would have no adverse effect on cooling. The impact on water supply of wastewater reuse for evaporative cooling make-up will depend on the source of water supply which is replaced. If a cooling tower utilizes a high-quality ground or surface water source which could serve as a municipal supply, the substitution of wastewater would have a beneficial effect. Certainly water conservation would in many cases be the strongest argument for wastewater use in cooling, and at present such use appears to be one of the most feasible means of wastewater recycling (20). Adverse environmental impacts include possible transmission of viruses in aerosols from cooling towers and spray ponds.

The application of treated wastewater to soil covered by plants has received much attention in recent studies (21-24). This disposal option is by no means new (25), and has seen wide application in various parts of the world for many years. A number of food processing plants and small U.S. cities dispose of wastewater by this means. Plant cover for disposal areas has included grass, forest, agricultural crops and natural vegetation. After renovation by a combination of biological, chemical and filtration processes within the plant/soil system, the wastewater may enter surface waters or recharge ground waters. The effects of a heated wastewater effluent on plant growth are somewhat uncertain. In the case of agricultural crops, studies have shown that in comparison with the use of water at ambient temperatures, warm water may offer better frost protection and atmospheric control, and has no adverse effects on plants when used for irrigation (26, 27).

Since the characteristics of a land application system are dependent on soils, climate and plant cover, it was necessary to select a particular location for the system in order to make an evaluation. Consistent with other aspects of the study, the Suffolk County area of Long Island was chosen. Suffolk County soils are sands and silt loams which are level, well-drained and easily tilled (28). They support a diversified agriculture with major crops being potatoes, cabbage, cauliflower and sod (29, 30). These crops require irrigation and high levels of fertilization on Long Island, and spray irrigation with wastewater could accomplish the same purposes. Based on mean freeze-free periods, an application period of 7 months can be anticipated. It would be necessary to store the wastewater flows in lagoons during the remaining months (November-March). Application rates were determined by a nitrogen mass balance for the irrigated crops. Based on a crop nitrogen removal of 180 lb/ac-yr for potatoes and sod and 150 lb/ac-yr for cabbage and cauliflower and a wastewater nitrogen content of 20 mg/1, irrigation rates were computed which would limit yearly average nitrogen concentration in seepage water (input from wastewater-crop removal divided by total seepage water volume) to 10 mg/1, the Public Health Service standard for nitrate nitrogen in drinking water. Computed rates were 2 in./wk in April and October for all crops and 3 in./wk and 4 in./wk, respectively, for potatoes or sod, and cabbage or cauliflower, during May-September. Required land areas and winter storage requirements are summarized in Table 6.

Table 6. Characteristics of a land application system for wastewater disposal in Suffolk County.

Average Wastewater Flow (mgd)	Winter Storage Required (million gallons)	Land Disposal Area (ac) Potatoes, Sod	Land Disposal Area (ac) Cabbage, Caulfilower
10	1500	1300	1600
20	3000	2600	3200
30	4500	3800	4900
40	6000	5100	6500
50	7500	6400	8100

A large portion of the applied wastewater will enter the ground water, since the excess of evapotranspiration (consumptive water used by growing plants) over precipitation during the growing season averages only 4.6 inches (31). Thus most of the wastewater would be conserved for a potential water supply use. Certain ecological impacts may result from the land application option. Insects, birds and small animals may be exposed to the wastewater sprays, and the possibility of adverse effects on fauna cannot be completely discounted. The unnaturally wet conditions in the irrigation area may encourage the growth of various insects, parasites, fungi and plant diseases which could depress crop yields. The economic feasibility of land application would be determined by the proximity of crop land to the wastewater source and the market acceptability of the irrigated crops. In this regard, it should be noted that several states expressly permit irrigation of secondary disinfected effluents on crops grown for human consumption (23).

EVALUATION OF INTEGRATED MANAGEMENT SYSTEMS

With the options of fresh and saline water use in evaporative cooling systems, and including the possible alternative of no heat transfer to wastewater, there are a total of 84 combinations of the system components shown in Figure 2. With the elimination of redundant or inconsistent combination (*e.g.*, a cooling/stabilization pond would not be used with open-cycle cooling), 62 possible integrated systems for power plant cooling and municipal wastewater management remain. As indicated in the previous discussions the level of detail in available data and the limitations of the study did not permit a complete quantification of system costs or environmental impacts. Moreover, the performance of many of the component alternatives is rather uncertain. With this in mind, a qualitative evaluation of selected systems was made. The evaluation was obviously a matter of judgment based on those characteristics of the system components outlined previously.

System Objectives and Relative Rankings

The four objectives of an integrated power plant cooling and wastewater management system are considered to be as follows:

1. Predictable or reliable *technical performance*.
2. Low monetary *cost*.
3. Minimization of adverse *environmental impacts*.
4. Enhancement of *water supply*.

The integrated systems which were evaluated with respect to these objectives are shown in Table 7. While somewhat arbitrary, the selection is intended to indicate a range of systems which could be considered for northwestern coastal areas. The performance of these systems is indicated in Table 8.

Table 7. Selected integrated cooling/wastewater management system.

System	Cooling	Heat Transfer	Wastewater Disposal
A	Open-Cycle	None	Ocean
B	Saline Tower	Barometric	Land
C	Saline Tower	Sectionalized	Land
D	Freshwater Tower	Sectionalized	Ocean
E	Freshwater Spray Pond	Sectionalized	Make-up
F	Cooling/ Stabilization Pond		Land

System A represents the conventional management option with no integration of power plant cooling and wastewater treatment. Technical performance is well understood and financial costs are most likely the least of any system evaluated. Water supply impact is unfavorable, since ocean wastewater discharge precludes further

Table 8. Evaluation of selected integrated systems.

System	Technical Performance	Costs	Water Supply	Environmental Impacts
A	+	+	−	−
B	−	−	+	0
C	−	−	+	0
D	0	+	−	0
E	0	+	0	+
F	−	0	+	+

KEY: + Relatively favorable effect on objective.
   0 Mid-range effect on objective; average with respect to other systems.
   − Relatively unfavorable effect on objective.

water supply use. Severe adverse environmental impacts are due to thermal and organic pollution of coastal waters.

System B has uncertain and potentially severe technical performance characteristics with regard to treatment efficiencies and radioactive contamination. Costs of saline cooling towers, boiler water make-up and winter wastewater storage make this system one of the most expensive considered. Since saltwater make-up would be used for the tower and land disposal would result in ground water recharge, net water supply impact would be positive. Environmental impacts are associated with drift, fogging, salt deposition and visual impact of cooling towers.

System C is similar to the previous system but would be somewhat less costly since boiler make-up is unnecessary. Technical uncertainties associated with radioactivity are also absent, but sectionalized condenser performance is unpredictable.

System D can be most usefully compared to System A. The costs of freshwater cooling towers are substantially more than open-cycle cooling, but thermal pollution of coastal waters by a cooling discharge would be avoided. Water supply impacts are severe since freshwater make-up is needed, and additional technical uncertainties are associated with the sectionalized condenser.

System E avoids environmental problems except for drift and fogging. Technical uncertainties are associated with the sectionalized condenser and any difficulties involved in the use of wastewater in the remainder of the cooling system. Spray ponds are less costly than saltwater towers and since make-up water is provided by wastewater, the system would result in no net consumption.

System F has fewer adverse environmental impacts than the other systems. Since a complete integration of cooling, wastewater, and winter wastewater storage is effected, cost savings may be achieved. Technical performance must be considered at least as uncertain as

Systems B and C. Water supply impact is favorable although evaporative losses from the pond will consume a portion of the wastewater volumes.

System Selection

The ratings of alternative systems as given in Table 8 indicate the degree to which any system satisfies four different and conflicting objectives. Since no system is superior in all categories, the selection of a "best" system is not obvious. The author's preference is for System F, which is indicated in more detail in Figure 3 and Table 9. This selection is based primarily on the system's favorable environmental impact, moderate cost and generally beneficial impact on water supply. An additional justification is perhaps the analogy which can be drawn between the system and the successful multipurpose water resource projects discussed in the introduction. These projects were based on a structure, *i.e.,* a reservoir, which was capable of being managed for a variety of objectives. A cooling/stabilization pond would be a similar multipurpose water body which would serve as a cooling water source, wastewater treatment process and storage lagoon for a land application-ground water recharge system.

CONCLUSIONS

The results of the study reported in this chapter are subject to several interpretations. An optimistic view would suggest that an application of multipurpose water resource management principles

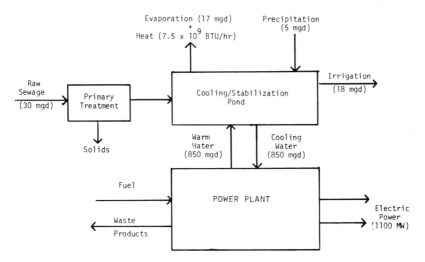

Figure 3. Major system balances integrated System F.

Table 9. Physical characteristics of an integrated cooling/wastewater management system—cooling/stabilization pond and land disposal.

Nuclear Power Plant Size:		1100 MW
Heat Rejection at Condensers:		7.5 10^9 Btu/hr
Terminal Steam Temperature:		120°F
Cooling Water Flow:		850 mgd
Cooling Water Temperature Range:		25°F
Pond Area:		1500 ac
Pond Depth:	Maximum	10 ft
	Minimum	3 ft
Pond Volume:	Maximum	4.9 billion gal
	Minimum	1.45 billion gal
Pond Temperature:		
	January — Cooling Intake	55°F
	Cooling Discharge	80°F
	Average	68°F
	July — Cooling Intake	90°F
	Cooling Discharge	115°F
	Average	102°F
Average Wastewater Flow:		30 mgd
Average Pond Overflow (Effluent for land application):		18 mgd
Wastewater Treatment:		Primary + Chlorination
Land Application Area		2300 ac
Crop:		Potatoes
Application Rates:		
	April, October	2 in/wk
	May - September	4 in/wk
Ground Water Recharge:		17 mgd

to the power plant cooling problem results in the evolution of several integrated management systems which are environmentally sound and which conserve the water resource. A more pessimistic outlook would hold that such systems are little more than conceptual frameworks of doubtful economic and technical feasibility. A reasoned midpoint between the two opposing opinions would be that at least a preliminary analysis indicates that the integration of power plant cooling and wastewater management is a possible means of avoiding some of the adverse effects of current management practices. Moreover, a mass and energy balance of the cooling-stabilization pond-land application system provides some evidence of feasibility.

It can be anticipated that even if further studies and demonstration indicate the economic feasibility and remove the technical uncertainties of integrated management systems, their actual implementation will not be easy. There are real institutional constraints on such management options. Regulatory agencies, power, wastewater, water supply and agricultural interests would all be involved in the venture and a viable management framework would have to evolve. Again, however, analogy to previous multipurpose water resource projects may be useful. Conflicting interests were reconciled in these projects, with federal funding and management playing key roles.

As a final note, the integrated management systems discussed in this paper represent applications of resource management policies expressed in the Federal Water Pollution Control Act Amendments of 1972, which call for management "which results in integrating facilities to treat, dispose of, or utilize other industrial and municipal wastes, including but not limited to solid waste and waste heat and thermal discharges (PL92-500, Sec. 201e)."

ACKNOWLEDGMENTS

Work reported herein was performed while the author was with Quirk, Lawler and Matusky Engineers (now Lawler, Matusky and Skelly Engineers), Tappan, N.Y. Stuart Bassell of that organization aided in preparation of cost estimates and system evaluations. The work was under contract to the North Atlantic Division of the U.S. Army Corps of Engineers.

REFERENCES

1. "Study of an Integrated Power, Water and Wastewater Utility Complex," New York State Atomic and Space Development Authority, New York, December, 1972.
2. Boersma, L. and K. A. Rykbost. "Integrated Systems for Utilizing Waste Heat from Steam Electric Plants," *Journal of Environmental Quality,* Vol. 2, No. 2, pp. 179-187 (1973).
3. Oswald, W. J. "Ecological Management of Cooling Basins," *Journal of Environmental Quality,* Vol. 2, pp. 203-206 (1973).
4. "Nuclear Waste Heat to Treat Municipal Sewage," *Water and Wastes Engineering,* November, 1971, pp. 46-48.
5. Gray, H. J., C. U. McGuigan, and H. W. Rowland. "Sewage Plant Effluent as Cooling Tower Makeup—A Case History," Presented at the Annual Meeting of the Cooling Tower Institute, Houston, January, 1973.
6. Hansen, N. S. "The Capacity of Thermal Electric Power Plants to Accept Municipal Wastewater," *Proceedings of the National Conference on Complete Water Reuse,* Washington, D.C., April 23-27, 1973.
7. Weddle, C. L. and H. N. Masri. "Reuse of Municipal Wastewater by Industry," *Industrial Water Engineering,* June/July, 1972, pp. 18-24.
8. Humenick, M. J., W. E. Morgan, and E. G. Fruh. "Wastewater Effluent for Power Plant Cooling," University of Texas Center for Research in Water Resources Report #CRW-95, Austin, October, 1972.
9. "An Opportunity for the Future: Integrated Water Supply—Power Generation—Wastewater Treatment Management—Land Use Control," Quirk, Lawler, and Matusky Engineers, Report to the U.S. Army Corps of Engineers, North Atlantic Div., December, 1973.
10. Sebald, J. F. "A Survey of Evaporative and Non-Evaporative Cooling Systems," Presented at the 74th National Meeting of the American Institute of Chemical Engineers, New Orleans, March, 1973.

11. Woodson, R. D. "Cooling Towers for Large Steam-Electric Generating Units," M. Eisenbud (ed.), *Electric Power and Thermal Discharges,* Gordon and Breach, New York, pp. 269-280 (1969).
12. Rainwater, F. H. "Thermal Waste Treatment and Control," M. Eisenbud (ed.), *Electric Power and Thermal Discharges,* Gordon and Breach, New York, pp. 189-212 (1969).
13. Baron, S. "Options in Power Generation and Transmission," *Energy, Environment and Planning,* Brookhaven National Laboratory, Upton, N. Y., October, 1972, pp. 129-144.
14. Shore, P. H. "Energy in the Long Island Sound Region," *Energy, Environment and Planning,* Brookhaven National Laboratories, Upton, N. Y., October, 1972, pp. 15-23.
15. Friedman, A. A. and E. D. Schroeder. "Temperature Effects on Growth and Yield of Activated Sludge," *Journal of the Water Pollution Control Federation,* Vol. 44, No. 7, pp. 1433-1442 (1972).
16. Mueller, A. C. "Heat Exchangers," W. M. Rohsenow, J. P. Hartnelt (ed.), *Handbook of Heat Transfer,* McGraw-Hill, New York, Chapter 18 (1973).
17. Fair, G. M., J. C. Geyer, and D. A. Okum. *Water and Wastewater Engineering,* Vol. 2, John Wiley & Sons, New York (1968).
18. Oswald, W. J. "Complete Waste Treatment in Ponds," 6th International Conference on Water Pollution Research, Jerusalem, June, 1972.
19. "Environmental Impact Statement on Wastewater Treatment Facilities Construction Grants for Nassau and Suffolk Counties, New York," U.S. Environmental Protection Agency, Region II, New York, July, 1972.
20. Gavis, J. "Wastewater Reuse," Staff report for the National Water Commission, Arlington, Va., July, 1971.
21. Thomas, R. E. "Land Disposal II: An Overview of Treatment Methods," *Journal of the Water Pollution Control Federation,* Vol. 45, No. 7, pp. 1476-1484 (1973).
22. Reed, S. *et al.* "Wastewater Management by Disposal on the Land," U.S. Army Corps of Engineers, Cold Regions Research and Engineering Laboratory, Hanover, N. H., May 1972.
23. Stevens, R. M. "Green Land Clean Streams," Temple University, Center for the Study of Federalism, Philadelphia, (1972).
24. Pound, C. E. and R. W. Crites. "Wastewater Treatment and Reuse by Land Application, Vol. I and II," Office of Research and Development, U.S. Environmental Protection Agency, Washington, August, 1973.
25. Metcalf, L. and H. P. Eddy. *American Sewerage Practice,* Vol. III, Chapter XVII, New York, McGraw-Hill, (1915).
26. "1969 Annual Report of the Thermal Water Horticultural Demonstration Project," Eugene Water and Electric Board, Eugene, Oregon.
27. "1970 Annual Report of the Thermal Water Demonstration Project," Eugene Water and Electric Board, Eugene, Oregon.
28. Cline, M. G. "Soils and Soil Associations of New York," Extension Bulletin 930, Cornell University, Ithaca, N. Y. (undated).
29. "1969 U.S. Census of Agriculture—New York," New York Crop Reporting Service, Albany, New York, June, 1972.
30. "Comparative Importance of Suffolk County, N. Y., Agriculture" (unpublished mimeo), Suffolk Co. Cooperative Extension Service, Riverhead, N. Y., June, 1973.
31. "Comprehensive Public Water Supply Study—Suffolk County, New York, Vol. II," Holzmacher, Mclendon, & Murrell—Consulting Engineers, Melville, N. Y. (1970).

17.

Manure Management Energy Consumption in Swine Confinement Systems

Herbert L. Brodie*

The management of waste products in a finishing swine confinement building can follow several methods, each requiring different amounts of investment and operating dollars, management, labor and energy. Any satisfactory waste management system must remove accumulated waste products from the animal confinement area such that all environmental constraints are satisfied. These constraints have generally included the maintenance of a proper animal environment and the protection of the air and water quality of the surrounding area. Within these constraints a low-level combination of dollars, labor, management and energy is desirable.

Two swine manure management methods practiced in Maryland satisfactorily remove animal waste within the existing constraints. These manure management methods are an oxidation ditch under a slotted floor and an open guttter flush with water recirculation through a lagoon. Both methods produce desirable environmental conditions and result in labor efficiency. However, there is a considerable difference between these methods with regard to management input, energy consumption and investment dollars.

Of primary interest in this report is the amount of energy consumed by each manure management method. A survey was conducted on two operating swine confinement systems to determine

* Extension Agricultural Engineer, Agricultural Engineering Department, University of Maryland, College Park, Maryland.

the energy utilized for the movement of manure from the feeding floor to land disposal.

THE SYSTEMS

Oxidation System

Animal waste from a partially slotted floor finishing building for 500 head is collected and treated in an oxidation ditch of conventional design. The finishing building is a partially enclosed block and metal-clad wood frame structure 7.3 meters wide by 76 meters long. The 1.8-meter wide by 1.2-meter deep oxidation ditch is aerated by two cage-type rotors, each powered by a 3.7 kilowatt electric motor.

Waste overflow from the oxidation ditch enters an underground concrete storage tank. The available storage volume is approximately 38 cubic meters. Accumulated waste is disposed once every seven days on four hectares of grassland. A big gun irrigation system operated by a tractor pto-powered progressing cavity pump is used to distribute the waste. An estimated 1.3 megaliters of waste are disposed annually.

Manure deposited on the slotted floors is pushed through the slots by animal activity. An occasional build-up of manure deposits adjacent to a wall area is allowed to exist until the animals are moved to market. The pens are then cleaned by hand labor prior to the introduction of new animals. The slotted floor is a cast-in-place reinforced concrete structure. No maintenance has been required in over four years of use.

Within the oxidation ditch the aerators are operated continuously resulting in a high energy demand. Each aerator requires maintenance or repair on a monthly basis. The maintenance of an aerator requires approximately four hours of labor and fifteen dollars worth of parts per month. Aerators are constructed of welded steel angle and fabricated in the farm shop. The aerator blades deteriorate sufficiently during use that annual replacement is required.

The biological oxidation process within the ditch becomes unbalanced on the average of twice annually. This occurs regardless of management and maintenance attempts to keep mechanical equipment continuously functioning. On the occasion of biological malfunction the entire content of the ditch is pumped to the disposal field. The ditch is then filled with water and the biological system is usually reestablished without additional expenditure of labor or energy.

The only function of the permanently sodded disposal field is to accept waste. The maintenance performed is limited to mowing twice annually. Each waste application occurs on a different location within the disposal field. A small amount of time and labor is required for relocating the spray nozzle.

The main lines of the irrigation system are buried PVC pipe. The system has been in operation for four years and maintenance has been negligible. Maintenance of the pump has been limited to periodic lubrication. A tractor is used for operation of the pump and maintenance is entirely charged to waste handling. The main energy input to the irrigation system is the fuel consumed by the tractor. The time required to set up the tractor and pump and to empty the storage tank is approximately one hour per week.

Flush System

Animal waste is flushed to a storage lagoon from a finishing building for 600 head. The finishing building is an open pole and block structure 9 meters wide by 73 meters long. The pen floors slope from one side of the building to an open gutter. The 1.8-meter wide by 0.1-meter deep gutters slope from each end toward the center of the building on the longer dimension.

The gutters are rectangular, constant width open channels. A gutter slope of 1% is used for the first half of the length with the remaining length at a slope of 2%. The gutters end at a catch basin from which waste is transported to a lagoon through a pipe approximately 20 cm in diameter. Flush tanks from which water is periodically released to develop a flushing action are located at the high end of each gutter. Self-dumping tanks requiring no external energy are used.

All wastewater is collected in a lagoon with surface dimensions of 95 by 91 meters. The depth ranges from 1.2 to 1.8 meters depending on the volume of storage. Ground water conditions prevent the use of a greater lagoon depth at this location.

Water from the lagoon is utilized as flush water in the gutters. Lagoon water is returned to the flush tanks by a 248-watt electric motor-driven centrifugal pump. The pump is operated continuously with a valved by-pass to allow control of the return flow rate. The return flow rate controls the rate of flush tank filling and the frequency of flushing. The major energy input is for continuous operation of this return flow pump. Maintenance of the pumping system has been negligible.

Animals become well trained to defecate in or near the flush gutter. The flushing activity reinforces animal behavior such that the majority of the pen area is kept clean. Any manure dropped in the pen is easily worked toward the gutter by animal traffic. The

6.25% floor slope prevents the accumulation of excess moisture in the pen and the floors are generally very dry.

On occasion a new group of animals may be slow to learn defecation patterns. These pens are scraped by hand. In addition, feed is spread on the floor to encourage a change of animal habit. Usually such a group of animals adjusts to the desired habits within two or three days. Other than this small amount of hand labor the entire system is operated automatically.

The disposal of accumulated water in the lagoon is performed annually. A portable big gun irrigation system operated by an engine-driven centrifugal pump is used to distribute lagoon water on 12 hectares of cropland. An estimated 6.4 megaliters of wastewater are disposed annually. Rainwater collected on the lagoon surface accounts for approximately 72% of this volume. Two big gun irrigation nozzles are manually moved to various locations within the disposal area. The actual pumping time is 40 hours. An additional 18 hours of labor are required for irrigation system operation and maintenance.

ENERGY INPUT

The energy input for these systems can be differentiated into several categories. These would include energy purchased for operation, energy associated with labor, and energy associated with the procurement or production of material components.

Operational Energy

Easiest to document is the energy purchased for system operation in the form of electricity and fuel. The purchased energy shown in Table 1 is of greatest concern because its cost and availability are

Table 1. Energy requirements for mechanical operations in two swine manure management systems.

Manure System		Oxidation	Flush
Animal Capacity, Head		500	600
Annual Energy Purchased			
Electric, Kilowatt-hours		63,824	2,172
Equivalent Gigajoules		230	7.8
Gasoline, U.S. Gallons		100	267
Equivalent Gigajoules		13	35
Total Energy Equivalent			
Megajoules per Head Capacity		486	71
Annual Energy Cost, Dollars			
Electric @ $.04/kWh		2,593	87
Gasoline @ $.45/U.S. gallon		45	120
	Total	2,638	207
Energy Cost per Head Capacity		$5.28	$.35

most responsible to the laws of supply and demand. High energy demanding production processes cannot profitably operate during periods of high energy prices or low energy availability.

The oxidation system is a high energy demand process. The treatment and disposal of manure from the oxidation system requires almost seven times the amount of energy required for the transport and disposal of manure from the flush system. The primary demand is electricity for operation of the oxidation wheel aerators. The dollar cost per head of capacity for this energy is fifteen times higher for the oxidation system than for the flush system.

This dollar cost reflects not only the difference in energy use, but also the relative expense of the different forms of energy purchased. One gigajoule of energy costs $3.42 as gasoline and $11.15 as electricity using the price schedule shown in Table 1. This cost is of the energy purchased and does not account for the efficiency of the energy conversion equipment.

If, for example, the oxidation system irrigation pump were to be powered by an electric motor rather than the internal combustion engine of the tractor, the energy cost would be reduced simply because of the greater efficiency of converting electrical energy to mechanical motion. In this example the annual cost of electricity would be approximately $31 in place of the $45 spent on gasoline.

Labor Energy

The hours of labor required for operation of the system are easily documented. However, assigning an equivalent energy demand on this labor is quite difficult. Various studies have attempted to develop human energy expenditures for various tasks. An attempt to equate these studies with the work patterns in the two swine systems would be unrealistic. For these two systems it is assumed that the greater labor requirement in terms of hours is also the greater energy requirement. The hours of labor for the operation of each system are shown in Table 2.

Table 2. Labor requirements for operations in two swine manure management systems.

Manure System	Oxidation	Flush
Animal Capacity, Head	500	600
Annual Labor, Hours		
Manure Handling	6	3
System Maintenance	134	1
Irrigation Operation	30	18
Irrigation Maintenance	13	2
Total	183	24
Annual Labor, Hours per Head Capacity	.37	0.04

The tasks within the two manure management systems can be considered somewhat equivalent in terms of physical exertion and, perhaps, in energy requirement. Both systems require a small amount of hand scraping, some floor washing with hoses, movement of irrigation equipment, and other light work associated with maintenance. Additional labor is required for the maintenance of the oxidation equipment including fabrication of new equipment each year.

The oxidation system requires 9¼ hours for each hour of labor per head capacity required of the flush system. This labor is primarily required for the maintenance of the oxidation equipment. The oxidation system also requires additional labor for more frequent irrigation equipment operation and for maintenance of the waste disposal field.

Investment Energy

The energy required for the procurement or production of material components is an important factor when considering an overall energy inventory for any particular system. For example, a system might be developed which requires very little operational energy to perform some function. Yet, this same system might require a costly initial energy input for construction. The total energy input over the life of the system could be greater than that for some other system performing the same function at a greater operational energy requirement but at a lesser initial energy input.

This concept has been well established in the area of dollar economics. A system selected for totally economic reasons is done so using a balance between investment costs and variable costs resulting in the lowest total cost over the life of the system.

The lack of adequate data prevents the development of an initial energy input comparison for the two manure mangement systems. However, if it is assumed that the dollars invested had equal energy buying power, a comparison of the capital investment in each system can be used for the desired energy comparison.

A comparison of capital investment for the two systems (Table 3) indicates that the flush system requires slightly over half the capital investment per head capacity required for the oxidation system. The construction of slotted floors and the oxidation ditch is responsible for the greater cost of the oxidation system.

Other Energy

There are additional energy costs that have not been included in the analysis. These costs are from such sources as replacement parts, electricity or fuel utilized for repair, and special equipment such as welding machines for repair. In addition there is some energy

Table 3. Capital investment requirements for two swine manure management systems.

Manure System		Oxidation	Flush
Animal Capacity, Head		500	600
Investment, Dollars			
Concrete Floor		11,400	5,651
Storage Structure		17,000	7,700
Equipment		1,600	988
Irrigation System		8,600	11,000
	Total	$39,100	$25,939
Investment per head Capacity		$78.20	$43.23

lost due to the utilization of land for a lagoon or waste disposal area rather than applying this land to future crop production. There is also some energy gained by utilizing lagoon effluent as a minor irrigation water source for cropland.

This list could be extended but the contribution to energy use becomes less defined with each addition. These additions, although necessary to develop a complete energy inventory, are not significant relative to the energy levels necessary for the construction and operation of these two manure management systems.

SUMMARY

If energy conservation is to be a major constraint for any process, then high energy-demanding activities must be altered to reduce the overall energy demand. Continued effort is necessary to find substitute activities within a process to allow energy conservation while any detrimental effect on the quality of production is minimized.

A desirable example of a substitute activity is the flush system for swine manure management. When compared to an oxidation system the flush system provides a considerable reduction in energy, dollars and labor while effectively performing the primary function of manure removal and disposal. Both systems perform this function within existing environmental constraints equally well.

Section III

Energy Reclamation From Agricultural Wastes

18.

From Biodung to Biogas– Historical Review of European Experience

Cord Tietjen*

BACKGROUND: MARSH GAS — MANURE GAS — SLUDGE GAS

"Biogas" results from the anaerobic decomposition of organic materials like marsh gas. "Marsh gas," so it is reported, was discovered by Shirley in 1667. Undoubtedly, it was known much earlier because of its universal appearance from decaying vegetation in swamps and lake beds (1). We may even blame Daniel Defoe (1719) for not letting Robinson Crusoe construct a "biogas plant" during his lonely stay on the isle where an abundance of plant material was available. As early as 1630, van Helmont mentioned, among 15 different kinds of gas, an inflammable gas that evolves during putrefaction and also is contained in intestinal gas (2). Priestley reported in 1790 of "air produced by substances putrefying in water" in his *Observations on Inflammable Air* (3). He confirmed the work of Volta, who in 1776 recognized a close relationship between the appearance of combustible gas and the decaying vegetation in the sediment of lakes and streams. Volta gave the first eudiometric analysis of methane, the constitution of which was first developed by Dalton in 1804 (2).

It was in 1808 that Humphrey Davy collected methane in his experiments with strawy cattle manure kept in a retort in a vacuum (4). This might be considered the beginning of manure gas research,

* Institut für Pflanzenbau und Saatgutforschung, Forschungsanstalt für Landwirtschaft, Braunschweig-Voelkenrode, West Germany.

but Davy was not interested in solving energy problems with natural fuel gas; his experiments were directed toward an evaluation of rotted and unrotted manure for crop production.

Gayon, a pupil of Pasteur, also recorded a success in his experiments with animal manure in 1883-84 (5,6). The volume of gas collected at 35°C was so great that Louis Pasteur concluded anaerobic manure fermentation might supply gas for heating and illumination under special circumstances. But the proposal, made in jest by the newspaper *Le Figaro* to improve the street illumination of Paris by manure fermentation from the numerous horses of the taxis and public works was not executed (7). It was only a few years later, in 1896, that sewage gas was used for lighting a street in Exeter, England (8).

In the following decades, experimental work with manure gas production never totally stopped; but to apply the results on a single-farm base was obviously considered an economical venture (9, 10). Municipalities using taxpayers' money were able to further the development. The next step was sewage treatment in a heated digester, which accelerated the fermentation process, improved the treatment efficiency, reduced the retention time and thus the digester volume requirement—the increase in release of sewage gas by heating was regarded only as a valuable by-product until the next critical state in energy supply. That occurred during and after World War II.

In 1951, 48 sewage treatment plants in West Germany provided more than 16 million m^3 of sewage gas, 3.4% of which was utilized for power production, 16.7% for digester heating, 28.5% was delivered into the municipal gas supply system, and 51.4% was converted to vehicle motor fuel (8). Today, 25 years later, flaring is prevailing.

Until now in this short review, methane bacteria microbiology has not been mentioned. Knowledge has developed slowly over a long period of time; the names of eminent scientists are like steps in the fascinating progress of this science from Béchamp, 1868, to Buswell, Barker, and Liebmann, 1950 (8).

Principal characteristics are briefly summarized: methane-producing bacteria is a specialized physiological group of strictly anaerobic organisms. They are much more sensitive to oxygen or other oxidizing agents, *i.e.*, nitrate, than are most other anaerobic bacteria. They can be grown more easily in liquid or semisolid media if fully protected from air. They appear to be restricted to the utilization of relatively simple organic and inorganic compounds. Complex mixtures of organic compounds need to establish a balanced population of bacteria for a multistage conversion process. Such a culture is capable of maintaining itself indefinitely when a

fresh supply of organic materials is added contim because
the gases escape from the medium leaving behind ve because
tially toxic by-products. These bacteria grow in med poten-
the carbon source, trace minerals, carbon dioxide, a retaining
and ammonium as a nitrogen source. The optimum agent
between 6.4 and 7.2. In terms of a group of bacteria, the is
optimum temperatures—mesophilic (32–37°C) and t two
(55°C).   ilic

BEFORE AND AFTER WORLD WAR II DEVELOPMEN

Knowledge of the biology, physiology, and biochei
methane bacteria will help to improve the process of ma
mentation. We soon realized that strawy animal manu
optimum substrate; nothing must be added when protecti
and constancy of temperature in the optimum range are wa

The impetus for the revival of manure gas research in G
after World War II was the breakdown of energy supply. Op
calculations led us to believe that at least every farm that pr
animal husbandry would be able to care for its demand (11).
this period had passed and the energy shortage was overcom
realized two facts: firstly, the input for a biogas plant under
tral European climatic conditions is so high that a small-
medium-sized farm can hardly achieve it alone (12, 13); and
ondly, the way to the liquid manure or guelle system was fe
with its benefits in labor savings and in a remarkably higher
tilizing efficiency of the manure (14).

The experimental work could be based on knowledge of the forn
tion of methane by bacteria to the extent indicated above, and
the experience collected elsewhere on the anaerobic fermentati
of farm wastes, *i.e.*, by Buswell and collaborators at Urbana, Illinoi
also by Jacobs at Ames, Iowa, where two decades later Taiganide
picked up the topic (15-17). In France, the attempt was made during
the war to solve the problem of fuel shortage on the farms by the
construction of several hundred, perhaps even a thousand small
installations (18, 19). The prototype had been developed by Ducellier
and Isman in North Africa in 1937 (6, 20-22).

In West Germany, three groups began nearly simultaneously to
develop their own procedures:

Strell, Goetz and Liebmann in Munich with a pilot plant at Grub:
Thirty days fermentation of solid manure with a natural moisture
content was followed by fill-and-draw of one-tenth of the mass at
intervals of at least six days; heated water or liquid manure was
circulated by spraying and withdrawing. No full scale plant was
built (8).

Reinhold and Noack at Darmstadt: Liquid and solid manure made

a passage gh a fermentation chamber in the form of an insulated co e (or iron) channel below ground level by the
action of a of agitator on a longitudinal shaft, which is operated
for a fev tes twice or three times a day. Spent solids are
forked o an uncovered portion at the end and are put on the
storage The chamber content is heated by steam injection
from a few plants have been built on small farms (23, 24).
Schr Eggersgluss at Allerhop: The procedure was aimed
from ning toward the reduction of labor requirement, an
impr of manurial value, and a high output of gas in a short
time he plant construction had to be adjusted carefully
to t anagement system (25-28). Short chopped straw was
use mall quantities in the stables as bedding material, to
re nount of water added to flush the manure through a
s es into a collection and mixing silo, where chopped
s er organic waste could be added before in a fill-and-
s e daily loading rate was pumped into the fermentation
sted to 3 kg organic matter per m^3 fermentation capac-
n circulation pump transported the manure from the
mixing silo, into the fermentation chamber, and later
age silo, also into the slurry tank spreader or through
o the fields. The pump also circulated the manure
ntation chamber by use of a rotating jet to break up
n 1956, Feraud, an advocate of manure gas in France,
rt, but striking description of the procedure and its
mechanization to the members of the Academy of
in France: "Summarized, one could nearly say, the
at Allerhop, that is a blower and a pump" (18).
osal of Schmidt and Eggersgluss was accepted by sev-
rms and also some smaller ones. At least 15 plants have
with fermentation capacities between 96 and 960 m^3 and
ling manure storage silos from 96 to 600 m^3. They were
o serve 25–220 animal units (500 kg); some were con-
h the municipal sewer system (29). One of these "bihugas
iological humus and gas plant) is still in operation with
ler of 200 m^3, a manure storage silo of 300 m^3, and 480 m^3
ion capacity in two digesters (Klostergut Benediktbeuren,
elz, Germany); some of the other plants have discontinued
ction only in the last few years (30).
a certain number of different procedures developed in the
years, I will mention three (8, 31, 32).
nheim": A horizontal, cylindrical steel fermentation vessel
east 10 m^3 is filled with strawy manure with a "natural"
re content. No additional fluid is necessary, also no chopping

of straw or other materials. The specific loading rate is high; a correspondingly high gas production rate has been claimed (23).

Gaertner and Ikonnomoff, "System Berlin:" Fermentation chamber and gas holder are one room, connected by a channel with a pressure chamber; all three are combined in a silo of about 65 m³, leveled with the stable floor. Loading is done through a tunnel by means of a winch; unloading is performed automatically by the differences in the gas pressure with a maximum value of 0.13 kg/cm². The prototype constructed in 1953 at Jungholz, Tirol, is still in operation (34).

Poetsch, in Hannover, with a pilot plant at Voelkenrode: An upright cylinder with a horizontal division, constructed of timber, 75 mm thick, to reduce heating and insulation input. The lower part contained the fermentation chamber and gas storage, the upper part was designed as fluid reservoir to be reached against a head of 7 m. Filling and emptying with strawy manure in "natural" moisture condition was still a problem; trap doors in the silo sides had to be opened. Heated fluid was circulated by spraying and withdrawing. This pilot plant of W. Poetsch was one of four different installations we used at Voelkenrode for our investigations (35).

The most important developments in East Germany in the same period are:

Kertscher and Poch at Jena: Chemical and technical investigations on manure gas production. Preference is given to the fermentation process at higher temperatures (52°C).

Rosegger and Neuling at Dresden and at Bornim: Intensive work on energy production and manure treatment. The aspect of energy balance was judged as unfavorable because of the climatic conditions in eastern countries.

Schmalfuss and Fiedler at Halle: Evaluation of the manure treated in biogas plants, in field experiments with crops.

GAS, DUNG, LABOR REQUIREMENTS

Although the impetus for the manure gas investigations immediately after the war was given because of the shortage in energy, another aspect should not be overlooked: Shortages of plant food and fertilizer were great and, in those years, of similar importance. When the first groups decided to begin with natural gas production, other groups observed anxiously that farm animals' excrements, the base of the conventional and now much more valuable manure, were not misused or even disappeared in the fermentation process. The question arose: How does the anaerobic treatment of the manure affect its quantity and its quality (41)? This may explain why we at the former Institute of Humus Husbandry of the

Research Center of Agriculture, Braunschweig-Voelkenrode, became members of that team of three groups than began research
into gas production. We constructed three different biogas
plants near our experimental fields. According to our task, we did
not try to develop technical matters or improvements, but made
our choice out of those installations available on the market, to
produce manure gas and manure varieties that were applied in field
and pot experiments with crops (8).

One biogas plant was set up according to Ducellier and Isman; we
obtained it from an Italian company: Three cylindrical containers,
2.2 m in height, 2.1 m in diameter, and with a capacity of 7.5 m³,
consisted of prefabricated concrete parts. Two were covered with
a sheet iron lid after loading, the third was completed by a gas
holder. No heating or insulation was provided for. In France,
farmers used to put manure around the fermentation vessels
and utilize the self-heating process under air access (8). We installed a water circulation heating system and glass wool insulation.

The second plant was built according to Reinhold in Darmstadt,
of normal size for a farm with about 10 animal units (500 kg) with
a fermentation chamber of 17 m³. The heating system by steam
injection was not satisfactory—water infiltrated into the slag insulation.

The third plant was constructed according to Schmidt and Eggersgluss, but on a reduced scale to allow substrate sampling for balance investigations (42). It consisted of a mixing chamber of 4 m³,
a fermentation chamber of 14 m³, a manure storage silo divided in
two compartments of 8 m³, and a gas container of about 12 m³. The
manure treatment installations were combined by a pipeline system;
a centrifugal pump allowed liquid manure transfer forward and
backward. Heating was done by steam injection.

Later a fourth plant was built by the Technical University of Hannover as a prototype, designed by W. Poetsch who was responsible
for the technical development (35).

The first three installations mentioned differed in several aspects,
as distinguished in Table 1. These indicated characteristics described the procedure sufficiently, and other biogas plants developed in the following years, differed little in general from one of
these three.

In field and pot experiments with different crops, the fermentation
residues and conventionally rotted stable manure were compared on
a balance base, so the results could be taken as an evaluation of
the treatment method. The conclusion was that the products of
liquid manure practices exercise a higher yield effect than solid
manure. There is no miracle effective in these results; the most
important is that loss of plant nutrients was, in practice, totally

Table 1

	Ducellier "batch"	Reinhold "channel"	Schmidt and Eggersglüss "fill-and-draw"
Condition of straw a.o.	Unchopped	Chopped	Short chopped
Loading	Periodically with prerotted manure	Daily fresh manure	Daily fresh manure
Sp loading rate	Batch	Undefined	3 kg
Scum break-up	No scum	Agitator	Rotating jet
Substrate consistency	Wet	Flowing	Pumpable
Unloading	Totally in long intervals	Daily	Partly in short intervals
Storage of fermented substrate	No storage	Stack behind the plant	Storage silo, air access hindered
Application of residues for manuring	Liquid and solids separately	Liquid and solids separately	Combined as guelle, slurry

excluded by the arrangement of the installation, and the liquid phase contains a remarkably higher plant-available nitrogen portion than does the solid manure (8, 14, 26, 43, 44).

These observations also affected the following development when there was no more need for manure gas, but labor costs or labor shortages could be partially overcome by liquid manure practices. In the case of the procedure according to Schmidt and Eggersgluss, the step from "bihugas to bihudung" was a short one, reached by leaving out the heating system, the insulation, and the gas holder. At the end of 1956 the ratio of bihugas plants to bihudung plants was already 1:4.

A few points from our manure gas investigations are mentioned here. Figure 1 shows sum curves obtained by gas collection at 30°C

Figure 1. Gas sum curves obtained from different organic materials at 30°C (8, p. 252).

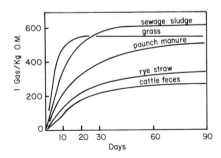

from different organic materials over a long period of time, 90 days (8.) It appears that such experiments were made again and again, even 200 years ago by Priestley. With regard to manure gas, the graph shows that the smallest quantity of gas to be expected from cattle feces and half the 90-day amount is produced in 15 days. So we can finally restrict our investigations in manure gas production on short observation periods. But we must define clearly the substrate to understand the differences, as shown in Figure 2: The gas

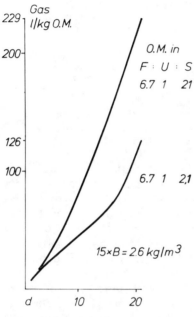

Figure 2: Gas production at 35°C influenced by straw. Cattle feces : urine : straw = 6.7 : 1 : 21 and 6.7 : 1 : 2.1 resp. (Base Organic Matter).

amount developed out of a cattle manure substrate with a great portion of straw was nearly twice the amount out of a similar substrate with less straw. Figure 3: It should be decided soon, for economic reasons, how important the specific loading rate is. Strawy animal manure is an optimum fermentation substrate, and in this respect very different from sewage. The loading rate can be much higher, and the CO_2 portion in the collected gas appears to decrease faster. But it should be realized also that the ratio of CH_4 to CO_2 in the gas collected from a fresh charge is shifting successively over a long period of time in favor of methane (Figure 4). For optimum utilization of the gas, conclusions are to be drawn with regard to loading rate, duration of fermentation, and capacity of gas storage facility. Finally, it is known that the fermentation temperature not only affects the total gas amount produced in a certain length of time, but also the ratio of CH_4 to CO_2 in a way that a

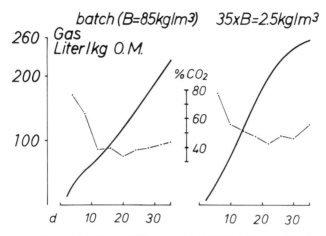

Figure 3. Gas production and CO_2 content at 35°C influenced by loading.

fermentation at 55°C yields much gas with a high methane portion in the first part of the period, contrary to fermentations at lower temperatures (Figure 5) (14). By admitting the progress in insulation material and methods and in equipment for adjustment and control, more attention should be given to higher temperature fermentations than previously.

This proposal not only refers to the field of manure gas production, but also touches the area of veterinary hygiene. In liquid manure practices, long storage periods are required for hygienic reasons. The thermophilic fermentation would essentially improve the contra-epidemic-hygienic effect, and thus reduce storage time requirement (45).

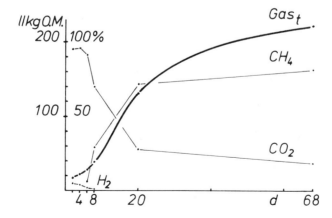

Figure 4: Change in composition of gas from straw at 35°C.

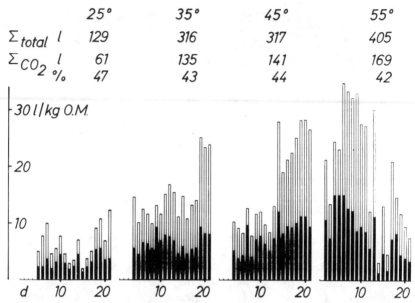

	25°	35°	45°	55°
Σ_{total} *l*	129	316	317	405
Σ_{CO_2} *l*	61	135	141	169
%	47	43	44	42

Figure 5: Temperature influence on gas production and CO_2 fraction (Black Column). Spec. loading rate 2.3 kg organic matter[2] on 15 successive days.

PERIOD'S END AND OUTLOOK

In 1958, a commission of French authorities made a visitation tour to study the installations of manure gas production in Germany. In their detailed report they concluded: ". . . In general, we think that the investigations on manure fermentation should be promoted to improve the value of the manure as well as to produce methane . . . But the interest in these installations lies beyond the rules of arithmetic and cannot be measured only in [economics]—it is of national [importance]. Consequently, the profitability is not of absolute importance . . . this does not mean that we shall neglect it . . ." (46).

In India, biogas plants have reached a prominent place in public discussions. The recent oil crisis and other reasons have caused decentralized energy production to become an attractive proposition. Today, plants of about 20 different sizes and capacities are available and 8000 plants are in operation. The Agricultural Ministry has plans to set up about 50,000 plants. However, the initial outlay on even the smallest plant is well beyond all except the affluent farmers. In 1973, the nationalized banks agreed to finance biogas plants in eight Indian states where the procedure has been proved very successful. Warning voices indicate the necessity not to rely only on subsidies and loans but to realize the importance of drastic cost reductions, *i.e.,* by developing alternatives to the use of cement and steel, and by changing the design of the plants according to new research on the fermentation rate, the gas yield and gas composition as a function of temperature, pH, viscosity, and agitation. In India, the conditions of manure gas production are different in several respects from those we are familiar wtih; it is difficult to judge the problems and the plans to master them (47-49).

In 1961, H. J. Nation, National Institute of Agricultural Engineering, England, made a visit to Germany to obtain information on "farm methane production." In his comprehensive, illustrated report to the British Society for Research in Agricultural Engineering, he formulated the following conclusions (50):

> "Different authorities have at various times put forward the views that the value of plants of the types studied . . . can be satisfactorily assessed under one, two or all of three headings:
> 1) reduction of labor requirement
> 2) improvement in manurial value
> and 3) power potential of gas produced.
> In fact, the value attached to these three yardsticks and their relationship to the capital investment varies according to the type of farming, size of farm, process employed, climate and local labour conditions. . .
> The chief conclusion arrived at after seeing all that has been described is that plants of any type for the production of methane from farmyard manure cannot be justified economically where other sources of power are available. In times of national emergency a significant contribution to the country's fuel requirements could only be made by many thousands of plants. . ."

Since Mr. Nation's report, 15 years have passed. Since then we have experienced the oil crisis and we have to deal with the requirements of environmental protection. Today, animal excrements have to be treated, in one way or other, to prevent public health hazards. Thermophilic fermentation which also excludes all unpleasant odor could be one way of treatment. Of course, facilities for animal excrement treatment have to observe labor reduction requirements, including the final disposal of the treated excrements, *i.e.,* by distribution upon cropland to reach an optimum utilization effect.

Methane from manure? Why not—but not as a priority—it might become available as a by-product of high value (51).

REFERENCES

1. Neumueller, O. A. Roempps Chemie-Lexikon. p. 2140. Stuttgart (1974).
2. Partington, J. R. A short history of chemistry. 3rd ed. London (1960).
3. Priestley, J. Experiments and observations on different kinds of air. Vol. I, p. 206. Birmingham (1790).
4. Davy, H. Elemente der Agrikulturchemie. Uebers. v. F. Wolff, p. 345. Berlin (1814).

5. Dubaquie, J. Le gaz de fumier. Compte-rendu de l'Academie d'Agriculture de France, séance de 1 déc. 1943.
6. LeSage, E. and P. Abiet. Gaz de fumier. 133 p. Soissons (1952).
7. Le Figaro. Un drôle éclairage. 5 mars 1884.
8. Muenchner Beitraege zur Abwasser-, Fischerei- und Flussbiologie, Bd. 3, Gewinnung und Verwertung von Methan aus Klaerschlamm und Mist. Hrsg. H. Liebmann. 343 p. Muenchen (1956).
9. Dubaquie, J. Modes actuels de fabriction des fumiers. Procès-verbaux de la Sociéte des Sciences physiques et naturelles de Bordeaux, séance de 4 juin 1931.
10. Gaesford, M. Muck power. The pros and cons of methane production. Farmers Weekly, May 31 (1974).
11. Hisserich, H. Ueber die Heiz- und Treibgasgewinnung und verlustlose Duengerbereitung auf jedem landwirtschaftlichen Betrieb. Land, Wald und Garten 2:131 (1947).
    Mehr Energie und Duenger aus der Landwirtschaft. Landtechnik 2: (1947).
12. Feldmann, F. Biogas—energiewirtschaftlich gesehen. Landtechn. Forschg. 4: 65 (1954).
13. Paulick, S. Beheizung von Trocknungsanlagen. Deutsche Landw. Presse 76: 12 (1953).
14. Sauerlandt, W. and C. Tietjen. Humuswirtschaft des Ackerbaues. DLG-Verlag, Frankfurt (Main), 240 p. (1970).
15. Buswell, A. M. and C. S. Boruff. Mechanical equipment for continuous fermentation of fibrous materials. Industr. and Engin. Chemistry 25: 147 (1933).
16. Buswell, A. M. and W. D. Hatfield. Anaerobic fermentations. State of Illinois, Div. State Water Survey, Bull. No. 32, 193 p., Urbana, Ill., (1939).
17. Jacobs, P. B. Heat and light on the farm.—The utilization of farm wastes. Architect Rec. 7: 358 (1934).
18. Feraud, L. Nouvelles observations sur le dévelopment des installations productrices de méthane biologique. Compte-rendu de l'Academie d'Agriculture de France 16 (1956).
19. Roussiaux, P. En France: 1000 fermiers cuisinent au gaz de fumier. J. France Agric. 6: 913 (1950).
20. Ducellier, G. and M. Isman. Procédé d'obtention d'un gaz combustible par fermentation de matière organique. Brevet d'invention gr. 15, cl. 3, nr. 893/767, Paris (1942).
21. Isman, M. Une etude sur les modes d'utilisation pratiques des appareils à gaz de fumier. Elevage et Culture 21 (1951).
22. Mignotte, F. Le gas de fumier à la ferme. La maison rustique, 85 p. Paris (1952).
23. Reinhold, F. Gasgewinnung in der Landwirtschaft nach dem System "Darmstadt". Landtechnik 7: 33 (1952).
24. Noack, W. Biogas in der Landwirtschaft. Elsner, Darmstadt, 111 p. (1955).
25. Kemmler, G. Ueber Veraenderungen in der stofflichen Zusammensetzung des Stallmistes bei der biologischen Gaserzeugung. Diss. Goettingen (1952).
26. Defu-Mitteilungen. Hrsg. F. Schmidt, Deutsche Futterkonservierungsges., Verden, 52 p. 9 (1951).
27. Schmidt, F. and W. Eggersgluss. Hochwertiges Gas aus landwirtschaftlichen Abfaellen und anderen organischen Stoffen. Brennstoff-Waerme-Kraft 7: 249 (1954).
28. Schmidt, F. and W. Eggersgluss. Method and apparatus for the processing of organic waste material. Canada 540, 755, filed Dec. 30, 1954, issued May 14, 1957, U.S. Env. Prot. Ag. 1973: Pat. Abstr., Internat., Sol. W. Man (SW-78c).

29. Tietjen, C. Bihundung aus Abwasserschlamm und Stroh. Landbauforschg Voelkenrode 6: 75 (1956).
30. Allgaeuer Rundschau 31. Aug. 1974.
31. Mueller, W. Untersuchungen ueber die biologische Gaserzeugung bei der Methangaerung von Stallmist und Stroh. Diss. Hohenheim (1955).
32. Wick, H. Diss. Hohenheim (1960).
33. Maschinenfabrik Adelsheim G. M. B. H. Biogasanlage System Hohenheim. Prosp. L 56, VIII, 56.
34. Gaertner, A. and S. D. Ikonomoff. Faulraum (Gasanlage) "System Muenchen". Staedtehygiene 5 (1956).
35. Poetsch, W. Entwicklung und Erprobung einer handwerklich gefertigten Biogas-Anlage aus Holz. Heinz-Piest-Institut f. Handwerkstechn., Techn. Hochschule Hannover, Arbeitsber. 13, 61 p. (1963).
36. Kertscher, F. Biogasgewinnung. Wiss. Z. Universitaet Rostock, math.-naturw. Reihe 2: 209 (1953).
37. Poch, M. Biogas. Deutscher Bauernverlag, Berlin, 45 p. (1953).
38. Rosegger, S. Der Entwicklungsstand von Biogasanlagen und Perspektiven fuer die landwirtschaftliche Praxis. Wiss. Z. Techn. Hochschule Dresden 6: 511 (1956/57).
39. Rosegger, S. and S. Neuling. Versuchsanlage zur Humus- und biologischen Gasgewinnung an der TH Dresden. Deutsche Agrartechn. 6: 147 (1956).
40. Schmalfuss, K. Fragen der organischen Duengung. Deutsche Akademie d. Landwirtschaftswiss., Berlin, Sitzungsber. VII, 3, 24 p. (1957).
41. Scheffer, F., E. Welte and G. Kemmler. Ueber Veraenderungen in der stofflichen Zusammensetzung des Stallmistes bei der biologischen Gaserzeugung. Landwirtsch. Forschg. 4. Sdh. (1953).
42. Tietjen, C. Bilanzuntersuchungen bei Stallmistaufbereitung in Biogasanlagen. Pflanzenernaehr., Duengung, Bodenkde. 77: 198 (1957).
43. Tietjen, C. Plant response to manure nutrients and processing of organic wastes. Proceed. Nation. Sympos. ASAE 136 (1966).
44. Tietjen, C. Anwendung von Fluessigmist im Ackerbau. Landw. Forschg. 25/II. Sdh., 25 (1970).
45. Poch, M. and L. Haenske. Die biologisch-thermische Desinfektion von Guelle mit Hilfe der thermophilen Methangaerung. Z. Hygiene u. Grenzgebiet 14: 553 (1968).
46. Brunotte, R. and A. Gac. Etude des installations de production de méthane biologique en République Fédérale Allemande. Bull. Techn. Génie Rural 45 (1958).
47. Gotaas, H. B. Composting. World Health Organization, Geneva, 205 p. (1956).
48. Ministry of Agriculture, New Delhi. The use of organic fertilizers in India. FAO Soils Bull. 27: 273 (1975).
49. Prasad, C. R., K. K. Prasad, and A K. N. Reddy. Biogas Plants-Prospects, Problems, Tasks. Indian Institute of Science, Bangalore, Mimeo 50 p. (1974).
50. Nation, H. J. Report on a visit to Germany and Holland, II. Farm methane production. Nation Inst. Agric. Engin., Brit. Soc. Res. Agric. Engin. 18 p. (1961).
51. Fairbank, W. C. Fuel from livestock wastes: an economic analysis. Agric. Engin. 20 (1974).

19.

Energy and Agricultural Biomass Production and Utilization in Canada

C. G. E. Downing*

Canada is one of the largest countries of the world with great extremes in climatic conditions for agricultural production. Wide variations in temperature exist from a high of over 90°F during part of the growing season in most areas of the country to a low of −30° to −40°F in most areas of the prairies and northern parts of eastern Canada during the winter. Precipitation varies from 11 inches per annum in the arid areas of the prairies to more than 35 inches per annum in the more humid areas of eastern Canada and the Atlantic provinces. It is estimated that more than three-quarters of the agricultural production in Canada comes from an area stretching from the Pacific to the Atlantic and less than 200 miles north of the U.S. border. Therefore east-west transport is a serious and an expensive aspect of Canadian agricultural distribution beyond the farm gate.

However, during the past three decades Canadian agriculture's greatest achievement has been an increased output of quality food at very modest increases in price. This has been possible by increased mechanization, increased use of fertilizer and improved crop varieties. During this period the value of implements and machinery has increased from one-half billion dollars to over 4.5 billion dollars. In 1971 farm operating expenses amounted to approximately

---

* Director, Agricultural Engineering, Research Branch, Agriculture Canada, Ottawa, Canada

3 billion dollars (one-half billion of which was spent directly on energy resources) to produce 4.5 billion dollars worth of farm output (1).

Agriculture is one of the largest users of petroleum products in Canada, consuming 7.9% of the gasoline and 12.2% of the diesel fuel. The range in utilization was 3% and 2% respectively for gasoline and diesel in the maritimes, and 34% and 45.6% in Saskatchewan. Direct farm expenditures on petroleum products represented approximately 10% of the operating costs of Canadian agriculture. In the prairie provinces it was between 15.4 and 17.5% while in the east and British Columbia, it was between 5.7 and 7.8%. Thus any increase in petroleum prices or reduction in supply could have a significant impact on the operating cost and/or the output of Canadian agriculture, particularly in the prairie region.

There are approximately 170 million acres of total farm land in Canada, of which 96 million are designated cropland with the balance as improved pasture and various types of unimproved land, giving a total of approximately 129 million acres of agricultural biomass area and about 41 million acres of nonproductive biomass area.

AGRICULTURAL PRODUCTION

Table 1 outlines some of the basic statistics of Canadian farms. All farms are grouped together under Canada, out of which has been separated those in Ontario and those in Saskatchewan (2). Saskatchewan has more than 40% of the cropland and is the main cereal and oilseed growing area of Canada, maintaining large farms using primarily a mulch tillage system with large tractors and equipment. Livestock production is extensive with large areas of rangeland and brushland. Ontario, with less than 10% of the cropland, has a much more intensive agricultural production, primarily related to corn and other cash crops, tobacco, fruit and vegetables, and an intensive livestock production, all quite highly mechanized It is to be noted that there are more tractors on farms in Ontario than in Saskatchewan, with a much smaller acreage per tractor

Table 1. Summary of Canada agriculture statistics.

	Canada	Ontario	Saskatchewan
Cropland (acres)	96×10^6	8×10^6	44×106
No. Farms	3£6,000	95,000	77,000
No. Tractors	600,000	1£6,000	133,000
Acres/Tractor	160	50	330
Fuel/Acre (gal)	4.2	9.2	2.5

but with almost four times the fuel per acre consumed in the production programs. Here plowing is the major primary tillage operation and, along with row crop operations and multiple spraying
operations, Ontario farm production is quite energy-intensive.

Table 2 outlines the energy value of the various crops included in
cropland in Table 1. These are given as thermal value of the usable
crop in its harvested form. The energy value used varies from 15
million Btu per ton for cereal grains, 17.5 million Btu for oilseeds,
14.4 for forage crops, 2.2 million for fruit crops, and 1.8 million for
vegetable crops (3). The energy value for the crop residue such as
straw is given as 12.5 million Btu per ton. The values for the animal
waste vary widely and are taken from the Canada Animal Waste
Management Guide (4).

Table 2. Energy value of crops.

	Canada	Ontario	Saskatchewan
Crop Energy (Btu)	109×10^{13}	23.5×10^{13}	35.4×10^{13}
Crop Energy (Btu/acre)	11.4×10^{6}	29.2×10^{6}	8.0×10^{6}
Available Crop Residue (Btu)	23.4×10^{13}	2.8×10^{13}	11.2×10^{13}
Available Animal Waste (Btu)	57.5×10^{13}	20.4×10^{13}	6.8×10^{13}

The determination of available crop residue is based on the assumption of ¾ ton per acre in western Canada under low rainfall
areas, where a high percentage of the residue is needed to maintain
humus in the soil and for mulch tillage purposes for wind erosion
control, and at the rate of 1½ tons per acre in eastern Canada where
unit crop production is much greater and where straw-to-grain ratio
is higher.

Available animal waste is determined on the basis that all poultry
and swine are housed, and therefore all waste is available. But
it is assumed in the West half of the cattle are on range or pasture
full time, and therefore their waste is left in the natural state,
and that the other half are housed for only half of the year in a manner in which their waste is accumulatable. In the East it is estimated
that most cattle are housed or are on feedlots for two-thirds of the
year. However, the major problems associated with the utilization
of these potentially available waste energies are the cost involved,
the energy input, the labor, transportation and storage of this material to an outlet where it might be consumed as a combustible fuel,
utilized in a chemical or digester system for the production of a
gaseous energy product, or recycled as feed through livestock. It
may be that the most appropriate utilization of this waste is still
back on the land under proper environmental quality conditions.

ENERGY INPUTS

Table 3 outlines the energy input into the crop production. The direct energy inputs are based on those used specifically on farm operations. The indirect energy used is based on equipment and materials used on the farm in the production of crops using Pimentel's values (5) for the original manufacture of these materials. The miscellaneous item covers pesticides, seed, irrigation, and a factor of approximately 20% of direct energy inputs to account for the energy required to produce this energy.

Table 3. Energy input in crop production.

	Canada	Ontario	Saskatchewan
Direct Energy Use			
Fuel—Tractors, Combine (Btu $\times$ 10^{12})	56.4	10.6	15.5
Fuel—Cars, Trucks (Btu $\times$ 10^{12})	47.0	10.0	12.7
Electricity (Btu $\times$ 10^{12})	8.4	2.9	1.2
Total Direct (Btu $\times$ 10^{12})	111.8	23.5	29.4
Indirect Energy Use			
Tractors, Machinery (Btu $\times$ 10^{12})	5.3	1.4	1.1
Cars and Trucks (Btu $\times$ 10^{12})	8.0	1.9	2.0
Fertilizer (Btu $\times$ 10^{12})	32.2	14.2	1.6
Miscellaneous (Btu $\times$ 10^{12})	30.0	7.6	6.6
Total Indirect (Btu $\times$ 10^{12})	75.5	25.1	11.3
Total D. & I.D. Input (Btu $\times$ 10^{12})	187.3	48.6	40.7
Energy Output/Input Ratio (Crops)	5.8:1	4.8:1	8.7:1
Energy O/I Ratio (Crop + Residue + Waste)	8.1:1	7.2:1	10.9:1

The crop energy output from Table 2 compared to the total energy input of Table 3 gives favorable energy ratios indicating that agriculture is a net producer of energy. There are, however, appreciable variations as indicated for Ontario and Saskatchewan. Although the unit production per acre is much higher in Ontario than in Saskatchewan, the high energy input in the form of fertilizer and higher input in field and mechanization makes crop production much more energy-intensive in Ontario than in Saskatchewan. When the energy output includes crop energy plus one-half the potential energy in the available crop residue and animal waste (allowing for energy input to gather, transport, process and convert), the energy rates are even much more favorable.

One factor of appreciable value in the Ontario production is the production of fruits and vegetables which have a very low energy value but require a high energy input. Thus an important factor to consider is that undue emphasis not be placed simply on the energy output to energy input ratio of all crops since the production and availability of good quality fruits and vegetables are of significant importance to the proper nutritional diets.

Table 4 brings the national and provincial analysis of Table 3 down to specific farm analyses (6). These five farm models are from a study made in Alberta under conditions not unlike the variation and cultural practices in Saskatchewan. The main difference between farm no. 1 and no. 2 is that farm no. 1 was over-mechanized and used more fertilizer per acre to maintain production on the longer rotation. These are the main factors accounting for the lower O/I ratio of farm no. 1. Farm no. 3, although the largest, is in the very low rainfall area in which half of the land is in summerfallow each year, registering the lowest ratio.

Table 4. Energy on specific farms

Farm No.	Acreage	Crops	Prec. Ins.	Yield Bus/Ac	O/I Ratio
1	620	B.O.R.S.	19	50,70,25	4.21.1
2	960	W.B.S.	15	32,50	5.50:1
3	1280	W.S.	13	20	4.19:1
4	300	Br-A.	19	2 tons	6.7 :1
5	300	A.	15	2T,4T	16.2 :1

N.B. Crop Rotation Abbreviations: B—Barley, O—Oats,R—Rapeseed, S—Summerfallow, W—Wheat, Br-A—Brome Alfalfa, A—Alfalfa.

It is an interesting factor that the practice of summerfallowing does not cause a very large reduction in overall energy efficiency. In the first place, summerfallowing requires a relatively small amount of energy in relation to that for total crop production. For instance, in farm no. 1, only 5.6% of the total energy was used in summerfallowing although it constitutes 20% of the farm acreage. In farm no. 2, 16.8% was for summerfallow with 33% of the acreage, and in farm no. 3, 26% for summerfallow in which 50% of the acreage was in fallow. Also, summerfallow is commonly considered a substitute for fertilizer because it aids the decomposition of residues from the previous crop with accompanying release of nutrients for the succeeding crop.

The main reason for the difference in O/I ratio in farms 4 and 5, both of which were forage production units, is the high fertilizer input in farm no. 4 primarily to support the brome production; in farm no. 5 no fertilizer was applied to the dryland alfalfa but with fertilizer applied to 20% of the land under irrigated alfalfa. The irrigation was carried out by flood irrigation; therefore, there was no high energy input for pumping and sprinkler requirements.

BIOMASS ANALYSIS

It is important to realize that all agricultural production energy comes from solar energy through biomass based on photosynthesis. An evaluation of biomass production and utilization is being undertaken and is outlined in Figure 1. The total solar influx falling on farm land in Canada averages about 140 watts per square meter, and this is shown separated into three types of farm land: 1) the basic cropland on which external energy is applied for production and from which crops are harvested in one form or another, 2) the sum-

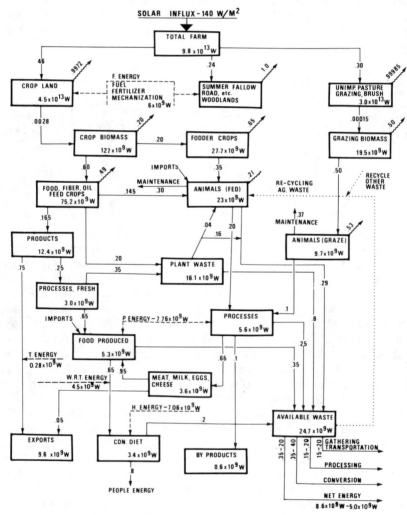

Figure 1. Biomass production and utilization.

merfallow or bare land from which no biomass is produced, and 3)
the unimproved pasture and grazing land on which no external energy is applied primarily in the rangeland of western Canada and
the brushland in the northern parts of most provinces where the
biomass is consumed directly by the grazing animals.

The biomass flow on an energy basis follows three specific patterns through the network system of Figure 1. One goes directly as
a crop products pattern either to export or through various degrees
of handling and processing to the Canadian diet. The second pattern
picks up the fodder crop from the crop biomass total and, with the
addition of feed crops and imports, follows through the animals
(fed) pattern through processing and handling to Canadian diet
and by-products. The third pattern is one directly from the grazing
land area and through animals (graze) to stages of processing and
again to Canadian diet and by-products.

The energy distribution from each of the rectangles of Figure 1
totals 1, indicating there is no storage build-up or deficiency in any
of the system components. The wavy arrows on the upper right-
hand corner from a number of the rectangles represents energy
losses from the system. These take different forms depending on
the type of product but include respiration losses, metabolic heat
losses from animals, the return of roots and crop waste to the soil,
and livestock manure deposited in rangeland and on pasture. Also
included in the outline are external energy amounts indicated by the
dashed lines which are based on actual farm energy inputs to the
farm gate with additional external energies applied to other areas
of the food chain to the consumer on a percentage basis as indicated
by Hirst (7). These are F energy, which is total farm energy to the
farm gate, P energy, which is processing energy applied by a variety of industries to the plant and animal products, and T energy,
which is transportation energy from the farm to the rail head for
export or other inland locations, WRT energy, which is wholesale,
retail and transport energy in the handling, packaging, etc. of the
food products after processing, and H energy which is home energy
used in the preparation, cooking and handling of food in the home.

Although appreciable amounts of the crop residue and animal
waste are returned to the soil in the process of agricultural production practices, there are many areas in the system from which
waste can be accumulated; these are shown in the rectangle, available waste. Estimates are shown indicating possible energy input
for gathering, transportation, processing and conversion of this
accumulated waste showing finally a potential net energy. It may
be noted that this net energy potential is about equal to the farm
energy that is directly applied to cropland production. Also indi-

cated by a dotted line is a possible secondary utilization of this waste as a recycling process back to animals (fed).

The general basis for the crop biomass produced from photosynthesis is that utilized by McHale (8) in which it is indicated that the distribution of photosynthetic material in plants is 20% as respiration loss, 25% as usable crops, 43% as leaves and stalks, and 12% as roots.

It is evident from the analysis that export of crops (wheat, barley, oilseeds), is a significant factor of Canadian agricultural production. The actual export of these crop products amounts to more than five times the amount of such products consumed in the Canadian diet and is equivalent to approximately three times the total energy equivalent of the total Canadian diet. However, approximately 45% of the crop products grown are consumed by livestock in one form or another, ending up in the Canadian diet, having been first transformed into livestock products.

The direct conversion of biomass through crop products to food of Canadian diet type is only about 40%. Although a considerable amount of the original biomass is returned to the soil, there is appreciable residue and waste developed at the different stages from production to consumption. However, this conversion of biomass to food products is still much more efficient than through animals in which it is only about 6% via the animals (fed system) and 4% via the animals (graze) system. However, there is considerable biomass utilized by animals and produced on land that is not suitable for producing food for people, such that the utilization of such products through animals, even at the low efficiency of conversion, is a plus factor for human food. There is a good potential for increasing production by this route.

The most significant factor in regard to energy consumption in making available to the Canadian diet the products of agricultural biomass is the large amount of energy utilized off the farm to process and handle the products to satisfy the diet needs. This amounts to almost 3½ times that utilized for on-the-farm production. This is quiet well illustrated in the following output-to-input energy ratios:

Crops (Export)	— at farm gate	— 5.4 :1
	— at rail head	— 4.6 :1
Crops (Food)	— at farm gate	— 5.4 :1
	— at home plate	— 1 :2.3
Animals	— at farm gate	— 1.52:1
	— at home plate	— 1 :10
Total Food	— at farm gate	— 3 :1
	— at home plate	— 1 :7

An evaluation of this analysis of biomass production and utilization indicates a number of areas for special study and further evaluation:

1. Reducing the field, storage and processing losses of crops and forages from production through livestock and human consumption;

2. Reducing the cereal, oilseed and other crops appropriate for human consumption presently being utilized in livestock production;

3. Exploiting land not suitable for efficiently growing crop products for human food to produce biomass for direct livestock feed or other forms of utilizable energy;

4. Exploiting ways of utilizing by-products and wastes for direct energy use such as methane gas or for indirect use such as livestock feed or soil fertility;

5. Studying ways of reducing and conserving energy inputs in the various aspects of agricultural production;

6. Improving photosynthesis efficiency of crops and management practices to increase unit production of biomass.

* * *

Contribution No. 512 from Engineering Research Service, Research Branch, Agriculture Canada, Ottawa, Ontario, Canada K1A 0C6.

REFERENCES

1. MacEachern, G. A. "The Impact of Energy Crisis on Canadian Agriculture," Agricultural Economics Research Council of Canada (1973).
2. Statistics Canada. Agricultural Statistics (1971).
3. National Academy of Science. *Atlas of Nutritional Data on U.S. and Canadian Feeds,* Washington, D.C. (1971).
4. Agriculture Canada. "Canada Animal Waste Management Guide," Publ. No. 1534 (1974).
5. Pimental, D. *et al.* "Food Production and the Energy Crisis," *Science,* Vol. 182 (November 1973).
6. Jensen, N. E. "Energy Budgets for Crops Grown in Western Canada," Agriculture Canada Contract, Ottawa (January 1975).
7. Hirst, E. "Food-related Energy Requirements," *Science,* Vol. 184 (1974).
8. McHale, J. *The Ecological Context* (George Braziller Inc., 1970).

20.

A Mobile Pyrolytic System — Agricultural and Forestry Waste Into Clean Fuels

J. W. Tatom, A. R. Colcord, J. A. Knight,
L. W. Elston and P. H. Har-Oz*

AGRICULTURAL AND FORESTRY WASTES AS A LOW-SULFUR FUEL

Each year prodigious amounts of waste are produced in the United States and, as is widely recognized, the disposition of this unwanted material is an increasingly difficult and expensive task. It is likewise widely known that the U.S. presently faces a severe shortage of fuel and that, in importing foreign oils to alleviate this shortage, the already serious problem of the balance of payments is aggravated. These two problems, as many have observed, are related since the chemical energy content of the wastes, by the most conservative estimates, is a very significant fraction of that of the imported oil. Thus, the desirability of converting these wastes into energy and thereby alleviating the problem of waste disposal is evident.

There are a number of problems with this idea, unfortunately, the most important being the costs of collecting, transporting and preparing wastes for use in either conventional or advanced thermal oxidation plants for ultimate production of mechanical or electric power. A considerable effort has been made in overcoming these

---

* Engineering Experiment Station, Georgia Tech, Atlanta, Georgia 30332.

problems as they relate to urban wastes, but a like effort in dealing with agricultural wastes has not been forthcoming. It is presumed that because urban wastes are already being collected and transported to some site of disposal, they have appeared as a better source of potential energy than agricultural wastes which are frequently located in generally more remote areas.

The fact that agricultural and forestry wastes outweigh urban wastes by more than an order of magnitude tends to be ignored or forgotten. The homogenous, lignocellulosic character of these wastes, which avoids the necessity for expensive metal sorting and/or other preparation prior to thermal oxidation, is apparently not widely appreciated. This fact is especially important when it is realized that, typically, in a modern urban waste disposal/utilization thermal oxidation system, 50% of the operating costs is spent on preparation of the wastes. There is also the very significant fact that the combustion products from these wastes have an extremely low sulfur content, thus allowing power production without further burdening the atmospheric SO_x problem. Alternatively, by combining these wastes with presently available high sulfur coals, acceptably low sulfur emissions can be achieved and in so doing the usable coal supplies effectively increased by a substantial margin. The extent of this coal-supply increase in the United States could be from 10 to 30%. There is also the advantage of low chlorine emissions; in contrast, the chlorine problem could seriously affect the successful utilization of urban wastes as a fuel.

To utilize economically agricultural and forestry wastes, though, still requires a practical technique for overcoming the transportation problem. This problem would not be so bad if the heating value of the raw waste were higher, if it did not typically contain 50% water, and if it were not so bulky. In addition, there is still the requirement that new burners, furnaces and pollution abatement facilities be built or old equipment, extensively modified before these wastes, as presently produced, can be burned. Clearly the mentioned difficulties must be overcome before these wastes can be successfully used as fuels.

PYROLYTIC CONVERSION OF
AGRICULTURAL AND FORESTRY WASTES

The Georgia Tech Engineering Experiment Station (EES) has long recognized these problems and has been working for more than eight years with the goal of converting waste materials into useful fuels. In the course of this work it has developed a unique, partial oxidation, steady-flow, low temperature pyrolysis process capable of converting these raw wastes into high energy fuels. This simple, self-sustaining, flexible process has developed from laboratory scale to pilot plant scale and finally to a large-scale demonstration facility capable of feed rates of more than 50 tons/day. (This system and the EES experience in waste utilization are described in a later section.)

In addition to the technical development of this process, a systems study of its application to agricultural[1] and forestry waste utilization has been conducted. One of the most promising systems studied involves a series of pyrolysis plants located at or near agricultural processing plants such as saw mills, cotton gins, and sugar mills, and in the vicinity of a thermoelectric power plant. The system would produce a char, an oil, and a gas from the wastes. The char would be crushed, mixed with the oil, and then transported to the power plant where it would be directly burned using existing facilities. Its form would closely resemble that of coal with a bulk density of 40 lb/ft^3 and its heating value would be typically in the range of 11,000 to 13,000 Btu/lb. Because of its higher heating value, higher bulk density and zero moisture content, the transportation costs per Btu would be greatly reduced compared with a system involving transportation of the raw, wet wastes. This basic fact, plus the ability to use existing equipment, gives such a system a significant advantage over other conventional and less thermodynamically efficient techniques which involve direct burning of the unprocessed waste. The pyrolysis gas generated at the production sites would be used to dry the wastes and to replace present sources of process heat; any excess could be sold to business or private consumers in the vicinity of the pyrolysis unit. Considering the rising costs and decreasing availability of natural gas, this approach has a valid economic basis.

The results of this study indicate that the concept has merit and that a highly profitable operation using either public or private monies is feasible. In this study, only economic factors were considered. No monetary value was assigned to the obvious but intrinsic advantages of: a) conservation of resources; b) alleviation of the problem of waste disposal with resulting air pollution; c) generation of a number of extra skilled job openings and a general upgrading of the technological capabilities of the area residents; and d) local retaining of money presently spent on coal. This could typically represent several hundred thousand dollars per year and would offer a significant stimulus to the local economy. Thus, when considered from a national perspective, these added intrinsic values, in addition to its economic merit make such a system highly desirable.

---

1. Manure, because of its very high moisture content, is not considered to be a good material for energy conversion by pyrolysis. For this reason, manure has been excluded in these studies, since digestion or fermentation processes appears to be far superior methods for converting this waste into energy.

In addition to the above study, considerable effort toward the development of a transportation version of the EES pyrolysis system has been made. This turns out in many cases to be an even more suitable system than the fixed pyrolysis plant concept. The need for transportability arises because many agricultural wastes are produced seasonally and in widely diverse locations. Economic considerations require that the system operate continuously rather than sporadically and that the transportation costs of the raw wastes be minimized. Thus a self-sustaining waste converter system that can be readily transported to the site of the waste production and then moved to another location when needed has many obvious advantages. The low temperature of the EES waste converter lends itself very nicely to this transportability concept since it results in a much lighter system, due to reduced insulation weight. In fact, preliminary work suggests that the insulation may be removed entirely for this type operation.

PRODUCTION AND AVAILABILITY OF SELECTED AGRICULTURAL AND FORESTRY WASTES APPLICABLE TO THE PYROLYTIC CONVERSION CONCEPT

There appear to be very considerable differences between various estimates of the production rates of agricultural and forestry wastes in the United States. And further, the availability of these wastes is also a source of some major differences of opinion. While the EES has not made an exhaustive study of this question, it has been in contact with several individuals and groups involved in such matters who have. The conclusions from these contacts, as they relate to the pyrolytic conversion concept, are offered here as interim estimates which hopefully will be improved in later more extensive studies. It is important to note that the various estimates encountered have differed by orders of magnitude, so perhaps the results described should not be taken to a high degree of precision. The purpose of offering these estimates is to demonstrate the potential amounts of waste that could be utilized for the production of energy. It is the orders of magnitude that are significant.

Concerning crop residues, Anderson (1) estimates that 390 million dry tons are produced annually with only 22.6 million tons being "readily collectable"[2] the remainder being left in the fields. At the

---

2. The expression "readily collectable" appears from a private communication with Anderson, to be a highly subjective term. The value of the "readily collectable" wastes is apparently based on the collective wisdom of a panel of experts. The concept of a mobile system, capable of moving to the site of the waste production, was not imagined when these estimates were made; only conventional transportation concepts were considered.

present time this estimate is probably not far off; however, the important point is that the available amounts of crop residue are relatively insignificant in comparison with the quantities of other available wastes. (The NSF-supported study by Inman, *et al.*, at Stanford Research Institute, when published should shed considerable light on the amounts of these residues.) For this reason, because of the large collection costs associated with these wastes, the high price of nitrogen fertilizer and the loss of organic material from our nation's soil, there is considerable question as to the practicality and wisdom of using these materials for energy production. Thus for purposes of this discussion, these wastes can be roughly disregarded.

Concerning manufacturing and process wastes, such as those produced at sawmills, cotton gins, sugar mills, etc., the total production is somewhat more impressive. Estimates from Forest Service contacts (2) indicate that almost 100 million dry tons of wood wastes are produced annually. The production of cotton gin wastes, peanut hulls, rice hulls, bagasse, etc., all together probably amounts to about 20 to 30 million tons/year. Since these wastes are produced in a concentrated form, on a regional basis they represent a significant source of energy and cannot be disregarded. For example, in parts of south Georgia and Alabama there are large areas which could become energy self-sufficient from the available cotton gin trash and peanut hulls. In Louisiana, bagasse could provide significant amounts of that state's energy needs. Likewise in the area surrounding Memphis, Tennessee, and Lubbock, Texas, there are potentially available for energy conversion perhaps two million tons/year of cotton gin trash. Correspondingly, there are concentrated areas in the midwest and far west where these type wastes represent a serious disposal problem and which offer the potential for considerable energy production.

Thus process and manufacturing wastes, especially wastes from wood, represent a very significant potential source of energy production. While there are some uses presently made for these wastes in the manufacture of other products, the proportion of the wastes used for these purposes is presently not large. Hence a very significant fraction, perhaps 70 to 80%, would be available using either the fixed or the portable pyrolysis concept.

In addition to the process wastes produced annually, there is available the accumulation of these wastes from previous years' operation. In the case of wood alone, this represents perhaps as much as 200 million dry tons. Indeed there are sawdust piles 60 feet deep and covering dozens of acres in the Southeast and corresponding deposits in Maine, Michigan, and Oregon. There are canyons filled with wood waste in Northern California. Thus these concentrated waste "deposits" literally represent a national resource that can be mined just as coal.

In passing it might be noted that these process wastes also represent an ideal material with which to initiate the development of the pyrolysis concept. Their present availability, their physical properties and the concentrated form in which they are produced all lend themselves favorably to the initial technology development required for the concept, especially the mobile pyrolysis unit.

It is in the forests themselves, however, where the real potential for energy production from wastes is evident. And it is here where the previous work of Anderson (1) and the more recent estimates by Taylor (2) and LaHaye (3) differ by more than order of magnitude. These differences may be explained partially by the fact that the earlier estimates were made from experience with "marketable" timber operations. Thus these estimates, rather than representing the entire forest situation nationally, were applicable only to locales where logging operations were being conducted. Since there are vast stretches of uncultivated forest where the ratio of cull trees to marketable trees is very large and where it is not practicable to conduct logging operations, at least part of the difference between these various estimates can be explained. Another explanation for the difference is the inclusion of diseased and dying trees in the more recent estimates. To illustrate, there are an estimated five million acres of trees in Maine presently dying due to infestation with the spruce-budworm. These trees because of the large ratio of culls to marketables and due to the type pulping operations in Maine do not represent a practical source of either timber or woodpulp. Therefore their conversion into energy is perhaps the only viable solution.

And finally it appears that previous estimates as to the total amounts of wood that can be produced annually per acre may be off by a factor of 20 or more. The commonly quoted figure of 0.2 to 0.3 chords/year-acre may more correctly be ten chords/year-acre (3). This altogether leads to the following conclusions:

1) On an annual basis the production of waste wood in our forests amounts to just over half a billion dry tons/year, not 55 million as reported in (1).

2) There is currently available in the form of culls, slash and logging residues something like three billion dry tons of wood wastes.

While clearly only a fraction of all these forest wastes is easily recovered, even with the portable pyrolysis concept, the amounts are nonetheless staggering and the potential of energy production from these wastes is clearly indicated.

EES EXPERIENCE IN PYROLYSIS

The process of pyrolysis, because it offers a means to efficiently convert raw lignocellulosic wastes into storable, clean-burning, high-energy fuels has recently received much attention. A number of organizations have developed different pyrolytic processes for the conversion of municipal wastes and some have developed them to the demonstration scale. Workers at the Engineering Experiment Station have also found that pyrolysis is readily adaptable for the conversion of cellulose and lignocellulosic wastes into useful fuels and other products. Involvement in the area of conversion of agricultural wastes by pyrolysis began with work over eight years ago to develop a means to dispose of peanut hulls without producing the pollution problems of incineration.

The first large prototype, developed from small-scale batch tests, was constructed in 1971. This system, shown in Figure 1, was built to operate continuously at an input feed rate of 4000 pounds per hour. The outside dimensions of this unit were approximately 8′ x 7′ x 3′ and the reaction chamber was mounted on top of a water-cooled collection chamber. This unit processed thousands of pounds of peanut hulls over a period of many months operation.

Based on the data and results from the experimental prototype,

Figure 1. First generation EES pilot plant pyrolytic converter.

an improved 6 ton/day pilot plant unit, was designed and built. In addition a large 50 ton/day demonstration plant has been built by the Tech-Air Corporation who is the licensee for the EES process.

The pilot plant, located at the EES on the Georgia Tech campus, has been modified and upgraded several times since its initial construction in 1972. Presently, the system includes a waste receiving bin, a belt conveyor to the waste converter, the converter and char handling system, an off-gas cyclone, a condenser and a gas burner. The present system will process 300 to 500 pounds waste/hour depending on the density of the feed material. A photograph of this pilot plant is shown in Figure 2. Types of wastes processed through the converter include peanut hulls, wood chips, pine bark and sawdust, automobile waste, municipal wastes, macadamia nut hulls, and cotton gin wastes.

Figure 2. Second generation 6 ton/day pilot plant pyrolytic converter.

The large 50 ton/day demonstration plant is located in a wood yard in Cordele, Georgia, and operates on wastes from the sawmill. This system has been in commercial operation now for more than two years and was field tested for two years prior to that. The char produced is sold on the commercial carbon market; the oil produced is used in an oil-fired kiln drièr; a portion of the gas is used to dry the feed, and the remaining gas is flared. Plans are to construct a process steam boiler which will utilize the remaining gases. The system is pictured in Figure 3. To get an idea of scale, note the control shed in the lower right hand corner of the figure. An attractive feature of this system is the cleanliness of its exhaust which is completely invisible to the eye. A recent analysis of the stack gases was made and the comparison of this analysis with the EPA exhaust standards reveals that the system easily meets all the federal standards.

Figure 3. Fifty ton/day commercial pyrolytic converter.

A new, larger 25 ton/day more sophisticated pilot plant, located at the EES has recently been constructed. This unit, similar to the 6 ton/day pilot plant is shown in Figure 4. This unit has processed many types of municipal plant and wood wastes and is currently being used in a study involving peanut hulls.

In the following paragraphs a description of experimental work done in the two EES pilot plants is presented. In addition, the results of a preliminary design study of the portable pyrolysis concept are described. The experimental programs have involved the processing of cotton gin trash, sawdust and peanut hulls and have demonstrated the wide versatility of the EES system to different types of waste materials.

Figure 4. Third Generation, 25 ton/day EES pilot plant pyrolytic converter.

TEST WORK

Cotton Gin Trash

Pyrolysis of cotton gin trash (supported under contract with Cotton, Incorporated) offers a method for converting this waste to useful energy forms and at the same time disposal of the waste in an acceptable manner. Cotton gin waste is a bulky, loose, non-free flowing material. The successful processing of cotton gin trash in the waste converter depended upon a method of handling the gin waste so that it would pass through the waste converter. Initially, pelletized gin waste was utilized in the waste converter pyrolysis runs. The use of pelletized or compacted gin waste was abandoned because studies showed that the high cost of compaction would make the process uneconomical. Following this, a mechanical agitator, which permitted successful processing of loose cotton gin trash, was developed. A maximum processing rate of 468 dry pounds per hour which scaled to 7500 dry pounds/hour or 90 ton/day in a Cordele size waste converter was achieved.

As a part of the program an investigation was made to determine if the char produced from the pyrolysis of the cotton gin trash would be suitable for producing charcoal briquets. A commercial briquet manufacturer successfully processed 1000 pounds of cotton gin trash char into clean-burning briquets. A limited economic analysis of utilization of the EES pyrolytic waste converter for cotton gin trash was made, and the general overall economics of the system appear favorable.

The results of the investigation have shown that cotton gin waste can be processed with mechanical agitation in a pyrolytic converter; that the process rate in the pilot waste converter is adequate for large-scale processing; that the char is of suitable quality for use in the manufacture of commercial briquets; that the economics appear to be favorable; and that pyrolysis offers a method of converting the waste material into useful products in a non-polluting manner and at the same time, the disposal problem would be eliminated.

Wood Wastes

The object of this test program (supported under EPA Contract 68-02-1485) was primarily to determine the optimum operating conditions for maximum production of char and oil and for minimum production of gas using the EES pyrolysis system. With this in mind, the test instrumentation, the test operation and the selection of the test parameters were made, within practical limits, to shed as much light as possible on those conditions most favorable for application to the portable converter concept. The primary test parameters were: the air/feed ratio, the depth of the porous bed, the geometry and number of process air cubes, and the use of mechanical agitation to increase throughput and prevent "bridging." The program included 14 tests. The testing was conducted in the six ton/day EES pilot plant.

Study of the data from the tests shows that with sawdust the effects of agitation on product yields are unimportant.[9] Analysis of the results also indicated that tube geometry probably has little effect on product yields, although a comprehensive study of this parameter was not made. Likewise, it appears that the effect of bed depth on char yields is unimportant; however, the deeper beds appear to result in greater oil yields and therefore would be more desirable in maximizing the production of the char-oil mixture.

The most dramatic effect and the one which overrides the importance of any of the other test parameters is the air/feed ratio. This is illustrated in Figure 5 which presents a graph of the percent of the available energy of the sawdust in the char-oil mixture as a

---

3. It is important to note that mechanical agitation has been shown in previous studies with cotton gin wastes to have a pronounced effect on the outcome of the testing. Thus the fact that in these tests with this feed and his agitator, the effects were unimportant should not be generalized to less free-flowing solids.

function of air/feed. This figure presents a plot of all the test data
at various bed depths, tube geometrics, feed rates, and with or
without agitation. Thus clearly the air/feed ratio is the dominant
parameter and small values are the more desirable. A complete
description of this work is presented in (4).

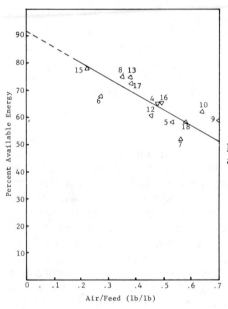

Figure 5. Percent theoretically available energy in Char-Oil mixture.

Peanut Hulls

General Description. The program (currently in progress under
EPA Grant R803-403-01-0) is directed toward determining the effects of feed characteristics, mechanical agitation and process air
system integration on the performance of the EES pyrolysis system.
As such, the study is almost completely experimental. However,
in the development of the mechanical agitation system and the
integrated process air system design and fabrication work is involved. A minimum of 14 tests is being conducted in the new 25
ton/day pyrolysis unit primarily using peanut hulls as the waste
feed. A total of 150 tons of hulls will be consumed in the program
which will last seven months. As in the study, with wood wastes,
the ultimate objective of this work is to provide the technical data
necessary for the development of the portable pyrolysis concept.

Feed Characteristic Tests. Previous experience has taught that
significant differences in the physical properties of the waste feed
can result in substantial changes in the material handling and
operating characteristics of the EES waste converter. The principal

influence is that of density on throughput since bulky materials, un-aided, do not flow well through the vertical bed which operates on gravity alone. Thus an investigation of the use of several different feed materials to determine quantitatively the EES pyrolysis system performance is an important element in this program. As such, investigations of the use of sawdust and peanut hulls are included in the study.

Mechanical Agitation Tests. The EES, in prior studies, has done considerable investigation of the use of agitation to enhance feed throughput. Therefore, there is already a substantial appreciation for the use of mechanical agitation as applied to various feeds. The approach, then, to the use of agitation is to apply this knowledge judiciously in this study where only the most desirable combinations of test parameters are investigated. Hence, the test matrix is being filled in with only the more important conditions; the conditions judged as providing only marginal improvement are being avoided.

Process Air-Agitation System Integration Tests. In this study an investigation of the possible benefits of integrating the mechanical agitation system with the process air system is being made. An integrated system has been designed, fabricated and tested.

There appears to be considerable advantage in the use of an integrated mechanical agitation—process air system, especially if the mechanical agitation system is a requirement, in any case to process bulky wastes. Such a system, conceptually, would avoid the hindrance to flow of fixed air tubes, to possibly allow processing of somewhat wetter feed than the present EES waste converter allows.

The approach in the study is to utilize as much as possible the hardware developed in the simple mechanical agitation investigations. Hence, relatively modest additions are being made.

To date only the first few tests have been conducted and results from this study are too preliminary to report. Therefore this section should be considered as a progress report and a more complete description of the results of this study will be presented at a later time.

MOBILE PYROLYSIS SYSTEM PRELIMINARY DESIGN AND ECONOMIC ANALYSIS

A preliminary design of 200 ton/day (assuming 50% moisture) mobile pyrolysis system is presented in this section. In addition to the design, the results of a simplified economic analysis are also shown. (Supported under EPA Contract 68-02-1485.)

As conceived, several mobile units would operate in zones with a large thermal conversion plant at their center. The several units would convert various agricultural and forestry wastes into the char-oil fuel which would be transported to the thermal conversion plant. As visualized, one char truck could service three or four pyrolysis units and likewise only one tractor for perhaps seven or eight complete systems would be required to move them from site to site. The char trucks would have a closed trailer which would also serve as a char-oil storage container and be exchanged twice daily.

The portable system was designed with certain basic ground rules which are listed below:

1) It must be completely self-sustaining.

2) It must produce no land, water or air pollution.

3) It must be transported readily and with no special highway permits, *i.e.,* its length should not be greater than 55 feet, its width is eight feet, its height is not more than 13 feet 6 inches and its weight is less than 73,000 pounds.

In addition to these basic limitations, it was assumed that:

1) The system would comprise two trailers.

2) The start-up would be accomplished using propane.

3) The oil condenser would operate at a temperature greater than the off-gas dew point to avoid moisture condensation.

4) A 50% derated gasoline engine operating on the low Btu pyrolysis gas would provide the power required to operate the system.

5) The design rated capacity of the system would be based on the Cordele, Georgia commercial unit. This unit has been operated at twice its rated capacity without agitation. Thus, the design capacity is highly conservative.

6) The feed, and the resulting char, because of this typically low thermal conductivity, would serve as an insulator on the sides of the unit. This would prevent not only the necessity for and associated weight of a ceramic insulator but also avoid the problems of transporting such a brittle, fragile system.

Presented in Figures 6 and 7 are preliminary design drawings of the two trailers which make-up the system. The drawings show the trailers in their stowed configurations to illustrate that space is available for all the required components. Deployment of the system would mainly involve moving the conveyors from their stowed to their operational locations, connecting up the flexible pipe between the off-gas system and the drier, and unloading the feed hopper by use of a collapsible ramp stowed underneath.

The economic analysis of the portable waste converter was not sophisticated and was limited to a study of only one system within an array of systems servicing wood yards. Assumptions were made,

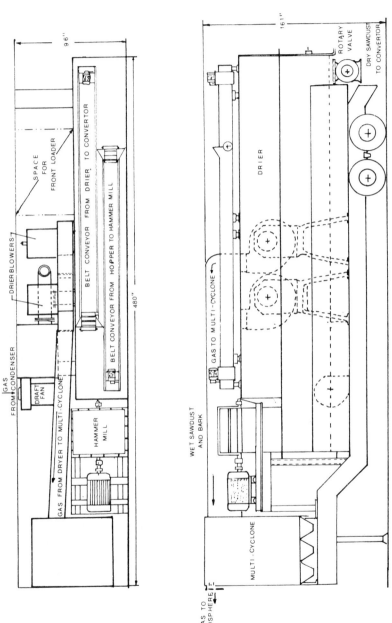

Figure 6. Plan and elevation views of drier trailer.

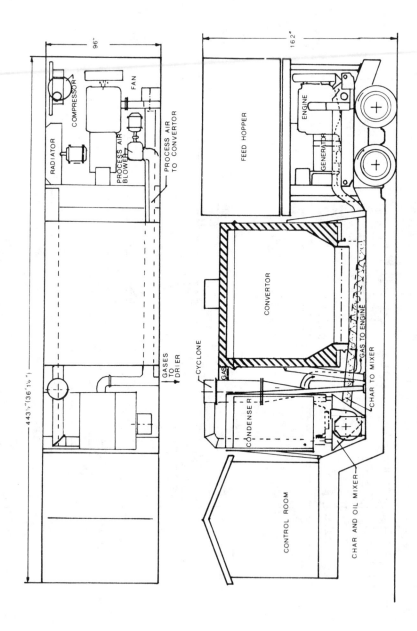

Figure 7. Plan and elevation views of converter trailer.

however, regarding the costs of sharing equipment among systems and the supervision costs of a group of systems. It has been also assumed that by proper scheduling the systems can be kept continuously operating except for down time between waste sites. Because so many coal prices are possible, depending upon whether "contract" or "spot market" values are used it was not practical to select just one. Therefore a range of different values were employed in the analysis.

Figure 8 represents a summary plot of the economic analysis for a 100 dry ton/day unit, with net income plotted as a function of coal prices. The figure shows the importance of three shift operation to the system economics and illustrates the influence on profit of a disposal credit. From the figure it appears that even for 16 hour/day

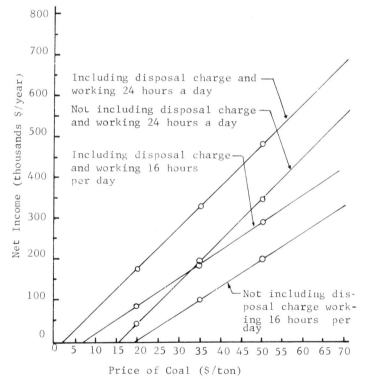

Figure 8. Net income of a 100 ton/day (dry) waste converter with varying prices.
Principal Assumptions:
1. Interest —9.5%
2. Depreciation—10 years
3. Raw wastes are 50% moisture
4. Disposal charge is $3.00/ton of wet sawdust

operation and without a disposal credit, the break-even point would be $20/ton. And clearly as the coal price increases, the net income goes up very rapidly.

From this work, it appears that on strictly economic grounds, a strong case can be made for the profitability of the portable waste conversion unit. When consideration is given to the more intrinsic advantages to society, our national economics and the environment, the case for the development of such a system is compelling.

CONCLUSIONS

The results of this paper should conclusively demonstrate the importance of agricultural and forestry wastes as a potential national source of significant quantities of clean fuels. The pyrolytic conversion concept using the EES process described provides a superior technical and economic means for utilizing these wastes as fuels. The techniques required for processing these wastes through the EES converter are available and methods for tailoring the operating parameters to the type of waste feed for maximum oil and char production are understood. The portable unit design appears technically feasible and its economics appear to be highly desirable.

REFERENCES

1. Anderson, L. L. Bureau of Mines Information Circular 8549 (1972).
2. Taylor, H. T. Private communication.
3. LaHaye, P. G. Hague International, S. Portland, Maine, Private communication.
4. Tatom, J. W. *et al.* "Utilization of Agricultural, Forestry and Animal Waste for the Production of Clean Fuels." Final Report under EPA Contract 68-02-1485. Engineering Experiment Station, Georgia Tech, Atlanta, Georgia, July 1975.

21.

Energy Recovery and Feed Production from Poultry Wastes

Awatif E. Hassan, H. Moustafa Hassan, and
Norman Smith*

The results of this study indicate that optimum solids concentration for anaerobic digestion of chicken manure was 7% at incubation temperature of 25.5°C. Operating the digester at this temperature implies that, for future application, the unit can be included in the broiler house. The presence or absence of exogenous carbon source did not affect the rate or the extent of methane production. Growing algae, *Scenedesmus*, on the digested effluent was a successful process. A yield of 20 gm of dried algae/liter of effluent was obtained. The protein content of the dried algae was 42-45%. This study also provided the preliminary design for a full-scale poultry manure digestion unit.

The living conditions in Maine and the New England States require a high consumption of energy per capita in the form of heating, transportation and handling of industrial products for long distance to rural areas. Due to present energy shortages, investigations concerning alternative energy sources are encouraged. However, the feasibility of some of these sources cannot be foreseen as long as fossil fuels are available even at much higher prices than today. Also the changeover to the new sources of energy is costly and requires considerable investment.

* Assistant Professor, Agricultural Engineering Dept.; Assistant Microbiologist, Dept. of Microbiology; and Professor and Chairman, Agricultural Engineering Dept., University of Maine, Orono, Maine 04473.

Maine's poultry industry produces annually over half a million tons of manure which represents a formidable waste disposal and environmental problem. Energy utilization from this large amount of manure is attractive to many farmers where the methane produced can be used directly on the farm for heating or perhaps operating electric generators. However, anaerobic digestion of waste requires great dilution of manure, resulting in much larger quantities of effluent to be disposed of. Hence, if anaerobic digestion of organic wastes for methane production is to be practiced, a means of utilizing the digested effluent must also be considered for the overall success of the project.

This paper deals with the anaerobic decomposition of poultry manure to produce methane, and the possible means for the utilization of the digested effluent. A model for efficient utilization and disposal of chicken manure is proposed.

LITERATURE REVIEW

The concept of generating methane from wastes is very popular with environmental groups as it appears to change a noxious by-product into a useful commodity. Many glowing reports of farmer- and home-operated digesters have been received but have been found to be much exaggerated.

Anaerobic digestion has been used for more than 100 years to stabilize sludge from municipal waste treatment systems (1). Many municipal sewage treatment plants in Europe and the United States are today operating lighting and heating systems with methane gas recovery from the digesters. Anaerobic digestion of animal wastes has received less attention. Recently, cow and chicken manure has been used for methane generation in India (2, 3). The application of this process to animal wastes in the United States is limited because of the lack of scientific and economic studies.

Energy production from organic wastes involves a biological reaction which requires a controlled environment. Changing any of the environmental factors will affect the quantity and quality of the methane gas produced (4). The most important factors in methane production from anaerobic digestion of wastes are: temperature, solids concentration, hydrogen ion concentration, volatile acids content, absence of oxygen, and absence of toxic materials.

The methane-forming bacteria currently implicated in anaerobic waste digestion are grouped within the mesophilic (18° to 45°C) and the thermophilic (above 45°C) groups of bacteria. Municipal sewage plants have been restricted to the mesophilic range (5). There is little information available on anaerobic digestion at lower temperatures, or on the effect of temperature on the kinetics of methane production from animal wastes.

Barker's work (6, 7) has increased significantly the understanding of the microbiology and biochemistry of methane production through fermentation.

Chicken feces contain gross energy from 3.2-4.5 Cal/gm dry manure and total nitrogen ranges from 30 to 70 mg/gm dry manure (8). Fresh manure averages 40% moisture, 1.3% nitrogen, 1.2% phosphorus, and 1.1% potassium (9). Other elements contained in manure are calcium, magnesium, copper, manganese, zinc, chlorine, sulfur, and boron (9). Most of these elements, N_2, K, and P, are not consumed during the anaerobic process for methane production. The digester effluent contains most of these elements which are essential for the growth of plants and algae, *i.e.,* the effluent could replace commercial fertilizers. Disposal of digested effluent is a major problem in anaerobic digestion systems. Several methods for effluent disposal can be considered such as aerated lagoons, disposal on forest land, use as a fertilizer on agricultural soils, or growing microorganisms such as algae (10), bacteria, and yeast for use in animal feed.

Single cell protein production for human food and animal food has been investigated by scientists in the fields of food industry, chemical engineering, animal nutrition, and economics (11-13). The advantage of using algae as a source of protein over other microorganisms (bacteria, yeast, and fungi) is that algae requires no carbohydrate in the media but is able to fix CO_2 from the atmosphere. However, the disadvantages are the enormous surface area, aeration, and light requirement (11).

Most of the studies related to algal production have been limited to raw sewage or animal wastes (14, 15). Algal production on effluent is the most successful process for effluent disposal and feed production. Oxygen, which is a product of algae photosynthesis on effluent, is essential for decomposition of soluble organic material by aerobic bacteria which in turn provide CO_2 required for algae growth. Po (10) has described a full system for energy recovery and feed production, however, no quantitative data were reported.

The study reported here was conducted with the following objectives in mind:

1. To determine the optimum conditions for methane production at temperature of 25.5 ± 3°C (78 ± 5°F) which is the average temperature for broiler houses;

2. To study the effects of depth, aeration, and effluent concentration on algal yield; and

3. To provide the preliminary design for a full-scale poultry manure digestion unit.

EXPERIMENTAL PROCEDURES

Methane Production

Fresh manure from caged layers from the University of Maine farm was used in this study. The solids content of the fresh manure varied from 25% to 35%. Sawdust was added at different rates to simulate the fluctuation that may be encountered when using broiler litter.

The fresh manure was mixed with water to the desired solids content and incubated in the laboratory digester (Figure 1). Sixteen

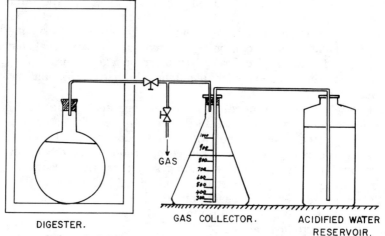

DIGESTER.   GAS COLLECTOR.   ACIDIFIED WATER
                                                          RESERVOIR.
Figure 1. Schematic diagram of a one-liter digestion unit.

one-liter boiling flask digesters were assembled to study the effects of solids content and sawdust percentage on methane production. Inoculum from previously digested effluent was added to all digesters at the rate of 50% by volume. The digesters were maintained at room temperature of 25.5 ±3°C (78 ± 5°F), and were mixed once daily by manual shaking.

The quantity (volume) and quality (composition) of the gases produced were recorded on a routine basis during the digestion process. The gas volumes were collected at atmospheric pressure and room temperature of about 25°C. The gases were analyzed for CO_2, O_2, N_2, CH_4 and CO using a Fisher-Hamilton Model 29 gas partitioner. A recorder (1mV) was used with the chromatograph to record the percentages of the respective gases in the collected samples.

Solids contents of the sludge before and after digestion were determined by gravimetric method.

Algal Growth

A green algae culture, *Scenedesmus,* was supplied by the Department of Botany and Plant Pathology. The algae strain was adapted to grow on the effluent. For the purpose of this study the effluent was clarified to allow maximum light penetration for growth under artificial light. The effluent was vacuum filtered through Whatman No. 40 filter paper.

Beakers of 2000- and 4000-ml capacity were used as laboratory ponds for the studies of the effect of effluent percentage, aeration, and depth of pond on algal growth. The algae required a dark period of at least 8 hours for cell division. A light source (intensity of 500-600 ft-candles) set on a time clock for 16 hours of light and 8 hours of darkness was provided. Aeration and mixing in the ponds are also required to supply CO_2 and to bring algae cells to the surface especially after the dark period. In the aerated treatments air was supplied during the light period by means of an air pump. The daily algal growth was measured by means of a spectrophotometer, at 600 nm. All ponds were mixed well before sampling to assure uniform results.

RESULTS & DISCUSSION

Energy Recovery

Kinetics of Methane Production: Figure 2 represents the accumulative methane production and the corresponding methane percentage in the evolved gas from one of the laboratory digesters, where solids content was 4.55%. All digesters exhibited the same kinetic pattern; however, the difference was in the rate of methane production (slopes of the curves) and the total amount of methane produced. The methane content of the evolved gas dropped slightly during the early stages (first week) of the digestion process, then continued to increase until it reached a maximum of 87% after 40 days, then slightly dropped again after 60 days of digestion (Figure 2). The initial drop in methane content is probably due to environmental shock of the methane bacteria present in the inoculum, as a result of fluctuation in anaerobiosis and temperature, caused by the introduction of fresh manure. The later drop in methane percentage was also accompanied by a drop in the rate of methane production indicating changes that might have occurred in the digester contents due to increase in pH, depletion of nutrients and/or ammonia accumulation.

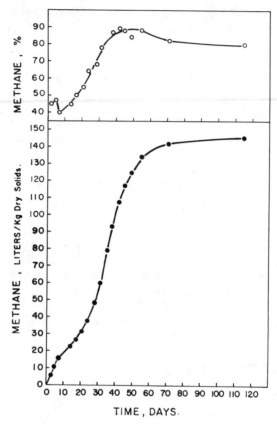

Figure 2. Kinetics of methane production from one of the laboratory digesters. Initial solids content was 4.55%.

In an attempt to study the effect of initial solids content on methane production, different laboratory digesters were used with initial solids ranging from 1.9-19.7%. It was found that the rate and extent of methane production were greatly dependent on initial solids content. However, with most concentrations the accumulated methane approached its maximum level after 70 days. The effect of initial solids contents on the methane produced after 70 days of digestion is shown in Figure 3. The data clearly show a plateau of 4-7% solids contents. Above or below that range a fast decline in rate and extent of methane production took place. A drop in methane production from 120 to 28 liter/kg solids was observed as a result of increasing solids contents from 8 to 19.7%, respectively. In other words the high solids concentration inhibits the growth and metabolic activity of the methane bacteria which is similar to that reported by Edwards (16) for bacterial inhibition by high substrate concentrations. Also this inhibition may have been due to the presence and accumulation of toxic compounds such as ammonia (17, 18). The low substrate concentration, below 4% solids content, may have been responsible for the reduction in methane production.

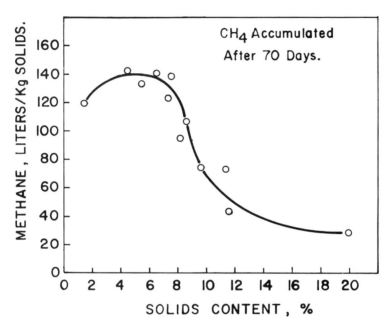

Figure 3. Relationship between initial solids content and methane production after 70 days of digestion.

Effect of the Percentage of Exogenous Carbon: The dried chicken manure used in this study was analyzed and found to contain from 7.0 to 7.12% total nitrogen. This high nitrogen content may result in an unbalance of C:N ratio required for methanogenic fermentation. Since sawdust or wood shavings are normally used in the broiler industry for litter, sawdust at room moisture equilibrium (~10% moisture) was added to the fresh manure at the rates of 2, 4, and 8% by weight of the fresh manure. From Figure 4 it is evident that the presence of sawdust increased the total production of methane after 35 days of digestion; however, the percentage appeared to have very little effect in the range tested. It is worth mentioning that broiler litter contains slightly more than the 8% sawdust used (19). Under a semi-continuous feeding operation, recharging will normally take place before the rate of methane decline, thus before the effect of sawdust becomes distinct (Figure 4).

Hence, it can be concluded that the presence or absence of sawdust will have no effect on the rate or extent of methane production for either broiler or cage layer operation.

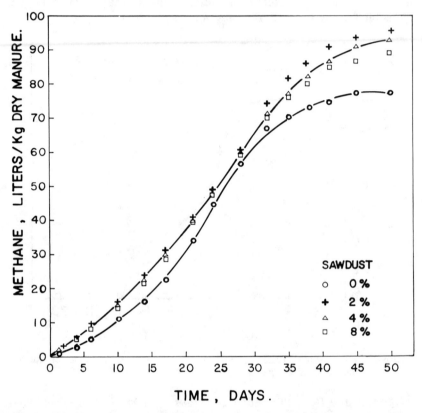

Figure 4. Effect of sawdust percentage on methane production. Rates of sawdust applications were 0–8% by weight of the fresh manure.

Effect of Solids Content on Recharging Intervals: The recharging schedule for optimum rates of methane production is of great interest for assuring a successful continuous operation. Figure 5 represents the effect of solids content on the recharging intervals and hence on digester capacity and also the frequency at which the digester may be recharged. The recharging interval is deduced from figures similar to Figure 2 for digesters operated at different solids contents, and defined as the time elapsed in days when the rate of methane production begins to decline, which was found to coincide with the time elapsed for 70-80% of total methane production. For example, from Figure 2 the recharging interval should be every 40 days. The recharging schedule of Figure 5 is based on an operation where 50% inoculum is used. Thus, the minimum recharging interval for a given digester operating at known solids content can be deduced directly from Figure 5. The recharging intervals increased with an increase in solids content beyond 8% and also increased slightly for solids content below 6%. Thus, optimum solids content for a minimum recharging interval is 7% (Figure 5) which also results in maximum methane production (Figure 3). The use of optimum concentration will have direct impact on the design and cost of operation.

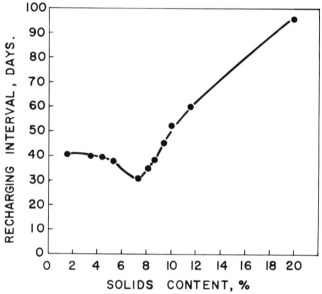

Figure 5. Effect of solids content on recharging schedule for optimum rate of methane production.

Feed Production

Effluent Composition: Chemical analyses of the digested effluent and reclaimed water were performed and results are listed in Table 1. The effluent was obtained from digesters which varied in.their initial solids concentrations (7.5 to 12.0%). Recalling that the total nitrogen content of the dry manure averaged 7.06%, therefore, the total nitrogen in the tested effluent should be .53 to 0.85%. Comparing these values with those listed for the digested effluent (whole) of Table 1 indicates that practically all the nitrogen content of the manure remains in the effluent and none is lost during the digestion process. The difference between the two values may have been due

Table 1. Chemical composition of digested effluent and reclaimed water.

	Total N_2 %	NH_4 %	K %	P %
Digested Effluent (whole)	0.60 to 0.87	NR*	NR	NR
Digested Effluent (Cheesecloth filtered)	0.48 to 0.66	0.42 to 0.53	0.25 to 0.31	0.056 to 0.100
Digested Effluent (filter paper Whatman No. 40)	0.4799	NR	NR	NR
Reclaimed Water after Algal Removal**	0.0028 to 0.0039	NR	0.0067 to 0.0083	0.00063 to 0.00078

* Not Recorded.

** The reclaimed water was obtained from an algal pond (133 liters) where the effluent was added at a daily rate of 0.5% for 8 days.

to variations in the nitrogen content of the fresh manure. It is evident from Table 1 that from 75 to 80% of the manure total nitrogen is retained in the cheesecloth filtrate of which 80 to 88% is in the form of ammonia.

Effect of Effluent Concentration: The nitrogen content of the effluent is very high. It is of interest to study the effect of effluent concentration on algal growth and to determine the optimum concentration which will provide maximum algal yield. Figure 6 shows the effects of effluent percentage on algal growth for both aerated and unaerated treatments. The algae cells were plasmolyzed at the higher concentrations (4 and 6%). However, better algal yield was obtained when the 4% effluent was added daily at the rate of 0.5% of the total pond volume. The 1% treatment provided the highest yield and also the aerated treatments gave better results than those which were unaerated. The aeration may have resulted in better gas exchange through the air-water interface and also may have brought the algae cells, which tend to settle to the bottom of the pond, to the surface.

It is recommended that the algae should be harvested after 3-4 days (10) as the cells will be easier to digest, even though the yield is only 20 gm dry wt of algae/liter of effluent.

Since algal growth took place in an open system (beakers) a wild culture was obtained. However, microscopic examination of the culture illustrated in *Scenedesmus* represented most of the cell population. *Chlorella* was also identified in the culture. A phase micrograph of the algae isolated from one of the ponds is shown in

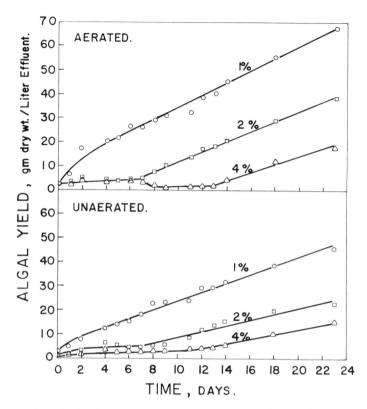

Figure 6. Effect of effluent concentration on algal production for aerated and unaerated treatments. Pond depth was 15 cm.

Figure 7. The protein content of dried algae flakes based on total nitrogen content was determined and varied from 42 to 45%.

Dabbah (11) reported a typical composition of freeze-dried chlorella to be 56% crude protein, 7.5% crude fat, 18% carbohydrates, 8% ash, 7% moisture, and 3.1% crude fiber. The difference in protein content between the reported data (11) and the results of this study (42-45%) may have been due to cultural differences and/or the presence of nonproteinaceous materials.

Effect of Pond Depth: Pond depth will normally affect light penetration and thus algal growth. Pond depths of 10, 15, and 20 cm were studied. Figure 8 represents the effects of pond depth on algae production for both aerated and unaerated treatments. The effluent concentration in these treatments was 2%. A depth of 20 cm (8″) provided the highest yield which is in agreement with the previous work of Oswald (15).

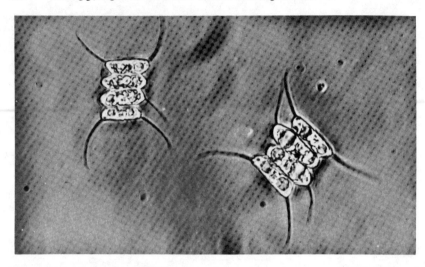

Figure 7. A phase micrograph of *Scenedesmus* algae grown on digested effluent (magnification = 1500 x).

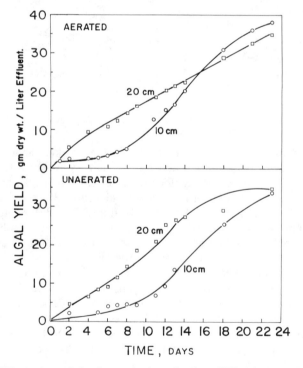

Figure 8. Effect of pond depth on algal production. Effluent concentration was 2%.

Algal Potential as Source of Protein Feed: The potential production of animal feed as indicated by Figures 6 and 8 is of great significance. The 20 gm of algae per liter of effluent represents a yield of 1 kg of protein per 10 kg of dry waste (manure). This means that the 500,000 tons (metric) of poultry manure produced in Maine each year could produce algae protein equivalent to 150,000 tons (metric) of soybeans, making a significant contribution to Maine farm income and easing transportation problems considerably.

A digestibility of 86% has been reported for algae protein (20). The use of algae, grown on sewage lagoons, as livestock feed additives has been studied on some feeding trials of swine by Hints and Heitman (21). They have reported that when mixed cultures of *Scenedesmus quadricaudy* and *Chlorella sp.* were fed as a replacement for meat and bone meal additives in barley rations swine grew and fattened in the normal manner.

The reclaimed water after algal harvesting contained traces of minerals (Table 1) and hence it may be recycled to the digester, algal pond, and the broiler house. Figure 9 illustrates the operation cycle for broiler production, energy recovery by means of anaerobic decomposition of organic wastes, possible combinations of effluent utilization, and high protein feed production.

Design Criteria for a Digestion Unit

If a net yield of 1666 kcal/kg (3000 Btu/lb) of dry manure can be obtained from a full-scale digester, a unit capable of digesting approximately 1225 kg (2700 lb) of broiler manure per day would be capable of supplying all the energy needs of an 85,000-bird broiler house under Maine conditions. This is based on a peak requirement of 3.785 liters (1 gallon) of oil per 20 birds per 53-day flock.

Net yields of 1666 kcal/kg (3000 Btu/lb) of dry manure should be possible if the digester is incorporated into the broiler house. A digester installed in this way would help to heat the house and would require much less insulation and supplementary heat than a separate unit. Its thermal capacity would also act as a "flywheel" and help to carry the house over extremely cold nights when heating system design conditions were reached, probably allowing a smaller heating system to be used in the house with consequent savings in first cost. Figure 10 illustrates a possible layout suitable for an 85,000-bird house. A digester for this purpose would need a capacity of approximately 750 m^3 (26,500 ft^3) to allow for a 7% solids concentration and some freeboard for foaming and gas collection. A tank 9.75m x 9.75m x 7.9m (32 ft x 32 ft x 26 ft) high, let into the ground 2.4m – 3m (8 ft – 10 ft) would use two stories within the house and provide the necessary capacity. The predigester could

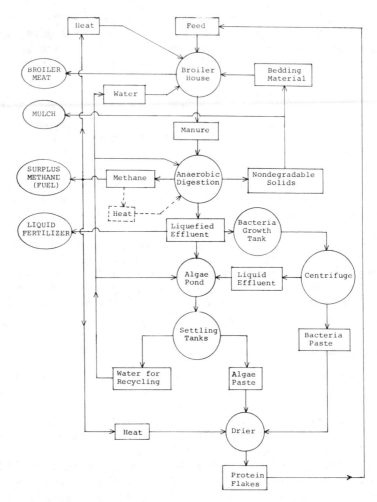

Figure 9. Proposed operation cycle for anaerobic decomposition of poultry manure and means of effluent utilization.

have approximately 25% of this capacity and would be used to recharge the digester every 21 days, along with a similar quantity of water. An open tank 7.3m x 7.3m x 3.7m (24ft x 24ft x 12ft) high would be suitable for this. Approximately 168 m³ (220 yards³) of reinforced concrete would be needed for these tanks. At an estimated cost of $196/m³ ($150/yd³) in place, this would mean a cost of $33,000.

Equipment to run the system would consist of a pump to charge the digester and provide agitation, a compressor to transfer the gas into an intermediate storage tank, and the services of a farm tractor with loader to charge the predigester every 21 days. If a 24-hour period were allowed for 50% recharging of the digester, a 265-liter/min (70 gpm) pump would be satisfactory. Such a unit with controls and plumbing would probably cost around $2000. Emptying of 50% of the digester contents is assumed to be possible by gravity, as the digester rises 4.9 meters (16 feet) into the broiler house.

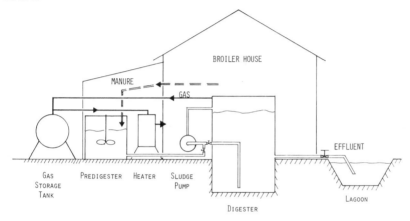

Figure 10. Schematic layout of a full-scale anaerobic digestion system.

A 113-liters/min (4-cfm) compressor and a storage tank sufficiently large to hold almost one day's gas production would be needed. At 3.5 kg/cm^2 (50 psi) a storage tank 3 meters (10 ft) diameter and 7.6 meters (25 ft) long would suffice. A cost of $5500 might be allowed for these units and necessary piping.

If the effluent from the digester were to be used for fertilizer-irrigation, a storage lagoon of 3058 m^3 (4000 yd^3) capacity might be necessary in northern climates. This would allow a 6-month storage period. Cost of excavating and partial lining for the digester excavation are estimated around $4500. Total cost of the system would therefore be approximately $45,000.

Operation might require 1 man hour of labor per day on average, plus 8 tractor hours every 21 days. This would approximate $6 per day in total or $2200 per year.

Assuming a 15-year life for the total unit, *i.e.*, a similar life to that of the broiler house; a 10% interest rate; taxes and insurance equal to 5% of the purchase price per year; and no salvage value at the end of the 15-year period, annual operating costs would be just short of $10,000. The methane produced would replace approximately 80,000 liters (21,000 gallons) of oil. If the fertilizer-irrigation value of the effluent could be regarded as just paying for the application costs, the methane cost would be equivalent to oil cost at 12¢/liter (46¢/gallon). This figure also assumes zero charge for the manure feed stock which would vary up or down depending on whether a negative or positive charge was placed against the manure. In many instances present disposal methods incur considerable transportation charges, indicating a negative cost for the manure if used in the anaerobic digester. The 46¢/gal (12¢/liter) figure is only slightly higher than present oil prices, indicating that a closer examination of the commercial feasibility is well worthwhile and that a detailed design study should be made.

It is worth noting that only about 12% of the broiler waste produced in the house is needed to provide the entire heating requirement if the 1666 kcal/kg (3000 Btu/lb) dry matter can in fact be obtained. This appears to be very likely, as yields of 1438 kcal/kg (2590 Btu/lb) have already been obtained in a rather elementary full-scale digester using the 21-day partial recharging technique previously described.

* * *

This investigation was supported by funds from the New England Regional Commission and Maine Agricultural Experiment Station.

REFERENCES

1. McCarty, P. L. 1973. Methane fermentation—Future promise or relic of the past? Proc. Bioconversion Energy Research Conference. Univ. of Massachusetts.
2. Patel, J. B. and R. B. Patel. 1971. Biological treatment of poultry manure reduces pollution. Compost Science, 12(5):18-21.
3. Singh, R. B. 1972. The Bio-gas plant—Generating methane from organic wastes. Compost Science, 13(1):20-25.
4. McCarty, P. L. 1964. Anaerobic waste treatment fundamentals II. Environmental requirement and control. Public Works, 95(10):123.
5. Smith, R. J. 1973. The anaerobic digestion of livestock wastes and the prospects for methane production. Agric. Eng. Dept. Iowa State University.
6. Barker, H. A. 1936. On the biochemistry of methane fermentation. Archiv für Mikrobiol. 7:404-419.
7. Barker, H. A. 1956. Biological fermentation of methane. In Bacterial Fermentations, CIBA Lectures in Microbial Biochemistry, Chapt. 1 p. 1-27. John Wiley and Sons Inc. New York.
8. Pryor, W. J. and J. K. Connor. 1964. A note on the utilization by chickens of energy from feces. Poultry Sci., 43:833-834.
9. Loehr, R. C. 1974. Agricultural waste management. Academic Press, New York.
10. Po, Chung. 1973. Production of methane gas from manure. Proceeding International Biomass Energy Conference. Biomass Energy Institute, P.O. Box 129, Postal Station C, Winnipeg, Manitoba R3M 3S7.
11. Dabbah, Roger. 1970. Protein from microorganisms, Food Technology Vol. 24: 659-664.

12. Humphrey, A. E. 1969. Engineering of single cell protein: state of the art. Chemical Engineering Progress Symposium Series No. 93 Vol. 65 p. 60-65.
13. Ward, C. H. 1973. Algae as food. Symposium on Teaching with Algae, Annual Meeting at the American Institute of Biological Science, Univ. of Massachusetts, Amherst, Mass.
14. Golueke, C. G. 1973. Bioconversion of energy studies at the University of California (Berkeley). Proceedings Bioconversion Energy Research Conference, University of Massachusetts.
15. Oswald, W. J. and C. G. Golueke. Harvesting and processing of waste-grown microalgae. Algae, Man and the Environment. p. 371-390. Edited by D. F. Jackson, Syracuse University Press.
16. Edwards, V. H. 1970. The influence of high substrate concentrations on microbial kinetics. Biotechnol. Bioeng. 12:679-712.
17. Albertson, O. E. 1961. Ammonia nitrogen and the anaerobic environment. J. Water Poll. Control Fed. 33:978-995.
18. Anthonisen, A. C. and E. A. Cassell. 1974. Methane recovery from poultry waste. North Atlantic Region of ASAE Paper No. NA74-108. 28p.
19. Gerry, R. W. 1968. Manure production by broilers, Poultry Science, Vol. XLVII No. 1 p. 339-340.
20. Lubitz, J. A. 1963. The protein quality digestibility and composition of algae *Chlorella* 71105. J. Food Science, 28:229-232.
21. Hintz, H. F. and H. Heitman Jr. 1965. Nutritive value of algae for swine, California Agriculture, 19(2):4-5.

22.

Anaerobic Digestion in Swine Wastes

J. R. Fischer, D. M. Sievers, and C. D. Fulhage*

Much publicity has been given to the work of Ram Bux Singh (1) in India on his small individual farmstead-type digester, and L. John Fry (2) in South Africa who operated two anaerobic digesters on his swine farm. In the U.S., there have been many laboratory-scale studies concerning the anaerobic digestion of swine manure but few larger studies. Even solar stills have been built anticipating the use of solar heat for the digestion operation.

In this study a pilot plant anaerobic digester, utilizing swine waste as a feed source, was monitored 1) to evaluate the effect of loading rate and maximum loading rate on gas production, 2) to determine the most stable chemical environment for gas production, and 3) to determine the most easily managed, best suited system for measuring digester stability.

EQUIPMENT

The digester was a 23 in. (58.4 cm) diameter x 55 in. (139.7 cm) long cylindrical container supported in a horizontal position (Figure 1). A heat tape was wrapped around the outside of the container to maintain the digester temperature at 95°F (35°C), and 1 in. (2.54 cm) of polyurethane insulation was then placed around the digester (Figure 1). The feed input into the digester was located at one side of the digester and three outputs were located on the

* Agricultural Engineer, U. S. Department of Agriculture, Agricultural Research Service, North Central Region; Assistant Professor, Agricultural Engineering Department; Associate Professor, Agricultural Engineering; University of Missouri, Columbia, Missouri 65201.

opposite end. One output for removing sludge from the digester was located near the bottom; the second output for removing circulation effluent was located midway along the side. Gas from the digester escaped via a third output (located above the circulation effluent output) to an inverted barrel and automatically discharged through a wet-test meter (Figure 1).

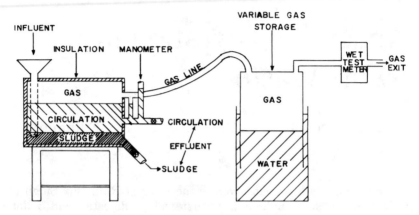

Figure 1. Schematic of anaerobic digestion system.

PROCEDURE

The loading-unloading procedure for the digester consisted of:

1. Recording the gas production for the previous 24 hr.

2. Recording the ambient air temperature in the room where the digester was located.

3. Removing circulation effluent from the digester. Enough circulation effluent was removed from the digester so that when 25 lb (11.34 kg) of influent was added, the digester maintained the same volume.

4. Removing sludge, instead of circulation, periodically when the solid content in the circulation increased.

5. Adding the manure-water mixture (influent) into the digester. A total of 25 lb (11.34 kg) of manure and water was added daily and retained for 15 days in the digester. The swine manure added to the digester was obtained from concrete floors of a swine barn where hogs were fed a typical 14% corn/soybean finishing ration. The manure was collected weekly from the swine barn, weighed, placed in plastic bags and frozen until needed.

Influent, circulation and sludge were sampled weekly, and more often when some parameter, like gas production, indicated a digester upset. The samples were analyzed for total solids (TS), volatile solids (VS), chemical oxygen demand (COD), ammonia nitrogen

(NH_3), total Kjeldahl nitrogen (TKN), pH, alkalinity and, for several months, methane (CH_4) and carbon dioxide (CO_2) content. Due to technical difficulties, gas samples were not analyzed for the complete experiment. Periodically, we analyzed samples for phosphorus (P) and potassium (K).

DIGESTER OPERATION

The digester began operation on January 10, 1974, when it was loaded with 16 gal (60.6 l) of digester sewage sludge obtained from Columbia, Missouri, activated sludge plant and 29 gal (110 l) of tap water. Influent feed was gradually increased until a loading rate of 0.04 lb VS/ft³ (0.64 kg/m³) was obtained within about a month. The digester was operated at this rate until February 22 when gas production unexplainably decreased 50%. On February 24, while gas production continued to decrease, influent loading was stopped. The gas production continued to decrease with time, and on March 1 no gas was produced from the digester. However, all parameters had remained essentially constant throughout this period.

A possible reason for the decreased gas production was that on February 14, without our knowledge, one hog from a pen of twelve ranging in weight from 120 to 140 lb (54 to 64 kg) from which manure was collected, was injected intramuscularly with two antibiotics (10 cc of 200-mg tylosin and 10 cc of 100-mg lyncomycin). On February 19 the manure from these hogs was collected and included with that from other hogs and placed in the digester on February 21. Later, after checking with a veterinarian, we discovered that tylosin has a residual time of approximately 15 days whereas that of lyncomysin is 7 days. On March 1 samples of the digester circulation and sludge were frozen, packed in dry ice, and sent to both antibiotic manufacturers for analysis. Neither company could find any residual of the chemicals. Although the residual life of these chemicals had been exceeded between the time the hogs were injected and the digester samples were obtained, possibly traces of these chemicals were excreted by the hogs and included with the digester influent feed when the chemicals were still active enough to destroy the bacteria in the digester. Some hog producers have reported that tylosin has caused lagoons to "go sour" on them. The only way they could reactivate the lagoon was to drain it and begin again. We found this also happened with our digester. Our attempts to revive the digester by reloading it with swine manure after it had not been loaded for a week were unsuccessful.

Finally, on April 1, 3 gal (11.4 l) of sludge from the Columbia municipal digester were again put into the digester. When the digester was reloaded, gas was produced on the following day. Figure 2 is the loading rate (lb VS/ft³ digester) of gas produced (ft³ of gas/ft³ of digester) and ammonia concentration, ppm, in the circulation for 1 year of digester operation from April to March. The four zones in Figure 2 represent four different digester operating characteristics.

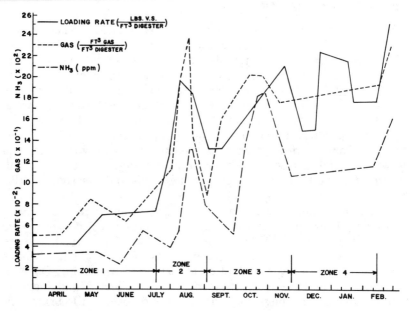

Figure 2. Loading rate and gas and ammonia production for one year of digester operation.

Zone I is a transitional zone where the digester becomes acclimated to swine manure as a source of nutrients. The loading rates varied from .04 to 0.09 lb VS/ft³ (0.64 to 1.43 kg/m³), gas production varied from approximately 0.5 to 0.9 ft³ (0.5 to 0.9 m³/m³) digester, and ammonia concentrations varied from 300 to 500 mg/1. Table 1 presents values for other parameters describing the digester conditions.

Zone II demonstrates the digester response to shock loading. When the loading rate was increased rapidly, gas production also increased rapidly but lagged by about 2 days behind the the loading rate. However, ammonia production lagged behind loading rate by 15 to 20 days. When the ammonia production reached a concentration of approximately 1200 ppm (1200 mg/1), gas production began to decrease. Again, average conditions for other parameters are shown in Table 1.

Table 1. Values representing the digester condition for each of the four zones.

	Zones			
	1	2	3	4
Loading Rate (lb VS/ft³ Dig)	0.083	0.181	0.163	0.195
Gas (ft³ gas/ft³ Dig)	0.84	1.8	1.75	1.92
Ammonia (ppm)	400	1000	800	1100
Alkalinity (ppm)	3500	6200	6000	7500
pH	7.02	7.3	7.25	7.33
Efficiency $\left(\dfrac{ft^3\ gas}{lb\ VS\ Des}\right)$	19	17	18	16.5
Gas Ratio (CO_2/CH_4)	*	0.67	0.63	0.64

* Data Not Available

For Zone III, the loading rate was again increased in an attempt to reproduce conditions of Zone II. But as the loading rate increased, gas production increased faster which is probably a result of residual organics in the digester. Ammonia began to increase reaching concentrations of approximately 1900 ppm (1900 mg/1). However, this did not substantially reduce gas production as it did in Zone II. After the increase, the ammonia concentration in the digester decreased and then began to level off.

In Zone IV, gas production and ammonia concentration stabilized even though the loading rate was varied significantly (0.15 to 0.24 lb VS/ft³, 2.38 to 3.81 kg/m³. Apparently the digester operation was stabilized; however, we were not sure about the reasons for this. Perhaps the bacteria became acclimated to the higher loading rate and the production of ammonia decreased, and the amount of ammonia in the digester decreased due to "washout."

EFFECT OF LOADING RATE

As loading rate increased, gas production increased up to a loading rate of approximately 0.18 to 0.19 lb VS/ft³ of digester (2.86 to 3.02 kg VS/m³). The maximum amount of gas produced was about 1.9 ft³ of gas/ft³ of digester (1.9 m³/m³ of digester) (Figure 3). Gas production seemed to be the best indicator of digester response since it was easiest to measure and responded more quickly to environmental changes in the digester. The gas production normally responded within 1 or 2 days after a new loading rate was introduced into the digester system. Ammonia increased as loading rate increased (Figure 4). We assumed that ammonia would continue to increase if loading rate was further increased until digester failure. However, we did not observe this reaction in this experiment because the digester was not overloaded enough for failure. As expected, when the loading rate increased, the TKN in the manure increased which also increased the TKN in the circulation.

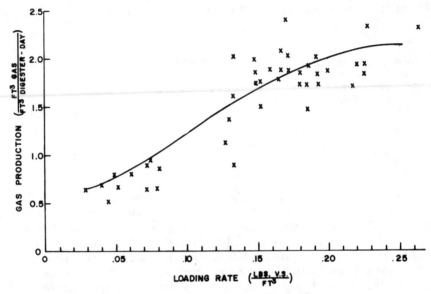

Figure 3. Gas production as affected by loading rate.

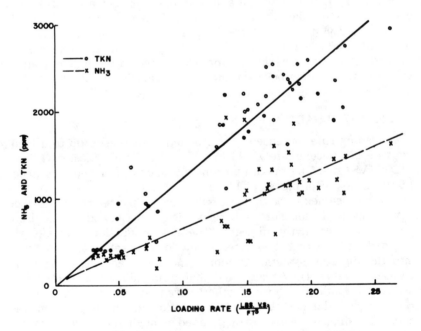

Figure 4. The effect of loading rate on ammonia and TKN.

The volume of gas produced per lb VS destroyed tended to decrease as loading rate increased. At a loading rate of 0.04 lb VS/ft³ (0.64 kg/m³), an average of 21 ft³ (0.59 m³) of gas was produced per lb of VS destroyed; whereas at a loading rate of 0.25 lb VS/ft³ (3.81 kg/m³), this figure was approximately 15 ft (0.42 m³) (Figure 5). Thus, as the digester was more heavily loaded, gas production efficiency seemed to decrease.

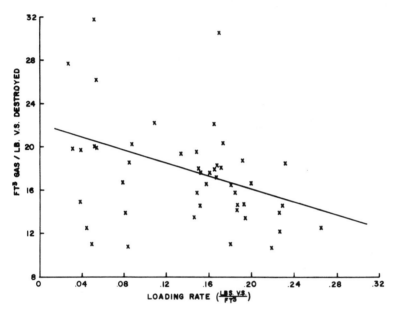

Figure 5. Cubic feet of gas produced per pound volatile solids destroyed as affected by loading rate.

The VS content in the circulation and sludge increased linearly with loading rate (Figure 6). Similar responses were obtained for total solids and COD.

The effects of alkalinity and pH are shown in Figure 7. Alkalinity increased linearly with loading rate from a concentration of 2000 to 9000 ppm (2000 to 9000 mg/ 1). The pH increased from 6.9 at the lowest loading rate to 7.3 at the highest. The CO_2/CH_4 content of the gas, which varied from 60 to 68%, was obtained only for loading rates in the range of 0.15 to 0.22 lb VS/ft³ (2.38 to 3.5 kg/m³).

The equations corresponding to the data in Figures 3 through 7 are given in Table 2.

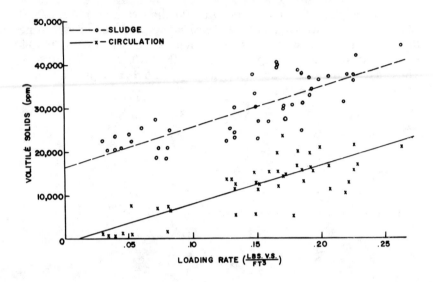

Figure 6. Volatile solids content in the circulation and sludge at various loading rates.

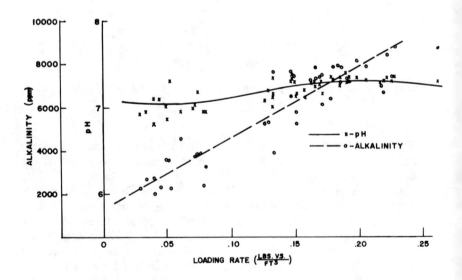

Figure 7. Effects of loading rate on alkalinity and pH.

Table 2. Equations which predict given variables for different digester loading rates. The R-Squared value for each equation is given.

Gas $\left(\dfrac{ft^3\ gas}{ft^3\ Dig}\right)$ $= 0.142 + 12.058\ (LR)* - 7.949\ (LR)^2 - 30.576\ (LR)^3$ R-Squared $= 0.83$

Ammonia (ppm) $= 85.9 + 6008.5\ (LR)$ R-Squared $= 0.55$

TKN (ppm) $= 87.8 + 11801.5\ (LR)$ R-Squared $= 0.85$

Alkalinity (ppm) $= 1453.7 + 31436.2\ (LR)$ R-Squared $= .85$

pH $= 6.86 + 3.344\ (LR) - 5.31\ (LR)^2$ R-Squared $= 0.62$

VS Sludge (ppm) $= 15632.4 + 97053.1\ (LR)$ R-Squared $= 0.55$

VS Circulation (ppm) $= -879.7 + 86287.6\ (LR)$ R-Squared $= 0.69$

Efficiency $\left(\dfrac{ft^3\ gas}{lb\ VS\ destroyed}\right)$ $= 73.1 - 325.6\ (LR)$ S-Squared $= 0.10$

* LR = Loading Rate $\left(\dfrac{lb\ VS}{ft^3\ digester}\right)$

PROPOSED DIGESTER OPERATION

Research from this study would indicate that 0.15 to 0.18 lb VS/ft³ (2.38 to 2.86 kg/m³) is the maximum loading rate attainable using swine manure. Values for the various parameters of digester operation corresponding to this maximum loading rate can be obtained from the above figures. Others have found similar maximum loading rates for swine manure (3, 4).

McCarty (5) proposed that good digestion requires an alkalinity ranging from 2500 to 5000 ppm (2500 to 5000 mg/1) and that ammonia ranging from 1500 to 3000 ppm (1500 to 3000 mg/1) would inhibit digestion. In our studies, ammonia inhibition depended on acclimation of the bacteria, but at a concentration of 2000 ppm (2000 mg/1) digester operation ceased. However, we recorded alkalinity as high as 9000 ppm (9000 mg/1). Schmid and Lipper (6) reported that a loading rate of 0.2 lb VS/ft³ (3.18 kg/m³) swine manure would produce more than 2000 ppm (2000 mg/1) of ammonia which could reduce the CH_4 in the gas produced to approximately 20%. The digester in this study could be operated for a short length of time at a loading rate higher than 0.2 lb VS/ft³ (3.18 kg/m³). In Figure 2, during the Zone IV time period, we varied the digester loading from 0.15 to 0.24 lb VS/ft³ (2.38 to 3.81 kg/m³) [0.18 lb VS/ft³ (2.86 kg/m³) average]. We observed no detrimental effects on the digester as long as the loading rate was not sustained above 0.18 lb VS/ft³ (2.86 kg/m³) for more than a week.

The amount of energy required to maintain the digestion temperature at 90°F (35°C) was obtained by calculating the heat loss through the insulation. For this digester, between 30 to 35% of the gas energy produced would have been required to maintain the digestion temperature.

Periodic analyses for phosphorus and potassium indicated that 14% of the K and 25% of the P remained in the sludge. Only 5% of the nitrogen, as measured by TKN, remained in the sludge, with 95% contained in the circulation effluent. This indicated that to make maximum use of the fertilizer nutrients of a digester, the circulation as well as the sludge must be applied to the land. However, when digester effluent is exposed to the air, a large percentage of the ammonia may be lost by volatilization which would decrease its value as a nitrogen fertilizer.

CONCLUSIONS

1. The manure of hogs given injections with antibiotics can disrupt digester performance.

2. Gas production rate is the best indicator of digester activity.

3. Stable digestion of swine waste can be obtained at loading rates ranging from 0.15 to 0.18 lb VS/ft³ (2.38 to 2.86 kg VS/m³).

4. At the loading rate stated above, approximately 16 ft³ (0.45 m³) of gas per lb VS destroyed is produced, the pH is 7.3, the ammonia is 1000 ppm (1000 mg/1) and the alkalinity is approximately 6000 ppm (6000 mg/1).

* * *

Approved by the director as a contribution from the Missouri Agricultural Experiment Station (Journal Series No. 7290).

REFERENCES

1. Singh, R. B. "Bio-Gas Plant," Gobar Gas Research Station, Ajitmal, Etawah (U.P.) India (1971).
2. Fry, L. J. *Practical Building of Methane Power Plants.* (Santa Barbara, California: Standard Printing, 1974).
3. Jeffrey, E. A., W. C. Blackman, Jr., and R. Rickets. "Treatment of Livestock Waste—A Laboratory Study." *Trans ASAE* 8:113-117 (1965).
4. Smith, R. J. "The Anaerobic Digestion of Livestock Wastes and the Prospects for Methane Production,'" Proceedings of Animal Waste, Iowa State University, Ames, Iowa (1973).
5. McCarty, P. L. "Anaerobic Waste Treatment Fundamentals. Part 2. Environmental requirements and control." *Public Works* 95(11):91-94 (1964).
6. Schmid, L. A., and R. I. Liper. "Swine Wastes, Characterization and Anaerobic digestion." In *Animal Waste Management,* Proceedings, Cornell Agricultural Waste Management Conference, pp. 50-57 (1969).

23.

Alternative Animal Waste Anaerobic Fermentation Designs and Their Costs

Gary R. Morris, William J. Jewell, and
George L. Casler*

The present and future trends in the U.S. toward extensive animal production facilities augment the problems associated with waste handling and disposal. Two complementary efforts are presently being directed at the animal waste management problem: the development of treatment alternatives to control environmental pollution and the search for beneficial uses for these highly concentrated organic wastes, not simply to contain and dispose of them. Anaerobically digesting these wastes in order to reclaim their intrinsic energy value while controlling pollution is one solution.

Some of the potential benefits in the utilization of the anaerobic fermentation process for animal waste management are listed in Table 1. The more important benefits of this technology include stabilization of the organic material, reduction of odor, and production of a useful by-product, methane, without decreasing the fertilizer value of the waste (1).

Studies have shown that animal wastes can be effectively digested anaerobically. But, the effects of several commonly referenced digester operational modes upon the anaerobic fermentation process and definitions of the limitations and potentials of these modes are

* Research Specialist and Associate Professor, Department of Agricultural Engineering; and Associate Professor, Department of Agricultural Economics, Cornell University, Ithaca, NY 14853.

Table 1. Potential benefits in the utilization of anaerobic fermentation for animal waste management.

---

1. Pollution Control
   —Stabilization and reduction of organic solids to a more acceptable form.
   —Liquid end-products are less offensive in odor and are homogenized to facilitate subsequent handling steps.
   —Rodents and flies are less attracted to the digested waste.
   —Destruction of pathogenic organisms.
2. Energy Generation
   —Production of a combustible gas—methane.
3. Nutrient Conservation
   —Conservation of the fertilizer constituents in the raw waste.

---

neither well documented in the available literature nor adequately understood. On this basis, it was decided that a knowledge of the fundamentals of anaerobic fermentation would allow the design and operation of various anaerobic digestion facilities in order to develop technically feasible system alternatives which would be compatible with current farm management practices. Thus, the specific objectives of this paper are:

1. To review the status of the available anaerobic fermentation technology and its application to animal wastes,

2. To apply this technology to develop technically feasible system alternatives which are compatible with current farm management practices, and

3. To assess the economic feasibility of incorporating the anaerobic fermentation systems into the farm both as an alternative energy source and a waste management practice.

ANAEROBIC FERMENTATION TECHNOLOGY

Anaerobic fermentation is one of the major biological waste treatment processes employed for municipal wastewater sludges. The anaerobic process can also be effectively applied to animal wastes because: a) the organic matter constituents are highly concentrated (*i.e.*, COD concentrations are greater than 4000 mg/1 (2) and total solids concentrations are greater than 1% (3); b) certain wastes such as those containing cellulose and lignin are more readily treated by the anaerobic process than by aerobic treatment (4); and c) high volumetric organic loading rates can be achieved.

The application of the anaerobic fermentation process to animal wastes requires an understanding of the environmental and operational factors that are involved. Factors influencing anaerobic fermentation are listed in Table 2. Some of these fundamentals will be reviewed in order to develop operational and general design criteria.

Table 2. Environmental and operational factors affecting anaerobic fermentation.

Environmental Factors	Operational Factors
1. pH	1. Composition of Organic Substrate
2. Alkalinity	2. Retention Time
3. Volatile Acid Concentration	3. Concentration of Substrate
4. Temperature	4. Organic Loading Rate
5. Nutrient Availability	5. Degree of Mixing
6. Toxic Materials	6. Heating and Heat Balance

Environmental Factors

Microbiology. Anaerobic fermentation is a highly sensitive microbial process, partly because it involves two major groups of bacteria which complete decomposition in two stages (4). First, complex organics are converted to low molecular weight alcohols and fatty acids by the "acid-forming" bacteria. In the second stage, the "methane-forming" bacteria utilize these low molecular weight compounds to form methane, carbon dioxide and other gases. Of the two, the methane-forming organisms are much more sensitive to various environmental conditions and will be inhibited whenever any of a number of factors are not suitable. Thus, the design criteria on which the anaerobic process is based must consider the environmental requirements of the methane formers.

pH-Alkalinity-Volatile Acid Concentration. The methane bacteria are affected by fractional pH changes. The pH is a function of the bicarbonate alkalinity of the system, the fraction of CO_2 in the digester gas, and the concentration of volatile acids (4). The alkalinity is a measure of the buffering capacity of the contents and acts as a safeguard against pH fluctuations due to accumulation of volatile acids. During normal operations, the pH is maintained naturally by the bicarbonate buffer system because of the carbon dioxide produced during anaerobic fermentation. The operational pH of municipal sewage digesters has ranged from 6.0 to 8.0 while effort has been directed toward maintaining an optimal pH of 7.0.

Temperature. The growth rate of microorganisms is a function of temperature which results in the assignment of digester operational temperatures. Fair and Moore (5) reported that several temperature ranges exist for the anaerobic fermentation process: a thermophilic zone (above 42°C), an intermediate or mesophilic zone (28-42°C) and a temperate zone (below 28°C). It has been common practice to digest municipal sludge in the mesophilic range where the temperature is 35°C. Investigators have revealed that reasonably effective and efficient anaerobic treatment can also be accomplished at both thermophilic and temperate temperatures (6, 7). Operation in the temperate zone would reduce or even eliminate the heating requirements, since the temperature of most raw wastes is less than 35°C. The major disadvantage to this operational temperature is that, with these slower reaction rates, longer waste retention periods are required. Operation in the thermophilic range requires an evaluation of whether thermophilic digestion is a sufficiently more efficient system than mesophilic digestion to warrant the additional external heat inputs to maintain the high operational temperatures. Although the energy inputs are higher, rapid microbial reaction rates may justify the use of smaller reactor volumes due to the shorter retention times required. Maly and Fadrus (7) concluded that the final degree of decomposition was virtually independent of the three temperature zones, although the reaction rates of decomposition were highest at the thermophilic range and lowest at the temperate range.

Nutrient Availability. The nutritional requirements are directly proportional to the synthesis of new microbial cells during fermentation. Ratios of available organic carbon:nitrogen:phosphorus of 100:2:0.5 are necessary to maintain a nutritionally balanced cell growth. The possibility of a limiting nutrient could inhibit the degradation of the substrate.

Toxic Materials. Toxicity can result from an excessive quantity of many organic or inorganic substances. Some of the inorganic substances that have caused toxicity have been summarized in an article by Kugelman and Chin (8) and will not be detailed here. The concentrations at which these materials begin to become toxic are difficult to define because they can be modified by complex interactions referred to as antagonism, synergism and/or acclimation of microbial systems to extreme conditions (8). The volumetric organic loading rate, in relation to the hydraulic retention time of a digester, can induce toxicity by increasing the concentration of the toxic substances. However, prediction of the operating conditions of a digester that will create concentrations which will inhibit methane production is not presently possible.

An example, relevant to fermenting animal wastes, is ammonia toxicity. McCarty (4) has suggested that anaerobic fermentation is inhibited by unionized ammonia concentrations in excess of 150 mg/l as NH_3-N. However, Kroeker *et al.* (9) has reported successful digestion of swine wastes at unionized ammonia concentrations of 500 mgl NH_3-N, possibly resulting from acclimation in combination with cation antagonism. This contrary reporting is just one example of the complexity involved with toxicity.

Operational Factors

Composition of Organic Substrate. The composition of the substrate is one of the major factors which determines the behavior of a digester. The various compounds and the form in which these compounds are present in complex substrates have a definite effect on their availability and subsequent microbial degradation to methane and carbon dioxide. The effects of complex substrates on anaerobic processes require a thorough knowledge of the composition of the substrate to be treated. Detailed information on the composition of animal wastes will not be given in this article.

Retention Time — Organic Loading Rate — Solids Concentration. The relationship between volumetric organic loading, retention time and solids concentration has been illustrated by Sawyer (10) for municipal sludges. Figure 1 shows organic loading as a function of the retention time and the solids concentration in the influent waste.

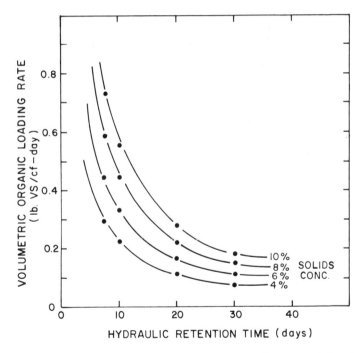

Figure 1. Relationship between volumetric organic loading rate, influent solids concentration and hydraulic retention time. [Adapted from Sawyer (10)].

The hydraulic retention time (HRT) is synonymous with the solids retention time (SRT) for a non-recycle system. The SRT is one of the most important operational factors controlling the design of an anaerobic fermentation system. It represents the average retention time of microbes in the system and can be determined by dividing the pounds of volatile solids in the system by the pounds of volatile solids leaving the system (2). Factors influencing the SRT are the temperature at which the process occurs, the volatile solids loading to the digester, the volatile percentage of the total solids, the total solids concentration in raw sludge, and the degree of stabilization desired. Minimum SRT values, where the methane bacteria will not wash out, are in the range of 7.5 days for 35°C, and 12.5 days for 25°C (6). Usually, SRT's of 10–30 days are employed to provide a safety factor for anaerobic fermentation of municipal sludges.

Although a high influent solids concentration is considered desirable, there are restrictions on the degree of concentration caused by the maintenance of adequate mixing in the digester and the difficulty of pumping the waste. The concentration of the substrate is dependent on the volatile solids of the raw substrate and the percent of solids breakdown. The optimum solids concentration in the digester should probably not exceed 6–8% (11, 12).

The anaerobic fermentation process can be designed by two different approaches. The first approach is called the rational method and is based on the kinetics of anaerobic fermentation and the theory of continuous culture of microorganisms (13, 14). This approach, illustrated by Lawrence (13) and O'Rourke (6), has been effectively applied to municipal sludges. The second approach is an empirical method using organic or solids volumetric loading rates. A knowledge of reported volumetric organic loading rates is used as a basis for digester design. This approach has been the most common method used in present agricultural waste practices (2).

Degree of Mixing. The degree of mixing influences the nature of a digester. Two extreme modes of digester operation are the plug flow reactor (no mixing) and the completely mixed reactor. In the past, only the completely mixed reactor was used.

Mixing of the digester contents offers several distinct advantages: (a) the substrate supply is kept in continuous contact with the microbes, (b) the substrate supply and temperature is uniformly distributed, (c) biological intermediates and end-products are uniformly distributed, and (d) scum layer formation is reduced to a minimum.

Application to Farm Animal Wastes

The information presented in Table 3 illustrates the relative quantities of manure emanating from different animals and the predicted quantities of gas generated from anaerobic fermentation. These estimates of gas production are subject to variations due to manure characteristics and efficiencies of digestion.

Table 3. Estimated manure and bio-gas production from animal wastes.*
(Estimated output per 1000 pounds live weight.)**

	Dairy Cattle	Beef Cattle	Swine	Poultry
Manure Production lb/day	85	58	50	59
Total Solids lb/day	10.6	7.4	7.2	17.4
Volatile Solids lb/day	8.7	5.9	5.9	12.9
Digestive Efficiency % of VS	35	50	55	65
Bio-Gas Production***				
cf/lb VS added	4.7	6.7	7.3	8.6
cf/1000 lb animal/day	40.9	39.5	43.1	110.9

* Adapted from Smith (15) and Loehr (2)

** Actual values may vary from these values due to differences in feed ration and management practices.

*** Based on theoretical gas production rate of 13.3 cf/lb VS destroyed [where 1 lb COD stabilized equals 5.62 of CH_4 (4) and assuminug the $CH_4:CO_2$ ratio is 60:40 and the conversion factor for VS to COD is 1.42]

Animal wastes vary in concentrations of organic and inorganic components, due to different housing and management conditions. Additional differences are caused by differences in types of animals, sizes, feed ration or diet and the animal's metabolism. With these factors in mind, manure characteristics should be carefully evaluated before proceeding with a design of the anaerobic system.

Information on the quantity of organics that are reported destroyed anaerobically with various animal wastes is limited due to the small number of investigations. Smith (15) and Loehr (2) have summarized the results of a number of laboratory-scale anaerobic digestion studies conducted utilizing livestock manure. The studies conducted thus far have been done on a case-by-case basis, and, while having demonstrated the amenability of animal manure to anaerobic fermentation, prior effort has not led to the establishment of a generally applicable design and operational model in terms of either the empirical or kinetic approach for predicting the performance of the anaerobic fermentation process.

ALTERNATIVE ANAEROBIC FERMENTATION DESIGNS

Anaerobic treatment can be adapted to several process configurations or modes of operation. Figure 2 illustrates how environmental and operational factors can be translated into a flow chart diagramming possible alternatives of digester operation and subsequent management and control. There are many possible pathways upon which to base the digester design. Four basic process schemes can be developed from the alternatives shown in Figure 2. These process schemes will be referred to as completely mixed, anaerobic contact, batch load and plug flow. The anaerobic contact process will not be considered here since it has not been shown to be an alternative for animal waste treatment.

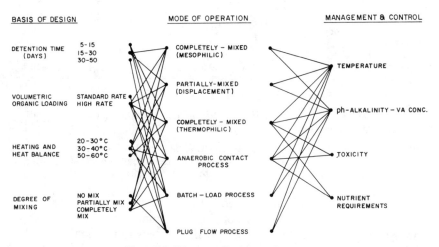

Figure 2. Digester design pathways.

Schematic representations of these various anaerobic processes are shown in Figure 3. The completely mixed process has been used most commonly in municipal sewage sludge digestion practice. In this process, the waste entering the reactor is immediately and uniformly distributed through the entire reactor. Digested effluent is withdrawn at a rate equal to the inflow rate to maintain the reactor at a constant volume. Variations of the completely mixed process relate to the degree of mixing attained and the operational temperature employed. These variations determine the retention time and volumetric loading rate needed for design purposes. For example, variations may include operating at either mesophilic or thermophilic temperatures and/or employing either a completely mixed or partially mixed reactor. In general, these digesters would employ relatively short retention times and high volumetric loading rates.

The batch load process operates as two completely mixed reactors where one unit is fed while the other continues to ferment as a full reactor. When both reactors become filled, one is emptied and it is loaded while the other continues to ferment. This operating procedure increases the efficiency of digestion. This design is recommended by the German consulting firm of Steinmann and Ittig for anaerobic treatment of animal wastes. Retention times of 30 days at 35°C are also recommended for operation.

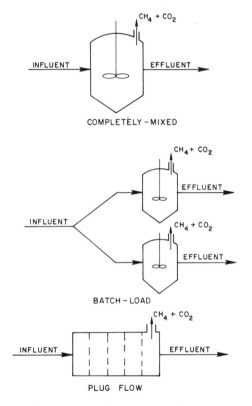

Figure 3. Schematic representation of anaerobic process configurations.

The plug flow operates as a longitudinal reactor where no intermixing occurs during the passage of the waste through the reactor. The biodegradable organics decrease through longitudinal distance while the microbial mass concentration increases. This is the general concept recommended for Monfort Feedlots, Inc. by Colorado

Biogas Co. A plug flow digester would be designed employing a long retention time and standard volumetric organic loading rate.

The next step in the design procedure is to incorporate these conceptual designs into operational fermentation systems. Figure 4 shows a schematic array of possible process trains which are available for anaerobic fermentation systems. The major components of this flow chart include the source, predigestion management, feeding component, digester design and by-product management.

Five alternatives were developed that were believed to be technically feasible. Brief descriptions of the five alternatives and their respective process groups follow. The partially mixed with liquid

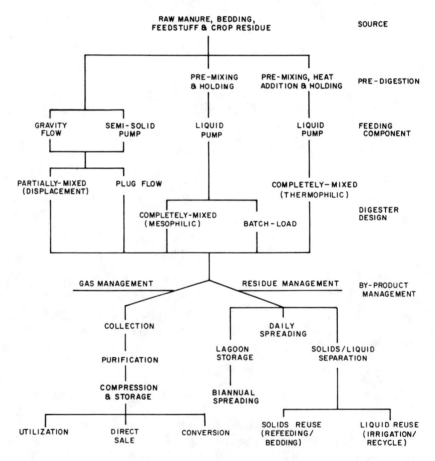

Figure 4. Flow chart of possible process trains for anaerobic fermentation systems.

displacement and plug flow systems are the simplest systems in terms of design and operation. They include gravity feeding and no premixing. The capital and operational costs are minimal, however, biodegradation and subsequent gas production are less than with the other three alternatives.

The completely mixed system, both mesophilically and thermophilically operated, and the batch load system have a higher degree of complexity. They include a predigestion holding tank for premixing and dilution water additions and are fed to the digester component by liquid manure pumps. The capital and operational costs are high, but in return, biodegradation and subsequent gas production are increased.

The gas and residue management components are common to all five systems differing in minor modifications for each system. Typical gas management components would include collection, purification and compression-storage facilities. The overriding factor in gas management design is safety. The practical solution to residue management would include temporary lagoon storage and periodic spreading, preferably biannually.

In summary, all design alternatives have been synthesized into five technically feasible alternatives which can be applied to animal wastes for stabilization of the biodegradable organics. These five possible alternative systems include: a) completely mixed digester operated in the mesophilic range, b) completely mixed digester operated in the thermophilic range, c) batch load digesters, d) partially mixed digester operated in the mesophilic range, and e) plug flow digester.

ECONOMIC EVALUATION OF ALTERNATIVE ANAEROBIC FERMENTATION SYSTEMS

In order to determine the overall feasibility of using one of the five design concepts suggested, an economic assessment of each of the five anaerobic fermentation systems was conducted. The feasibility analysis includes the evaluation of capital costs, operation and maintenance costs and the cost of generating energy in relation to current energy costs. In addition, focus is placed on the effect of adopting these systems as waste management practices in relation to other available practices. All of the economic considerations are based on 1975 dollars.

Included in the cost analysis is a typical off-the-shelf municipal sewage treatment plant digester. The reason for this inclusion is for cost comparison of a highly equipment oriented digester in relation to "modified" farm digesters.

Systems Design

A hypothetical case of a 100-cow free-stall dairy operation represents the basis of the feasibility analysis. Manure production emanating from the milking herd was calculated using data given in Table 3. Dilution water was necessary to obtain an influent solids concentration of 10%. The predigestion storage tanks were designed for a two-day holding capacity. Hydraulic retention time was the basis of design for digester sizing. Twenty-day retention times were chosen for the partially mixed and completely mixed digesters, operating at 35°C. A retention time of twenty days was selected for each of the batch load digesters, also operated at 35°C. The completely mixed digester, operated at 60°C, was designed for a five-day retention time. A forty-day retention time was chosen for the plug flow digester operating at 25°C. Construction materials included concrete and polyurethane with fabricated steel covers for all the digesters except the plug flow reactor. Plastic liner and styrofoam were selected as the construction materials for the trench-like plug flow digester.

The gas and residue management component designs were common to all five systems differing in minor modifications for each system. Gas management components included collection, purification and compression-storage equipment. Residue management component designs included six-month earthen lagoon storage and spreading facilities.

Gas production rates (Table 3) necessitated some minor adjustments based on the estimated digestive capabilities of each system. These adjustments were made because gas production is a function of volumetric loading rate, temperature and retention time, all of which varied for each alternative system.

The estimated heat required to operate the systems includes the amount of energy necessary to maintain the desired operational temperature divided by the efficiency heat transfer. Energy requirements include digester heat losses and influent manure heating requirements. It does not include the energy required for pumping and mixing. The estimated total heat requirement is a function of ground, air and manure temperatures, operational temperature of the digester and constructon material employed. In this chapter, the total heat requirements were calculated using climatological conditions equivalent to northern New York.

Cost Analysis

Capital investment costs for the five anaerobic fermentation systems were developed by identifying the components of each system, estimating design requirements and assigning costs to each component. Prices were obtained from equipment manufacturers and contractors. Equipment prices included installation; construction costs incorporated materials and labor. Table 4 presents the total capital investment cost and investment per cow for each of the five alternatives.

Annual costs for each system represent the yearly amortization of the initial capital investment and the necessary operational and maintenance costs. Capital costs were amortized at a rate of 9% by assuming a 20-year life for structures and a 10-year life for equipment. Maintenance and operational costs consisted of equipment maintenance and repairs, labor, power, water, taxes and insurance. Maintenance costs were assumed to be 2% of the initial investment per year. Taxes and insurance were based on a rate of $3\frac{1}{2}\%$ of the capital cost per year. Labor, power and water costs were estimated by determining their respective requirements for each system. Table 4 shows the annual total cost and cost per cow for each of the alternative systems.

The total investment and annual costs, shown in Table 4, clearly indicate that there are variations depending on the degree of complexity of the systems. The capital investment varied from a low of $200/cow for the simpler systems to a high of $550/cow for the off-the-shelf municipal digester system. The annual costs range from a low of $45 to a high of $150 per cow. It is evident from this table that the economic feasibility of methane production depends heavily upon the system design. These high costs were estimated on the conservative side and they might be reduced by making use of cheaper equipment, different construction materials and other modifications.

Energy Production Cost Comparison

The economic feasibility of incorporating the anaerobic fermentation systems into the farm as an alternative energy source can be evaluated by determining the unit cost of the net available energy. Energy production costs were calculated by dividing the net available energy by the annual costs. The costs for producing energy are presented in Table 4 for the five design alternatives. The unit costs of net available energy range from a low of $10 per million Btu to a high of $25 per million Btu. The higher costs of energy were a result of the excessive amounts of heat input necessary to maintain the desired digester operating temperature.

The significance of these energy costs can only be evaluated by comparing them with present prices of conventional fuels. Figure 5 compares the cost of energy production in relation to current costs of propane, natural gas, gasoline and electricity (16, 17). This figure

Table 4. Feasibility analysis summary of alternative anaerobic fermentation systems. (Based on a 100 cow free stall dairy operation.)

	Typical Municipal Digester	Completely Mixed (Mesophilic)	Partially Mixed (Displacement)	Completely Mixed (Thermophilic)	Batch Load	Plug Flow
Capital Investment						
Total $	55,000	27,000	20,000	21,000	31,000	22,000
$ per cow	550	270	200	210	310	220
Annual Costs						
Total $	15,000	8,000	4,500	7,000	9,000	5,000
$ per cow	150	80	45	70	90	50
Net Annual Gas (10^6 Btu/yr)	600	600	450	325	755	250
Energy Production Costs ($/$10^6$ Btu)	25	13	10	22	12	20

shows that the cost of producing methane would be more than double the cost of propane which is the most expensive fuel used on farms.

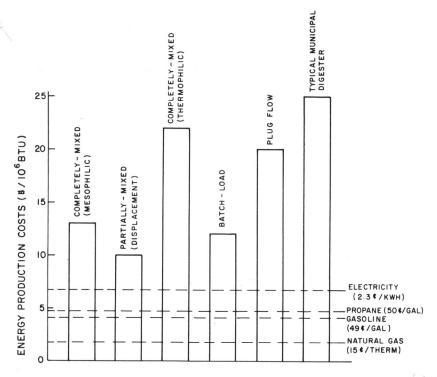

Figure 5. Cost of energy production in relation to current energy costs.

Energy generated by anaerobic fermentation does not presently compete on an economic basis with other conventional sources. However, in the near future, simpler and more efficient designs may lead to the development of systems that produce energy at comparative costs. Preliminary calculations show that a herd size of 1000 cows has the potential of competing with present energy costs. For this size operation, the methane production costs are approximately $2 per million Btu. On this basis, gas production from farms of 1000 head or more would be economically competitive with current energy costs, assuming that there is a use for the gas produced.

Waste Management Cost Comparison

Although energy production may not pay for the use of this process, the decision to use the process should take into account its value as a pollution control method, and in specific cases, its fertilizer conservation properties. Evaluation of the fertilizer value is beyond the scope of this paper. Thus, an alternate approach is to economically assess the fermentation systems not as an alternative energy source but as a waste management practice.

Other viable animal waste pollution control alternatives include mechanical stacking of manure and liquid manure handling, each with six months' storage. The economic assessment of alternate manure handling systems and their subsequent impact on the dairy farm have been examined by several researchers (18-22). Annual costs for these alternate manure handling systems are shown in Table 5.

Table 5. Annual costs of alternative manure disposal systems.

Annual Costs ($/cow)		
Mechanical Stacking System	Liquid Manure Handling System	Reference
18*	58*	Buxton and Ziegler (18)
19	44	Ashraf *et al.* (19)
20*	48	Jacobs and Casler (20)
16*	—	Johnson *et al.* (21)

* Extrapolated from referenced data.

Figure 6 presents a bar graph comparison of the annual cost per cow of the five alternative systems in relation to the ranges of annual costs per cow obtained for the two alternate manure handling systems. As can be concluded from this figure, the anaerobic fermentation systems are not competitive with mechanical stacking systems, but the simpler fermentation systems can be economically competitive with current costs of liquid manure handling systems. In addition, the value of the by-product gas can be treated as a potential income.

Economic Summary

An important conclusion that can be drawn from this economic analysis is that the use of the anaerobic fermentation process cannot be justified on the sole basis of being an alternative energy source. Whereas, it might be concluded that the anaerobic fermentation process can be an economically feasible waste management practice when liquid manure handling is an alternative.

The use of the fermentation process as a waste management practice offers additional benefits over the other manure disposal systems. Some of these benefits cannot be reduced to income credits. The one major credit is the production of a useful by-product, methane. The benefit of organic stabilization and subsequent odor reduction enables the farmer to increase his spreading period and the amount spread per acre. In addition, the digested manure has been homogenized, which facilitates subsequent handling steps. The fertilizer constituents, particularly nitrogen, in the raw manure are conserved, which significantly adds to the value of the digested manure as a fertilizer. Intangible benefits include fly and pest control and the destruction of pathogenic organisms.

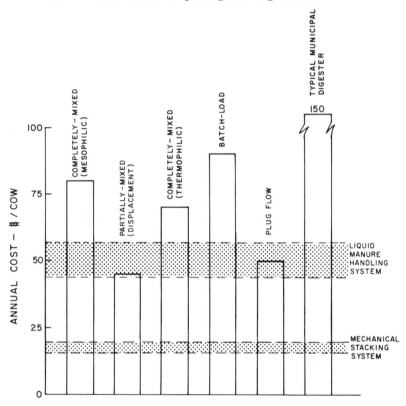

Figure 6. Annual cost comparisons in relation to waste management practices.

A second conclusion that can be drawn from this economic study is that the feasibility of anaerobic fermentation is dependent upon system design. The designs described in this chapter utilize current practices employed for sewage sludge digestion. The major problem of this approach involved finding reliable, but low-cost solutions to the design of system components. The authors feel that a sound research and development effort in the areas of system technology and design would lead to substantial cost savings. Areas that may have the potential for cost savings include a) development of prefabricated units or detailed design plans so the facilities can be self-erected, b) investigation of fermentation fundamentals to better understand, and thus, optimize the process, c) improvement in equipment (pumps, heat exchangers and gas handling facilities) in order that they may be reliable and available at a reasonable cost, and d) investigation of additional sources of increasing gas production such as manure pretreatment and crop refuse digestion.

CONCLUSIONS

In regard to the generation of methane from the anaerobic fermentation of animal wastes, the following is concluded:

1. Anaerobic fermentation technology for animal waste treatment is presently not documented in detail. More data is needed to provide the engineer with information necessary to design workable alternative systems.

2. Efforts to modify conventional municipal sewage digester designs are necessary in order to provide systems that are economically feasible for animal production operations.

3. Under present market conditions, the generation of methane as an alternative energy source for average sized dairies is not economically competitive with other fuels.

4. With larger animal operations, for example a 1000-head beef herd, methane gas production costs might be competitive with the present market value of natural gas.

5. As a pollution control device, anaerobic fermentation processes have the potential of being competitive with present costs of liquid manure handling and storage systems.

6. Anaerobic fermentation technology has the potential of playing a significant, future role in agricultural waste management.

* * *

This chapter is part of an investigation supported by funds from the National Science Foundation under RANN Grant No. GI 43099.

REFERENCES

1. Jewell, W. J., G. R. Morris, D. R. Price, W. W. Gunkel, D. W. Williams, and R. C. Loehr. "Methane Generation from Agricultural Wastes: Review of Concept and Future Applications." Amer. Soc. Agr. Eng. Paper No. NA74-107, 1974.
2. Loehr, R. C. *Agricultural Waste Management.* Academic Press, Inc., New York, 1974.
3. Lawrence, A. W. "Anaerobic Biological Waste Treatment Systems." Agr. Wastes: Princ. and Guidelines for Pract. Solutions, Proc. Cornell Univ. Conf. Agr. Waste Mgmt., 1971.
4. McCarty, P. L. "Anaerobic Waste Treatment Fundamentals: I Chemistry and Microbiology; II Environmental Requirements and Control; III Toxic Materials and Their Control; IV Process Design." Public Works, Vol. 95, No's. 9-12, 1964.

5. Fair, G. M. and E. W. Moore. "Time and Rate of Sludge Digestion and Their Variation with Temperature." Sewage Works Journal, 6, 3, 1934.
6. O'Rourke, J. T. "Kinetics of Anaerobic Waste Treatment at Reduced Temperatures." Ph.D. Dissertation, Stanford Univ., Stanford, CA, 1968.
7. Malý, J. and H. Fadrus. "Influence of Temperature on Anaerobic Digestion." Jour. Water Poll. Control Fed., 43, 4, 1971.
8. Kugelman, I. J. and K. K. Chin. "Toxicity, Synergism and Antagonism in Anaerobic Waste Treatment Processes." Anaerobic Biological Treatment Processes, F. G. Pohland, ed., Advances in Chemistry Series No. 105, American Chemical Society, 1971.
9. Kroeker, E. J., H. M. Lapp; D. D. Schulte and A. B. Sparling. "Cold Weather Energy Recovery from Anaerobic Digestion of Swine Manure." Energy, Agriculture and Waste Management, Proc. Cornell Univ. Conf. Agr. Waste Mgmt., 1975.
10. Sawyer, C. N. "Anaerobic Units." Proceedings, Symposium on Advances in Sewage Treatment Design, SED, ASCE, New York, 1961.
11. Sawyer, C. N. and J. S. Grumbling. "Fundamental Considerations in High Rate Digestion." SED, ASCE, 86(SA2), 1960.
12. Malina, Jr., J. F. and E. M. Miholits. "New Developments in the Anaerobic Digestion of Sludges." Advances in Water Quality Improvement, Water Resources Symposium No. 1, edited by E. F. Gloyna and W. W. Eckenfelder, Jr., Univ. of Texas Press, 1968.
13. Lawrence, A. W. "Application of Process Kinetics to Design of Anaerobic Processes." Anaerobic Biological Treatment Processes, F. G. Pohland, ed., Advances in Chemistry Series No. 105, American Chemical Society, 1971.
14. Gaddy, J. L., E. L. Park, and E. B. Rapp. "Kinetics and Economics of Anaerobic Digestion of Animal Waste." Water, Air and Soil Pollution, 3, 1974.
15. Smith, R. J. "The Anaerobic Digestion of Livestock Wastes and the Prospects for Methane Production." Unpublished manuscript, Iowa State Univ., 1973.
16. Agway, Inc., Syracuse, N.Y. Private communication on propane and gasoline prices, 1975.
17. New York State Gas and Electric, Ithaca, NY. Private communication on natural gas and electricity prices, 1975.
18. Buxton, B. M. and S. J. Ziegler. "Economic Impact of Controlling Surface Water Runoff from U.S. Dairy Farms." Agricultural Economic Report No. 260, U.S. Dept. of Agriculture, Washington, DC, 1974.
19. Ashraf, M., R. L. Christensen and G. E. Frick. "The Impact on Dairy Farm Organization of Alternative Manure Disposal Systems. A Method of Assessing the Cost of Environmental Regulation." Research Bulletin No. 608, Mass. Agricultural Experiment Station, University of Mass., 1974.
20. Jacobs, J. J. and G. L. Casler. "Economic and Environmental Aspects of Systems for Handling Dairy Manure." Unpublished manuscript, Cornell Univ., 1972.
21. Johnson, J. B., C. R. Hoglund, and B. Buxton. "An Economic Appraisal of Alternative Dairy Waste Management Systems Designed for Pollution Control." Journal of Dairy Science, Vol. 56, No. 10, 1973.
22. Casler, G. L. and E. L. LaDue. "Environmental, Economic and Physical Considerations in Liquid Handling of Dairy Cattle Manure." Food and Life Sciences Bulletin No. 20, Cornell Univ., 1972.

24.

Cold Weather Energy Recovery from Anaerobic Digestion of Swine Manure

E. J. Kroeker,* H. M. Lapp,* D. D. Schulte,*
A. B. Sparling**

Concern that known fossil fuel reserves are being rapidly depleted has focused attention on the feasibility of economical recovery of energy and plant nutrients from livestock manures through anaerobic digestion. The large concentration of organic matter contained in animal manures may, if broken down by anaerobic bacteria in a controlled environment, yield significant quantities of methane gas. This potential for energy recovery is enhanced in North America by modern systems of confinement housing where large quantities of manure are rapidly accumulated.

The anaerobic digestion process discovered in the late 1600's by Robert Boyle, has been used for many years as a sludge stabilization process in municipal sewage treatment systems. Farm-scale digesters for gas production have been operated in Europe, Asia, Africa and India (1-3). However, the technical and economic feasibility of large-scale anaerobic digestion of animal manure in cold climates has not been fully assessed, and there is speculation as to whether the process has the potential for net energy recovery or whether it consumes energy.

* Research Engineer, Professor and Assistant Professor, respectively, Department of Agricultural Engineering, University of Manitoba; Associate Professor, Department of Civil Engineering, University of Manitoba, Winnipeg, Manitoba R3T 2N2.

Pilot-scale experiments on anaerobic digestion of swine manure were started at the University of Manitoba in late fall, 1974. The general objective of the project was to assess the technical and economical feasibility of energy recovery from livestock manure while specific goals included: 1) determination of design parameters for methane production from livestock manure in cold climates; 2) determination of the efficiency of the anaerobic digestion process in recovering energy from livestock manure; 3) development of safe and economical methods of collecting, scrubbing, storing and utilizing digester gas on farms; 4) analysis of the effluent from anaerobic digestion and assessment of its value as a fertilizer; and 5) assessment of the environmental impact, if any, of the anaerobic digestion process.

The purpose of this chapter is to compare experimental results with a theoretical energy balance for the pilot-scale digesters. Discussion centers on energy consumption and production data and the implications of these results for large-scale anaerobic digestion for energy recovery.

EXPERIMENTAL PROCEDURE

Pilot Plant

The pilot plant used to conduct the experiments is located adjacent to a four-building swine complex housing approximately 800 hogs at the University of Manitoba's Glenlea Research Station. Components are housed in a 7.3 x 7.9 m (24 x 26 ft) insulated steel building equipped with explosion-proof electric lighting, heating and ventilation. Two electrically heated hot water tanks used for digester heating and a series of watt-hour meters used to monitor electrical energy consumption in the pilot plant are housed in a non-explosion-proof building adjacent to the pilot plant.

The pilot plant, shown schematically in Figure 1, consists of four cylindrical single-stage 3194-1 (700-Imp gal) fiberglass digestion tanks. It is serviced by a 91.3-m³ (20,000 Imp gal) raw manure holding tank equipped with mixing and transfer pumps to facilitate the delivery of raw manure at relatively uniform consistency into the pilot plant digesters.

Two duplicate digester heating systems, each consisting of a domestic hot water tank, water circulation pump, insulated 3.18-cm (1.25-in.) copper pipe and stainless steel heat exchange piping have been installed. Mixed-liquor temperatures have been controlled individually for each digester with thermoelectrically-controlled solenoids in the hot water circulation lines. The circulation pumps were operated continuously. Standard watt-hour meters on the hot

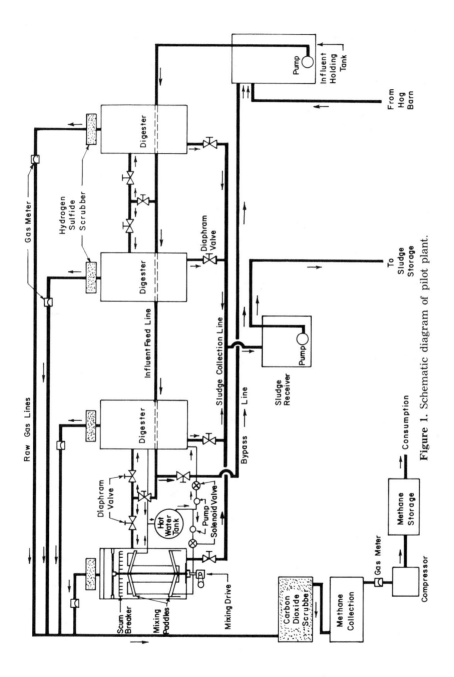

Figure 1. Schematic diagram of pilot plant.

water heaters and circulation pumps have been used to monitor heating requirements for each pair of digesters (A-C and B-D as reported in this paper).

Continuous mixed-liquor agitation in each digester was provided by rotating paddles (Figure 1) powered by a 0.56-kw (.75-hp) electric motor and a 40:1 gear reduction drive unit. Watt-hour meters were also used to monitor the energy requirements for mixing the contents of each digester.

Initial Experiments

The digesters were initially seeded with anaerobic digester mixed liquor from the Winnipeg North End Municipal Sewage Treatment Plant. Operations volume of each digester was 2300 liter (505 Imp gal). Following an accelerated period (20 days) of acclimation to the swine manure substrate, conditions characteristic of digester failure were encountered. Gas production rates were decreasing rapidly and volatile acids concentrations rose to between 7350 and 8550 mg/1 as HAc. Loading was then stopped for a 40-day period to allow digester recovery. At the end of this period, feeding of the raw swine manure was resumed. After reacclimation over a time interval (37 days) much longer than initial acclimation, digesters A and B were being operated at a 15-day solids retention time (SRT) and digesters C and D were being operated at a 30-day SRT. Table 1 summarizes the process performance data after reacclimation.

Table 1. Performance indicators at end of reacclimation period.

Parameters	Digester			
	A	B	C	D
Volatile Acids, mg/1 HAc	1,330	1,330	1,500	1,450
Alkalinity, mg/1 $CaCO_3$	14,480	14,960	14,640	14,340
pH	8.4	8.4	8.2	·8.5
Ammonia, mg/1 N	2,930	2,930	2,870	2,840
CO_2/CH_4	34/66	35/65	33/67	39/61

Data reported in the paper are indicative of digester performance at SRT's of 15 and 30 days. Minor fluctuations in volatile solids loading rates occurred (Figure 2) due to difficulty in maintaining homogeneity of suspended solids in the storage pit. The average organic loading rates to the following reacclimation were 1.3 g/1-day (0.08 lb/ft³-day) for digesters C and D and 2.6 g/1-day (0.17 lb/ft³-day) for digesters A and B. Mixed-liquor withdrawal and raw waste feeding were accomplished on a once daily basis. Mixed-liquor temperatures in the four digesters were maintained as nearly as possible at 35°C (Figure 3). Air temperatures outside the digesters were maintained at 7°C (45°F) ±3°C (6°F). Raw manure influent temperature was 0°C (32°F) during the period. Continuous mixed-liquor agitation was used to maintain high-rate anaerobic digestion.

Analytical Procedures

Data were collected on a daily, bi-weekly and weekly basis in order to determine rates of energy consumption and production

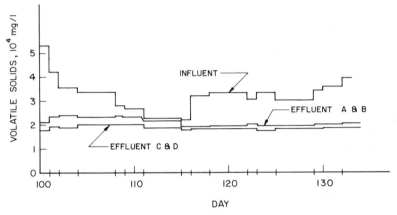

Figure 2. Influent and effluent volatile solids concentrations.

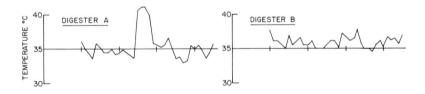

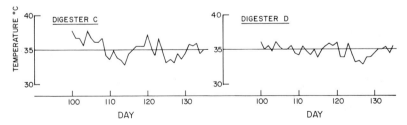

Figure 3. Digester temperature.

and to monitor biological and chemical characteristics of the digester influent and effluent.

Standard watt-hour meters were used to monitor daily rates of electrical energy consumption by hot water heaters, circulation pumps, mixed-liquor agitators and raw manure feed and effluent removal pumps. Thermocouples were used to monitor digester, outside air and raw manure influent temperatures. Precision wet-test meters were used to monitor daily gas production and gas analyses were performed by gas chromatography on a Fisher-Hamilton gas partitioner using a dual-column arrangement of 30% DEHS on 60/80-mesh Chromosorb P and a 40/60-mesh Molecular Sieve (14X).

Digester pH was monitored daily in the pilot plant and samples were brought to the laboratory to determine total and volatile solids, alkalinity, ammonia and total nitrogen, COD and volatile acids. Alkalinity was determined potentiometrically according to Standard Methods (1971). Ammonia and total nitrogen were also determined according to Standard Methods (1971). Volatile acids were determined spectrophotometrically at 490 mμ using procedures outlined by the Hach Chemical Corporation (1971).

Digester Operating Characteristics

Although fluctuations in operating parameters such as volatile acids, pH and gas composition occurred during the experimental period reported in this paper, steady-state operation had been achieved because long-term trend changes were not present. Steady-state gas production rates during the last 15 days of the 35-day period are shown in Figures 4 and 5. Gas composition data are also illustrated in Figures 4 and 5. Summary data of digester operating characteristics are presented in Table 2. The estimated energy value of gas produced in each digester during the steady-state operation has been summarized in Table 3.

Table 2. Summary data of steady-state operating characteristics.

Parameter	Digester			
	A	B	C	D
Loading rate, g VS/litre-day	2.10	2.10	1.05	1.05
Solids retention time, days	15	15	30	30
Gas production rate,				
l/g Vs added	0.68	0.62	0.82	0.71
l/g VS destroyed	1.98	1.82	1.85	1.62
$CH_4:CO_2$	2.04	2.13	2.02	2.00
Volatile acids, (mg/l as HAc)	2,130	2,050	1,950	1,550
Alkalinity (mg/l as $CaCO_3$)	16,000	16,100	15,600	15,400
NH_3-N (mg/l as N)	3,450	3,430	3,430	3,390
pH	8.0	8.0	8.0	8.0
Volatile solids destruction, %	36		44	

Table 3. Energy value of digester gas.

Heat value of pure methane gas — 10.38×10^{-3} kWh/1 (1012 Btu/ft^3)
Approximate heat value of digester gas $= 10.38 \times 0.67 = 6.95 \times 10^{-2}$ kWh/1
(678 Btu/ft^3)

Digester	Steady-State Gas Production Rate (liter/day)	Steady-State Energy Recovery Rate (kWh/day)
A	3273	22.75
B	3015	20.95
C	1972	13.71
D	1724	11.98

Total energy consumption data for each pair of digesters in the pilot plant are illustrated in Figures 6 and 7. As mentioned previously, each pair of digesters was serviced by one hot water heating system (hot water tank and circulation pump). Digesters were arranged in the manner outlined in Table 4.

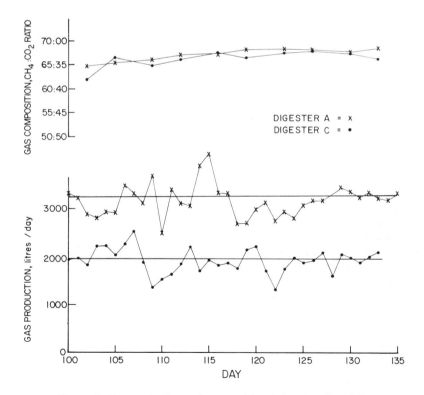

Figure 4. Gas production and composition (Digesters A and C).

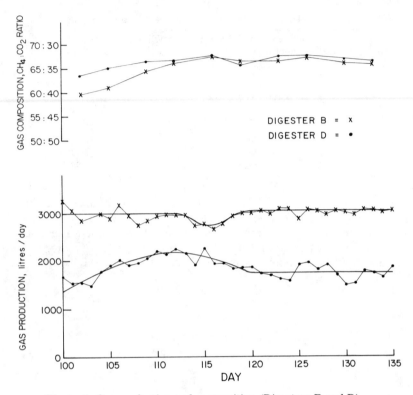

Figure 5. Gas production and composition (Digesters B and D).

Although energy consumption by the influent feed and effluent removal pumps was recorded, the total consumption was insignificant and was therefore not included in Figures 6 and 7.

High levels of methane gas production continued during steady-state operation in spite of ammonia nitrogen levels in excess of 3400 mg/1. In spite of pH levels of 8.0 in all digesters free-ammonia toxicity did not occur. Acclimation of the "methane formers" to high free-ammonia concentrations combined with cation antagonism effects reported by Kugelman and Chin (4) may have accounted for operation success.

Trade offs between the mean cell residence time (SRT), the or-

Table 4. Digester grouping.

Parameter	Digester			
	A	B	C	D
SRT, days	15	15	30	30
Loading rate, g VS/1-day	2.6	2.6	1.3	1.3
Hot water heating system	1	2	1	2

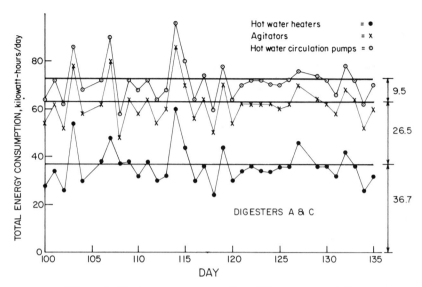

Figure 6. Total energy consumption (Digesters A and C).

ganic loading rates and the efficiency of gas production appear to exist as indicated by Table 2 and Figures 4 and 5. Although total gas production was higher in digesters A and B than in C and D, the fraction of volatile solids destroyed was higher at the lower organic loading rates and the higher SRT. It is not clear from the data gathered to date how much the SRT could be reduced and/or the organic loading rate increased without adversely affecting methane gas production rates.

THEORETICAL HEATING REQUIREMENTS

In the high-rate anaerobic digestion process, energy is required to maintain complete mixing, to maintain the process temperature, to remove the effluent and to premix and pump the raw waste influent.

During cold-weather digester operation, the mixed-liquor heating requirements will be significant, especially at the elevated process temperatures used in these experiments. The total amount of energy required to maintain the mixed liquor at the desired operating temperature is the sum of 1) heat losses through the digester walls, roof and floor, and 2) heat required to raise the temperature of the digester influent to the desired operating temperature.

In order to assess the accuracy of using heat-transfer theory to estimate digester heating requirements, the heating requirements necessary to replace wall, roof and floor losses and to raise the

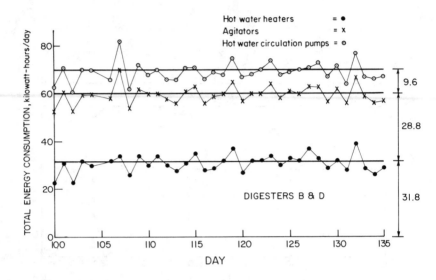

Figure 7. Total energy consumption (Digesters B and D).

temperature of the raw manure effluent were calculated. These theoretical values were then compared with actual energy consumption data.

The total digester heating requirements can be represented by the following equation:

$$Q_T = Q_L + Q_I \qquad (1)$$

in which Q_T = rate at which heat energy must be supplied to the digester mixed liquor (energy/time)

Q_L = rate of heat loss through digester walls, floor and roof, energy/time)

Q_I = rate of heat transfer to raw manure influent, (energy/time).

Digester Heat Losses

All heat losses from the digester were assumed to be by conductive heat transfer. The general equation for steady-state one-dimensional conductive heat transfer is:

$$Q = U \cdot A \cdot (T_2 - T_1) \qquad (2)$$

in which Q = rate of heat loss, (energy/time)

U = overall coefficient of thermal conductivity, (energy/time-area-temperature)

A = area normal to the direction of heat flow

T_2 = air temperature outside the digester

T_1 = mixed-liquor temperature

For composite wall, roof and floor materials made of structural material and a layer of insulation the overall coefficient of thermal conductivity, U, is a function of the unit-surface conductances inside and outside plus the thermal resistances of each of the materials as described by:

$$\frac{1}{U} = \frac{1}{h_i} + \frac{x_1}{k_1} + \frac{x_2}{k_2} + ... + \frac{1}{h_o} + \frac{1}{K_a} \quad (3)$$

in which x = material thickness, (length)

k = material coefficient of thermal conductivity, (energy/time-length-temperature)

h_i and h_o = inside and outside unit — surface conductances, (energy/time-area-temperature)

K_a = coefficient of conductivity of air and gas, (energy/time-area-temperature)

The walls and roof of the pilot-plant digesters consist of 0.64-cm (0.25-in.) rigid fiberglass covered with 3.80 cm (1.51 in.) of urethane foam insulation. The fiberglass bottom is supported by 1.9-cm (0.75-in.) plywood also covered with 3.80 cm of urethane. Numerical constants used in the calculations are summarized in Table 6. Using the equations summarized in Table 5, the theoretical heat loss, Q_L, from each digester was computed to be 9.7 kWh/day (33,000 Btu/day).

Influent Heating

The heat required to raise the temperature of the raw manure influent to the digester operating temperature can be calculated by:

$$Q_I = W \cdot C \cdot (T_1 - T_i) \quad (4)$$

in which W = weight of influent added

C = specific heat of influent, (energy/weight-temperature)

T_1 = mixed-liquor temperature

T_i = influent temperature

The data used to calculate influent heating requirements are summarized in Table 6.

The theoretical energy requirements for influent heating were computed to be 6.2 kWh/day (21,231 Btu/day) at the 15-day SRT and 3.1 kWh/day (10,584 Btu/day) at the 30-day SRT.

DISCUSSION AND RESULTS

Pilot Plant Energy Balance

Total energy consumption in the pilot plant was much greater than total energy recovery. The total energy recovery rate from digesters

A and C was 36.5 kWh/day (Table 3). This was approximately 50% of average total energy consumption rate by these digesters (Figure 6). Similarly the total daily energy recovery rate from digesters B and D (32.9 kWh/day) was 47% of the total consumption rate (Table 3 and Figure 7). The energy required to heat the incoming manure and to maintain digester temperature was approximately equal to energy rate of the digesters.

Table 5. Digester heat loss equations.

(a) *Heat Loss Through Roof*

$$Q_r = A_r (T_1 - T_2) / \frac{1}{h_i} + \frac{1}{K_a} + \frac{x_1}{k_1} + \frac{x_3}{k_3} + \frac{1}{h_o}$$

in which Q_r = rate of total heat loss through the roof, (Btu/hr).
$\quad A_r$ = roof area, (ft³).

(b) *Heat Loss Through Walls Below Liquid Level*

$$Q_{w1} = A_{w1} (T_1 - T_2) / \frac{x_1}{k_1} + \frac{x_3}{k_3} + \frac{1}{h_o}$$

in which Q_{w1} = rate of heat loss through the walls below liquid level, (Btu/hr).
$\quad A_{w1}$ = wall area below liquid level, (ft³).

(c) *Heat Loss Through Walls Above Liquid Level*

$$Q_{w2} = A_{w2} (T_1 - T_2) / \frac{1}{h_i} + \frac{1}{K_a} + \frac{x_1}{k_1} + \frac{x_3}{k_3} + \frac{1}{h_o}$$

in which Q_{w2} = rate of heat loss through the walls above liquid level, (Btu/hr).
$\quad A_{w2}$ = area of wall above liquid level (ft³).

(d) *Heat Loss Through Floor*

$$Q_f = A_f (T_1 - T_2) / \frac{x_1}{k_1} + \frac{x_2}{k_2} + \frac{x_3}{k_3} + \frac{1}{h_o}$$

in which Q_f = rate of heat loss through the floor, (Btu/hr).
$\quad A_f$ = area of floor, (ft³).

(e) *Total Digester Heat Loss*
$$Q_L = Q_r + Q_{w1} + Q_{w2} + Q_f$$

The organic loading rates used in these experiments probably represent the low range of rates realistically achievable in order to maximize energy recovery. Gramms *et al.*, (7) reported successful digestion of swine wastes at loading rates of 3.84 g VS/1-day (0.24 lb VS/ft³-day) and a 10-day SRT. The suggested loading range for high-rate municipal sewage sludge digestion is 1.5 to 6.5 g VS/1-day at an SRT of 10 to 20 days. Organic loading rates twice as large as those used in the operation of digesters A and B may be achievable. If the same efficiency of methane production were assumed, rates of energy recovery would then have exceeded total energy consumption for each pair of digesters. Further investigation is needed to determine maximum achievable organic loading rates to maximize energy recovery.

Very little empirical data exist with which to quantify power or energy requirements of mixing devices for anaerobic digestion. Metcalf and Eddy (8) stated that typical power requirements for maintaining a completely mixed flow regime in aeration tanks vary from 13.2 to 26.4 watts/m^3 (0.5 to 1.0 hp/1000 ft^3). Energy requirements for mixed-liquor agitation proved to be a significant component of total energy consumption in the pilot plant (Figures 6 and 7). Cumulative power consumption for mixed liquor agitation in the pilot plant was 239 watts/m^3 for digesters A and C and 261 watts m/^{3} for digesters B and D. Consequently the data reported here probably represent maximum requirements and not necessarily the minimum energy requirements to maintain the same gas production efficiency. More information is needed to determine the minimum power requirements to maintain high-rate anaerobic digestion.

Table 6. Digester heating requirement data.

Materials Dimensions

Floor and roof diameter	=	5 ft
Total digester height	=	75 in
Liquid depth	=	54 in
Thickness of fiberglass, x_1	=	0.25 in
Thickness of plywood, x_2	=	0.75 in
Thickness of urethane, x_3	=	1.50 in

Conductivity Constants

Thermal conductivity of fiberglass, k_1 = 0.42 Btu/hr-ft-°F
Thermal conductivity of plywood, k_2 = 0.06 Btu/hr-ft-°F
Thermal conductivity of urethane, k_3 = 0.03 Btu/hr-ft-°F
Unit surface conductance of air inside the digester, h_i = 1.6 Btu/hr-ft^2-°F
Unit surface conductance of air outside the digester, h_o = 5 Btu/hr-ft^2-°F
Coefficient of conductivity of air and gas, K_a = 1.1 Btu/hr-ft^2-°F

Temperature Data

$T_2 = 95°F$
$T_1 = 45°F$

Influent Heating Data

$W = 337$ lb/day at a 15-day SRT $\left(\dfrac{505 \text{ gal}}{15 \text{ days}} \times \dfrac{10 \text{ lb}}{\text{gal}} \right)$

$W = 168$ lb/day at a 30-day SRT $\left(\dfrac{505 \text{ gal}}{15 \text{ days}} \times \dfrac{10 \text{ lb}}{\text{gal}} \right)$

$C = 1.0$ Btu/lb-°F
$T_1 = 95°F$
$T^1 = 32°F$

Theoretical and Actual Heating Requirements

Calculation of theoretical digester heating requirements indicated that energy requirements to replace heat losses from the digesters were much higher than energy requirements to heat the raw waste influent. At the 15-day SRT, theoretical raw waste influent heating requirements were 39% of total heating requirements while at the 30-day SRT they were 24% of the total. Minimization of digester heat losses with adequate insulation was known to be important during cold-weather operation, but in relation to heat loss due to cold-liquid influent, no experimental evidence exists on which to base design decisions.

Unfortunately, the empirical data on influent heating requirements could not be separated from measured energy consumption to replace heat losses. Therefore, the data for hot water heater energy consumption (Figures 6 and 7) reflect the total heating requirement, Q_T.

Theoretical heating requirements, Q_T, for each digester pair (A-C and B-D) were 28.7 kWh/day. This compares favorably with measured digester heating requirements of 36.7 kWh/day for the digester pair A-C and 31.8 kWh/day for the digester pair B-D. In addition to the heat loss from the digesters, wall-heat losses occurred from the hot water circulating pipes (a total of approximately 50 m in length) and from the hot water tanks. These losses were not considered in the theoretical calculations; this may account for the fact that theoretical heat losses were slightly lower than measured heat losses in both digester pairs. Heat-transfer theory appears to have provided satisfactory prediction of the digestion heating requirements.

IMPLICATIONS FOR LARGE-SCALE DIGESTION

If controlled anaerobic digestion is to be a viable energy-recovery process, the total energy value of the biogas produced must be at least greater than the energy consumed by the process. In addition, the economic value of this surplus biogas and the digester effluent must be greater than the capital and operating costs of the process. Preliminary results from this pilot-plant study have indicated that net-energy recovery did not occur. However, the digester operating characteristics indicated that the upper limit of biogas production had not been reached. The magnitude of net-energy recovery obviously depends upon the insulating quality of the digester walls during cold-weather operation. Although the raw-waste influent temperature (0°C) was probably a realistic minimum during the cold-weather conditions characteristic of Manitoba, the temperature outside the digesters would normally be lower than 7°C, at least for digesters built above ground level.

The effects of lower digester operational temperatures on process stability and net-energy recovery need to be investigated especially for cold-weather operation. Cassell *et al.* (9) studied the theoretical effects of several different tank materials and digester operational temepratures on the net-annual energy recovery from the digestion of dairy cattle wastes in the state of Vermont. They concluded that even well-insulated anaerobic digestion systems may not produce enough biogas from dairy manure to maintain the required digester operating temperature during the winter months.

A comparison of theoretical and actual digester heating requirements has indicated the validity of using rational conductive heat transfer theory to predict digester heating requirements. Thermal conductivity data, together with a knowledge of temperature changes in the air or soil surrounding the digester and in the raw-waste influent may be used to establish a realistic heat balance for a large-scale system. Empirical conductivity coefficients for digesters built both above and below ground level have been quantified (10).

Because of the marginal nature of energy recovery from anaerobic digestion, especially during cold-weather operation, process mixing requirements must be minimized. Further investigation is needed in order to quantify these minimum requirements.

SUMMARY AND CONCLUSIONS

Pilot-plant anaerobic digestion of swine manure began at the University of Manitoba in the late fall, 1974. Initial experiments conducted during winter operation of the four digesters have indicated the following:

1) Process stability was achieved despite adverse environmental conditions within the digesters including high pH (8.0) and high ammonia concentrations (3390-3450 mg/1);

2) Despite relatively high rates of methane gas production, only 50% of the energy expended was recovered through gas production at the loading rates used in these initial experiments; and

3) Rational conductive heat-transfer theory accurately predicted energy requirements for digester heating.

Although not all of the empirical data collected from pilot-plant or laboratory studies may be quantitatively applied to an assessment of large-scale digestion of swine wastes, empirical biogas production data together with heat-transfer theory may be used to assess the economic feasibility of the anaerobic digestion process for cold-weather energy recovery.

ACKNOWLEDGMENTS

The authors acknowledge the Biomass Energy Institute Inc., the Faculty of Agriculture at the University of Manitoba, Agriculture Canada and Shell Canada Limited for financial support of the project. Appreciation is also expressed to N. Parreno and J. D. Haliburton for their technical assistance during the experiments and to L. C. Buchanan, B. H. Topnik and R. H. Mogan for their technical assistance during the design and construction of the pilot plant.

REFERENCES

1. Imhoff, K. Digester gas for automobiles. Sewage Works J. 18 17-25 (1946).
2. Fry, J. L. Methane digester for fuel gas and fertilizer. The New Alchemy Inst., Woods Hole, Mass. (1973).
3. Singh, R. B. Bio-gas plant generating methane from organic wastes, Gobar Gas Research Station, Ajitmal, Etawah (U.P.), India (1973).
4. Standard Methods for the Examination of Water and Wastewater, 13th ed., Amer. Public Health Assoc., New York, N.Y. (1971).
5. Hach "Colorimeter" Methods Manual, 7th ed., Hach Chemical Corporation, Ames, Iowa. p. 124-125 (1971).
6. Kugelman, I. J. and K. K. Chin. Toxicity, synergism, and antagonism in anaerobic waste treatment processes. pp. 55-90. In "Anaerobic Biological Treatment Processes", F. G. Pohland, Ed., Advances in Chemistry Series 105, Amer. Chem. Soc., Washington, D.C. (1971).
7. Gramms, L. C., L. B. Polkowski, and S. A. Witzel. Anaerobic digestion of farm animal wastes (dairy bull, swine and poultry). Trans. Amer. Soc. Agric. Engrs. 14: 7-11, 13 (1971).
8. Metcalfe and Eddy Inc. Wastewater Engineering. McGraw-Hill, New York, N.Y. p. 519 (1972).
9. Cassel, E. A., R. N. Downer, J. C. Oppenlander, and J. H. Corazzini. Energy analysis of anaerobic digestion of dairy cow manure. Symp. on Use of Agric. Wastes, Regina, Sask. (in press). (1974).
10. Sewage Treatment Plant Design. WPCF Manual of Practice No. 8. Water Pollution Control Fed., Washington, D.C. pp. 222-226 (1961).

25.

Energy and Economic Analysis of Anaerobic Digesters

Tom P. Abeles*

The concept of anaerobic digestion for the stabilization of manure and the production of methane has been well documented by a number of sources (1, 2, 3). Most of the full-scale operations have been either in Germany or India (2). A variety of "bench top" digesters have been constructed in the United States which tend to confirm the feasibility of these systems for use on both farms and feedlots. One U.S. manufacturer is testing a prototype system for feedlots (4) and there is an operational digester for 350 cattle in Ludington, Michigan (5).

Because of the close agreement of other researchers with the result obtained by us on "bench top" models, it was decided that the time was appropriate to build a small field model (3-4 cows) system which could serve as a prototype for a large-scale system. Co-incidental with our decision, a local stable decided to build a full scale system which would handle the output of approximately 100 thoroughbred horses. The construction of both systems was begun in the summer of 1974. Because of the recent parts shortage full operation of both plants has been held up until the writing of this chapter. The full-scale system is currently operational and design work is being carried out for a dairy farm and a turkey raising operation.

---

* College of Environmental Sciences, University of Wisconsin-Green Bay, Green Bay, Wisconsin.

The close proximity of the two systems has permitted us to develop a research program with the following basic objectives:

1) A total systems study to determine the optimum operating system for use on midwest dairy farms.
   a) mechanics of operation
   b) energy optimization
   c) environmental impact

2) A sociological study to determine the acceptability of anaerobic digesters as a viable farm management technique.

3) The feasibility of the establishment of new types of agribusiness operation.

4) The impact of anaerobic digesters on the utility companies. This includes not only anaerobic digesters but combinations of other alternate energy systems such as solar, wind and fuel cells.

This chapter will present some of the work in progress and tentative conclusions which are under further investigation.

SYSTEMS STUDY

There are a significant number of small-scale farms in the state of Wisconsin. Many of these operations are marginal in the economic sense. As energy and pollution restrictions become tighter, these operations may be forced out of operation at worst. At best one will see consolidation which is also filled with unforeseen consequences.

Therefore, the first priority of our study was to determine the economic feasibility of a digester system for a typical farm of 200 dairy units (1 dairy unit = 1000 lb). The costs are based on the design of our prototype model. This includes a construction cost and energy balance study. We have not as yet completed a cost/benefit ratio for fertilizer but all our preliminary results coincide with the study done by ECOTOPE Group on the Monroe Honor Farm in Washington (6).

The ECOTOPE Group estimates the cost of a 175-head digester at $45,000. This falls within the range of the cost of our 100-horse digester (approximately $40,000) which was a prototype with some research and other proof-of-concept expenses. The actual price of our unit would be closer to $30,000, or less. This digester uses recycled brewery tanks (epoxy-lined steel). The cost would be comparable if a "Controliner" (Environetics Corporation, Bridgeview, Ill.) system were used as a primary digester. This system which uses "vinyl" type liners for both the bottom and top has been under study here for two years and is similar to the operation in Ludington, Michigan (5).

We find that there are several recurring base costs for all systems such as the tank or container for digestion, pumps and generators. Thus increased capacity for larger operations increases only modestly over smaller systems until more than one digestion tank is added. This again is borne out by the theoretical study by ECOTOPE.

When one starts to figure the cost/benefit ratio of the digester as an energy-producing system, the break-even-point appears to be about 250 animal units. In designing the 100-horse operation we found that we could not carry out our systems analysis on this premise alone. Because of land, zoning and health restrictions, the stable needed to install some form of waste management system such as a lagoon or silo-type storage container. This could have been either anaerobic or aerobic. If waste handlers are deducted from the cost of the digester a more favorable energy cost results. In fact, the digester has cut substantially the waste handling costs both in time and dollars for the horse stable.

As of this writing studies are being initiated to use the effluent from the digester as a nutrient base for aquaculture of carp (7) and/or algal production such as that proposed by Boersma at this conference or Dor (8). Also our small digester is being readied for retrofitting with solar panels. When enough data is in it may show that the energy return is only an incidental profitable by-product of a "well managed" farm waste management system.

One other point needs serious consideration in the energy cost/ benefit analysis. This is the rising price of fuels and the rising needs of energy-intensive farming operations as documented by several state studies and the USDA (9-12). Even with projected fuel cost increases, the digester as just an energy-producing system is expensive. As a "well managed" waste system it may prove even more economical. Even more pressing is the prospect of the loss of one of our fossil fuels, natural gas. Industries in Minnesota and elsewhere are being given warnings of cut-offs from this fuel in the near future.

Mechanics of Digester Operation

Interviews with local farmers indicate that farmers are short of help and do not have either the time or monies to become or to hire a sewage treatment plant operator nor do they wish currently to develop new alternatives to the use of spent sludge. Currently, we are working with a local engineering firm and irrigation company. Their analysis shows that the biogas plants can be fully automated and equipped with appropriate safety devices so that routine operations similar to automated free-stall operations are possible within the prices utilized above.

As in the case of the horse farm, many agricultural operations will be forced to go to some form of waste management system. This is particularly true in colder climates which prohibit daily disposal. There is a growing agribusiness which sells and services such systems routinely. The addition of the digester operation appears to be a neutral extension of a technology-oriented farm operation.

We are currently looking at two schemes whereby tractor operation on biogas may be feasible. The first of these involves the use of a gas compressor which operates in the range of commercial gas cylinders (approximately 240 psi). This would be used with quick connect tanks on large tractors. The other system would be low pressure and would deliver the biogas to select points on the farm via pipelines much like an irrigation system. This would allow in-field refueling.

Energy Optimization

A true energy balance study would involve not only the energy balance of the digester but also the energy units required to manufacture the system, to produce the feed for the animals and the return to the land of the spent slurry. We are developing a conceptual model to handle this type of analysis. Currently, the following are under consideration:

a) Addition of solar heating—Digester operation requires the use of high-grade biogas to maintain the optimum operating temperature of 95 degrees Fahrenheit. By the end of this summer our model digester will be equipped with solar panels to test the feasibility of reducing the quantity of biofuel to maintain the digester operation. This is in conjunction with a solar project now underway which will determine the feasibility of reducing energy consumption in agriculture by the use of solar energy for both heating and drying of agricultural products.

Theoretically, an electrical generator using the biogas could utilize the waste heat to maintain proper digester temperature. This has been verified by Maelstrom (5) which implies that proper sizing and design are crucial to the energy optimization.

b) Use of spent slurry—our preliminary results show that even though the digested manure has converted the nitrogen to the more usable nutrient form, its value as a fertilizer is not substantially enhanced unless it can be applied at the appropriate time and by the proper management technique which permits immediate use by the field crop. Thus the increased value of anaerobically digested manure seems doubtful.

The real return from the digested waste appears to be its conversion into a source for protein. Hamilton Standard (4) has recommended the drying of the waste and using it directly as a feed for cattle. Others, as mentioned above, are looking toward enhanced use of solar energy through growing of algae or fish on the slurry.

The algae production has another interesting side. Certain enzymes have the capability of producing hydrogen. By proper combination of enzymes with algae it appears possible to generate hydrogen. This has several side effects. First, the production of a protein rich in biomass and secondly the production of an energy source, hydrogen. There is serious talk of a hydrogen economy and there is quite a bit of both chemical and biological research on hydrogen production. Large-scale fuel cells with an overall efficiency of greater than 40% which run on hydrogen are a real possibility. Thus, new technologies can, in the foreseeable future, greatly change the systems analysis of digesters.

Environmental Impact

Both the preservation of our dwindling fossil fuels and the saving of fertilizers are the most obvious environmental impacts. What is not clear yet, though, is the effect of returning of digested slurry on improving soil tilth and thus cutting energy consumption by restoring organic matter to the earth. The restoration of a substantial organic matrix still needs study as does the restoration of micronutrients.

SOCIOLOGICAL STUDY

An important element in the development of anaerobic digesters for farm use is the public acceptance of a novel system. This implies not only that they understand the concept but that they can see that there is some return (in most cases immediate).

The summer of 1974 was used to survey a group of 100 farmers in Brown County, Wisconsin. The survey included a distribution of farms of various sizes and concerned itself primarily with the attitudes of farmers toward manure handling practices. The majority of the farmers were older which is typical of the northeastern Wisconsin region. The survey was done on the farm and was preceded by a letter explaining the purpose and the letter also had a small brochure describing the anaerobic process.

A summary of the results indicates that the farmers were quite independent and adamant about change. Most of the farmers did not see the need to change their manure handling procedures unless forced to by pollution abatement laws or an energy crisis which would make biogas essential to their survival. It is interesting to note that over 10% were actively interested in trying the process but only if they could see an operational system and could be assured that there would be adequate service and trouble-free operation.

When questioned about cost most farmers seemed unconcerned about a $40,000 price tag since silos and large tractor costs are similar.

As the survey proceeded the energy climate changed in the state in that farmers were becoming more concerned about rising fuel costs. Also, there had been increasing coverage in the local papers on the work on alternate energy sources being done at the University and across the country. As this public awareness grew, the interest in biogas also grew. Since the survey there has been an increasing interest on the part of the farmers. Many have called and asked to be kept abreast of the project. That this is the result of increased publicity seems obvious because most of the farmers have become even more interested in solar energy as a result of the national publicity. A follow-up study is planned in the near future to see if there is a substantial attitudinal shift.

ESTABLISHMENT OF NEW AGRIBUSINESSES

The study on the horse stable has shown that the digester does not get rid of the waste but in fact increases its volume due to dilution of the manure with water. We are currently exploring with a local fish farm the possibility of a leasing operation where the fish farmer would manage several aquaculture ponds utilizing digester effluent as a nutrient base. This points out a real need for a well established, diversified service industry which can support digesters. Not only must the farmer have readily available knowledge and service for his digester but since each farm is unique there must be a flexible service system which can optimize the digester by integrating it into the operation with minimal problems for the operator. In some cases the problem may be solved with a simple irrigation system; in many cases it may mean the addition of an auxiliary operation either by the farmer operator or as in the case of the horse stable, by a contractual relationship with another industry. The entrance of utility companies (next section) is a real and viable alternative.

UTILITIES AND BIOFUELS

We are currently working with two utility companies to try to assess the impact on various alternate energy systems on power company operations. The development of numerous biogas plants has a large impact on the natural gas and electricity consumption.

Solar energy also has a similar impact. One of the major issues is that of load leveling of power. If a number of biogas and solar plants are in operation and then fail, the power companies must supply the energy which, in the case of electricity, is not instantaneous.

The use of fuel cells and bioproduction of hydrogen have similar impacts on the energy industries. This leads to the serious concern of the extent of involvement of utilities in alternate energy projects.

One concept which would service several farms and possibly a small farm-community is the development of intermediate-scale biogas plants. The slurry could be pumped to a central plant for digestion. The plant could be solar-heated and possibly have some wind power also. The output of gas and electricity could be distributed as needed and the slurry returned to the fields by similar lines in an irrigation type of a system. This would be very appropriate in conjunction with fuel cells in the megawatt region which worked from biogas (methane) or hydrogen. Affiliated with this intermediate plant could be an algae industry, hydroponics or aquaculture. The advantages of medium-scale integration in terms of efficiency and ease of maintenance and monitoring are obvious.

Another route which is being explored is the construction of biogas plants, solar and wind systems by the utilities and a direct leasing of these systems to either the farmer or homeowner in much the same fashion that the phone company leases phones. Monitoring systems would enable the company to anticipate peak loads and system failures.

CONCLUSIONS

There appears to be little doubt that biogas plants are not only economically feasible for the average farmer but are feasible from an operational point of view. Work still needs to be conducted on standardization of system design and components and on the development of a viable service industry.

The possibility that utility companies will play an active role in this area is a real issue. There is a growing interest in the use of biogas from feedlots for neighboring communities. Several utilities are actively trying to enter this field as well as looking seriously at involving themselves in solar heating and cooling and fuel cells.

Because of the imminence of these alternate energy systems one is being confronted not only with a developing technology but also a real legal, political and social dilemma. It is time we started to consider the social implications before we find that technology has once again outstripped society and unknown potentials are unleashed without proper thought about the future of mankind. Energy is just one part of a complex ecosystem which includes both human and natural resources and we definitely need a systems study of the potentials for mankind both here and abroad.

* * *

This work has been supported in part by the National Science Foundation (GY 11445), the Office of Environmental Education (HEW - OEG-0-73-5447) and the Ford Foundation.

REFERENCES

1. Fry, J. L. Practical Building of Methane Power Plants for Rural Energy Independence. Standard Printing, Santa Barbara, CA. (1974).
2. Singh, R. B. Bio-Gas Plant Generating Methane From Organic Wastes. Bobar Gas Research Station, Ajitmal Etawah (U.P.), India. (1973).
3. Merrill, R. Methane Digestors For Fuel Gas and Fertilizer. New Alchemy Institute Newsletter, No. 3 (1973).
4. Pionzio, J. Personal Communication (1975).
5. Maelstrom, J. Personal Communication (1975).
6. ECOTOPE Group, Process Feasibility Study: The Anaerobic Digestion of Dairy Cow Manure . . . ECOTOPE Group, Seattle, WA. (1975).
7. Krishnamoorthi, K. P., et al. Production of Sewage Fertilized Fish Ponds, Water Research, 9:269 (1975).
8. Dor, I. High Density, Dialysis Culture of Algae on Sewage, Water Research 9:251 (1975).
9. Cervinka, V. et al. Energy Requirements for Agriculture in California, California Department of Food and Agriculture, January 1974.
10. Candill, C. E., et al. 1974 Texas Farm Fuel and Fertilizer Survey, Texas Crop and Livestock Reporting Service, Bulletin 116 (November 1974).
11. Gunkel, W. W., et al. Energy Requirements for New York State Agriculture, Part I: Food Production, Agricultural Engineering Extension Bulletin 405 (November 1974).
12. "The U.S. Food and Fiber Sector: Energy Use and Outlook", prepared by the Economics Research Service, USDA, G.P.O., Washington, (September 20, 1974).

26.
Dry Anaerobic Digestion

G. M. Wong-Chong*

In 1776, Volta (1) discovered the formation of a combustible gas which he related to the quantity of plant material in bottom sediments in lakes, ponds and streams. It was not until 1868 that Béchamp (1) defined the reaction as a microbiological process. This process, the biological decomposition of organic matter with the liberation of methane gas, is the anaerobic digestion process.

Despite the production of methane, with its recognizable value, the utilization of the anaerobic digestion process in waste treatment has been mainly justified on the following: a) sludge volume reduction, b) stabilization, and c) the production of a non-noxious, more acceptable product for final disposal. However, with the increasing demands for fuel, heretofore wastes are being re-examined as sources of energy and anaerobic digestion as a processing alternative.

Anaerobic digestion as of sewage sludge does not achieve complete decomposition; about 60-75% reduction in volatile solids is generally achieved. Sludge is fed to these digesters at about 5.0 to 10% solids, thus the residual sludge is about 3 to 6% solids which may undergo some treatment prior to disposal. In applying this technology to animal wastes the handling and disposal of the residual sludge could be the greatest deterring factor. This investigation examined the anaerobic digestion of animal wastes under relatively dry (solids concentrations greater than 20%) conditions in an effort to circumvent the problems of postdigestion treatment of digester supernatant, solids concentration and dewatering prior to disposal, and thus enhance the economics of the overall process.

* Assistant Professor of Agricultural Engineering, University of Hawaii, Honolulu, Hawaii 96822.

PROCESS FUNDAMENTALS

The anaerobic digestion process is the result of the activities of two groups of microorganisms—acid-forming and methane-forming bacteria. The acid-forming organisms, a mixture of facultative and obligate anaerobes, decompose organic matter with the production of short-chain fatty (volatile) acids). The methane bacteria, obligate anaerobes, metabolize these acids, CO_2 and hydrogen to produce methane. Equation 1 represents the overall anaerobic digestion reaction; details of the microbiology and the biochemistry of this sequence have been reported (1-5).

$$\text{Organic Matter} \xrightarrow{\text{Acid Formers}} \begin{array}{c}\text{Organic Acids} \\ +CO_2+H_2 \\ +H_2O\end{array} \xrightarrow[\text{Bacteria}]{\text{Methane}} \begin{array}{c}CH_4+CO_2 \\ +H_2O\end{array} \qquad (1)$$

The technology of anaerobic sludge digestion is essentially centered around a conventional and high-rate process (Figure 1), although several modifications of the latter exist. In the conventional process the charged material undergoes decomposition in a quiescent reactor where a highly liquid feed tends to stratify (Figure 1). This stratification reduces the contact between the active

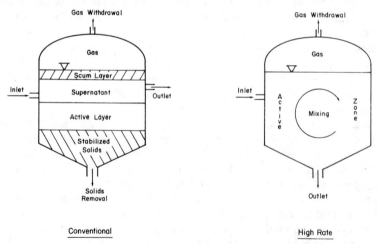

Conventional   High Rate

Figure 1. Digestion systems for wastewater processing.

organisms and the material to be decomposed, thus by agitation more effective digestion is achieved. This latter process modification is the high-rate process and is capable of performing at higher loading rates.

The proper operation of this process depends on maintaining a proper balance between the activity of the acid formers and that of the methane bacteria and any sudden increase in the formation of acid will disrupt this balance which may result in a decrease in the activity of the methane bacteria. To insure this balance, anaerobic digesters have been designed on loading rates, 0.04 to 0.10 lb volatile solids/ft³/day for conventional reactors and 0.15 to 0.40 lb volatile solids/ft³/day for high-rate reactors. Other design methods include a) volumetric loading (8) and b) the "rational approach" based on microbial growth kinetics in continuous culture (5, 9-11).

In the anaerobic digestion of sewage sludge there is a limit to the solids concentration in the feed material and this limit is the result of two factors: a) the ability to concentrate the sludge, and b) the handling of concentrated sludge. Conventional sludge thickening processes are capable of producing sludge concentrations of about 4 to 10% depending on the nature of the sludge (primary or activated). Sawyer *et al.* (12) reported that in high-rate digesters mixing becomes a problem when the solids content approaches 6%. Similarly at high solids concentrations there could be pumping problems in sludge transportation. However in applying anaerobic sludge digestion technology to municipal solid and agricultural wastes, having to dilute the feed material to solids concentrations around 5% results in a greater residual sludge volume which is generally a greater problem than the starting material and the economics of postdigestion sludge handling could make this mode of energy recovery uneconomical. The technology for sludge dewatering is well known (6, 13) and guideline economies have also been reported (6), showing this operation to be fairly expensive. On the other hand, the recovery of methane gas from municipal solid waste landfills (13-16) strongly indicates the possibility of a dry anaerobic digestion process. However, the reports of non-decomposition in landfills (17) and that of Schulze (18) strongly suggest the great need for further examination of the process before the benefits of omitting residual sludge dewatering and handling and supernatant treatment could be realized or the dilution of wastes with high solids concentration becomes an unnecessary step.

MATERIALS AND METHODS

All experiments were the batch type, carried out in 1.0 to 4.0-1 reaction flasks fitted with tubing to inverted graduated cylinders filled with CO_2 saturated water. The only exception was the batch fed reactor which was a 3.0 ft³ steel vessel. The gas from this reactor flowed through a wet test meter, Figure 2. After about 100 days of operation, this reactor was placed in a temperature control (85°F) chamber.

Both reactor feed and processed manure were analyzed for nitro-

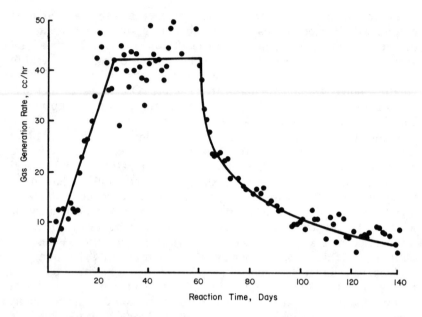

Figure 2. Schematic diagram of 3.0 ft³ batch fed reactor.

gen (Kjeldahl and ammonium), solids and pH. The analytical procedures were those described in Standard Methods (19) and
Prakasam *et al.* (20). Gas analysis was limited by CO_2 adsorption
in sodium hydroxide and combustion.

Inoculum

An initial inoculum was prepared by adding 200 ml of digester
supernatant (digesting dairy manure) to 6 lb manure. This initial
batch of digesting manure was subsequently (after 30 days) used as
inoculum for the large reactor (3 ft³) and other experiments.

RESULTS AND DISCUSSION

Methane potential in dairy manure

A batch reactor with 4.38 lb of fresh dairy manure inoculated with
5.0% weight of an active anaerobic sludge was observed. The initial dry matter (DM) content of the reactor was 20.8%; the final
DM was 16.5%. Figure 3 presents the gas production characteristics
of this reactor, showing an initial (almost linear increase in gas
production rate during the first 20 days, then a somewhat constant
rate for the next 40 days and a final (almost exponential) decline
in rate. A material balance on this reactor showed that over the

140 days 36.3% of the volatile solids (VS) were degraded and the gas generation rate was 11.3 ft³/lb VS destroyed with a composition of approximately 60 to 65% methane. There was an overall 8.5% loss in mass, an a temperature fluctuation from 70 to 82°F over the test period. The reactor pH was 7.4.

Halderson *et al.* (21) reported 55% reduction in VS and a gas production rate of 6.1 ft³/lb VS added (11.1 ft³/lb VS destroyed) for dairy manure. The differences in percentages reduction in VS could be due to differences in feed rations. A ration high in lignin will produce a waste similarly high in lignin. This lignin will constitute a portion of the VS and because of its very slow decomposition, under the best of conditions, there will be a corresponding low reduction in VS after digestion. Thus the VS characteristics are not a very definitive measure of the methane potential of a waste since lignin and many other inert materials will volatilize at the conditions (600°C) of the VS test.

Fresh vs. aged manure

Experiments were conducted to examine the relative potentials of fresh and aged dairy manure. The aged manure was partially composted. The reaction flasks were made up with a) 90 g fresh manure, and b) 35 g aged manure in 350 ml of digester liquor.

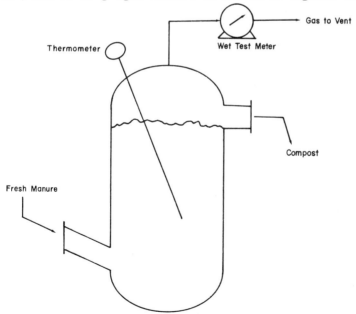

Figure 3. Gas generation characteristic of batch reactor, dairy manure at ambient temperature (75-82°F).

The fresh manure flask had an initial 5.0% solids content and the aged manure was 7.0% solids. In both reactors, pH remained about 7.3 to 7.4 throughout the test. Figure 4 presents the respective gas production characteristics. The fresh manure reactor produced 2.3 times more gas than the aged manure reactor.

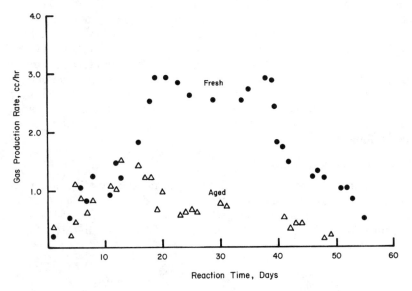

Figure 4. Comparison of gas production from aged and fresh dairy manure. Temperature 75-83°F.

The aged manure would have a high content of humus-like material, mostly lignin, and again, despite its high VS, the fraction available for bioconversion to methane was small.

Poultry manure

All attempts to digest poultry manure failed because of high ammonium concentrations. Ammonium levels of about 1.0% greatly reduced the gas production. In other experiments with dairy manure the ammonium nitrogen levels were generally low, about 0.1 to 0.2%. The effects observed at these levels of ammonium concentration coincide with those reported by McCarty (5).

Batch Fed Reactor

This reactor was a 3.0 ft³ steel vessel, Figure 2, and was operated to test operational procedures. This experiment was started with 71 lb of fresh manure seeded with a 5% (weight) sludge inoculum.

Figure 5 presents the gas production profile for this reactor. The initial 55 days was essentially a larger-scale check of the methane potential of dairy manure. In this initial stage, the gas production rate was 12.9 ft^3/lb VS destroyed.

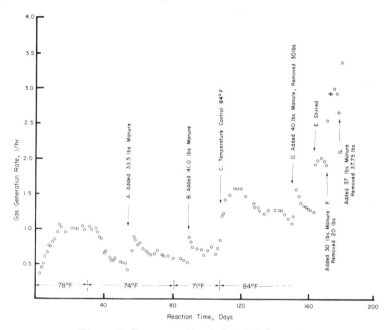

Figure 5. Gas generation in batch fed reactor.

The loading rates at Points D, F and G were about 0.25 to 0.40 lb VS ft^3/day and assuming a plug flow characteristic within the reactor, the effective residence time for loadings of 30 to 35 lb manure once per week would be about 28 to 35 days.

In earlier experiments free moisture was observed to accumulate in the bottom of the reactor and the top portion of the reacting mass appeared to be very porous, probably due to the fluffing action of the escaping gas. It was hoped that in the batch fed reactor this condition would be achieved, thus effectively providing *in situ* inoculation by the percolating moisture. Points A, B and D, Figure 5, are points where fresh manure was added to the reactor without prior inoculation. From all appearances significant inoculation was not achieved despite the instantaneous increases in gas production rate and further increase in gas production on "mixing" the reactor. Figure 5, Point E. Also observed was moisture accumulation at the top of the reaction mass when the reactor was opened (Figure 5, Point E), suggesting that the water formed was not percolating down the reactor. However, on subsequent openings, Points F and G, the reaction mixture was fluffy and no moisture accumulated at the top of the reactor, suggesting that with the improved activity in the bottom section of the reactory the reacting mass was fluffed enough to permit percolation of any free moisture down the reactor. The fresh manure added at Points F and G (Figure 5) were "stirred" into the reacting mass at the bottom of the reactor, rather than the plug addition at Points A, B, and D.

The sludge removed from the reactor, Points D, F and G, was essentially odorless with a hint of a petroleum odor. The nitrogen (TKN) content of this waste sludge was about 0.5 to 0.7%. The fresh manure fed to the reactor had nitrogen content of 0.5 to 1.0%. On oven drying of this sludge from 15-18% DM to 50% DM, the nitrogen content was improved to about 1.5-2.0% suggesting no significant loss in the course of drying, which is unlike aerobically composed manure where a significant amount of the nitrogen would be lost by ammonia volatilization. Further, when allowed to dry under ambient conditions, flies were not attracted to this sludge.

PROCESS COMPONENTS AND ECONOMICS

The major concern in designing a process for anaerobic digestion at high solids content, about 20%, is materials handling. This materials handling is essentially the input and removal of materials from the reactor and possibly some mixing of the fresh feed material with the reactor contents. Conceptually, the materials movement into and out of the reactor could be achieved by screw conveyors (augers) or a solids handling pump like the Moyno. The mixing could be achieved by a nonconfined screw auger. Figure 6 is a conceptual diagram of the process for dry anaerobic digestion. The feed-stream is shown entering the bottom of the reactor with the feed auger extended into the reactor, thus providing simultaneous feeding and internal mixing of the feed material with the reactor content. The reactor discharge could also be a screw type conveyor or a Moyno type sludge pump.

The process presented here is conceptual and I recognize that the area of materials handling and temperature controls needs further examination. However, the process offers numerous economies, the major ones being:

(a) possible energy saving in the use of low speed screw conveyors for material transport and mixing;

(b) a smaller digester volume, about 50% smaller than conventional with 15 days' retention;

(c) no digester supernatant for postdigestion treatment.

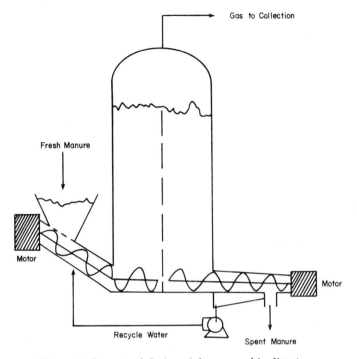

Figure 6. Conceptual design of dry anaerobic digester.

SUMMARY AND CONCLUSIONS

The existing anaerobic digestion technology was developed for municipal wastewaters, where the stabilization and reduction of waste sludge were the principal objectives. The digestion process for these sludges was operated at solids concentrations about 5%; the major limitations being a) the ability to economically concentrate the sludge, and b) the energy demands for continuous mixing and pumping of the sludge. However, in anaerobic digestion of high solids wastes such as animal manures for energy recovery, the problems associated with diluting the waste would be deterring economic factors. Some of the problems associated with diluting high solids wastes are a) postdigestion sludge dewatering, and b) treatment of digester supernatant.

This investigation examined the anaerobic digestion of animal wastes (dairy and poultry) at relatively high solids concentrations (>20%) in both batch and batch fed reactors at ambient temperatures (70-85°F). From the information gathered the following conclusions were made:

(a) From fresh dairy manure 11.3 to 13.0 ft^3 of digester gas was generated per pound of volatile solids destroyed. Methane composition of the gas was 60 to 65%.

(b) There is a limit to the amount of volatile solids in a waste which is convertible to gas.

(c) Fresh manure has the greater gas potential than aged manure.

(d) Ammonia inhibition to methanogenesis would be significant with highly nitrogenous wastes such as poultry manure.

(e) Anaerobic digestion of wastes with high solids concentrations is feasible and offers economies in reduced reactor volume, digester sludge handling, and avoids treatment of digester supernatant.

ACKNOWLEDGMENT

The author wishes to acknowledge Miss Mae Takemura for her able assistance in carrying out the chemical analysis in the course of this study.

* * *

Published with the approval of the Director of the Hawaii Agricultural Experiment Station as Journal Series Paper No. 1873.

REFERENCES

1. Barker, H. A. 1956. Biological formation of methane. *In* Bacterial Fermentation, John Wiley & Sons, New York.
2. Buswell, A. M. and H. F. Mueller. 1952. Mechanism of methane fermentation. Industrial and Engineering Chemistry, 44(3):550-552.
3. Burbank, N. C. Jr., J. T. Cookson, J. Goeppner, and D. Brooman. 1956. Isolation and identification of anaerobic and facultative bacteria present in the digestion process. Int. Jour. Air and Water Pollution, Vol. 10, pp. 327-342, Pergamon Press, New York.
4. Kortze, J. P., P. G. Thiel, and W. H. J. Hattingh. 1969. Anaerobic digestion II. The characterization and control of anaerobic digestion. Water Research, Vol. 3, pp. 459-494, Pergamon Press, New York.
5. McCarty, P. C. 1964. Anaerobic waste treatment fundamentals. Public Works Journal, Vol. 95, No. 9, pp. 107-112; No. 10, pp. 123-126; No. 11, pp. 91-94; No. 12, pp. 95-99.
6. Burd, R. S. 1968. A study of sludge handling and disposal. U.S. Dept. of the Interior; Fed. Water Pollution Control Administration, WP-20-4.
7. Anaerobic sludge digestion, Water Pollution Control Federation Manual of Practice, No. 16, 1968.
8. Rich, L. G. 1963. Anaerobic digestion. *In* Unit Processes of Sanitary Engineering, John Wiley & Sons, New York.
9. Lawrence, A. W. 1971. Anaerobic biological waste treatment systems. Proc. Cornell Univ. Agricultural Wastes Management Conf., Syracuse, New York.
10. Design of facilities for treatment and disposal of sludge. In Wastewater Engineering, Metcalf & Eddy, Inc., McGraw-Hill, 1972.
11. Andrews, J. F. and E. A. Pearson. 1965. Kinetics and characteristics of volatile acid production in anaerobic fermentation processes. Int. J. Air and Water Pollution, Pergamon Press, Vol. 9, pp. 439-461.

12. Sawyer, C. N. and J. S. Grumbling. 1960. Fundamental considerations in high-rate digestion. J. San. Eng. Division, Am. Soc. Civil Engrs. SA2, pp. 49-63.
13. Rovers, F. A. and G. J. Farquhar. 1973 Infiltration and landfill behavior. Jour. Environ. Eng. Div., Am. Soc. Civil Engrs., EES, pp. 671-690.
14. Dair, F. R. and R. E. Schwegler. 1974. Energy recovery from landfills Waste Age, 6-10, April/May.
15. Stone, R. 1974. Methane reclaimed from an old landfill. Public Works, pp. 70, May.
16. Eliassen, R. 1942. Decomposition in landfill. Amer. Jour. Public Health, Vol. 32, p. 1029.
17. Herig, R. and S. A. Greeley. 1921. Collection and disposal of municipal refuse. McGraw-Hill Co., New York.
18. Schulze, K. L. 1958. Studies on sludge digestion and methane fermentation. I. Sludge digestion at increased solids concentration. Sewage and Industrial Wastes, 30(1)28-45.
19. Standard Methods for the Examination of Water and Wastewater. 12th Ed. APHA, AWWA and WPCF.
20. Prakasam, T.B.S., E. G. Srinath, P. Y. Yang, and R. C. Loehr. 1972. Evaluation of methods for the analysis of physical, chemical and biochemical properties of poultry wastes. American Soc. Agricul. Eng. Meeting, Chicago, Illinois.
21. Halderson, J. L., A. C. Dale, and E. J. Kirsch. 1973. Anaerobic digester response with dairy cow manure substrate. Amer. Soc. Agricul. Engrs. Paper No. 73-4532.

27.

Technologies Suitable for Recovery of Energy From Livestock Manure

C. N. Ifeadi and J. B. Brown, Jr.*

The quantity and character of livestock wastes that can be collected and processed for the recovery of synthetic fuel are important factors in assessing the technical and economic suitability of any waste-to-fuel process.

Efforts to recover fuel from manure at a facility must be based on good operational practices that will generate sufficient quantities of manure with minimal inert-material contamination and pretreatment requirements. However, to date, typical farming practices in the United States have not resulted in the generation of quantities of manure from an economically viable energy recovery process. The collection of manure from open fields even for large livestock operation is not economically feasible, and the quantity from small farms, with few animals, is too small for any energy recovery process, even when the animals are in confined areas. Furthermore, some farming operations use unpaved lots where the manure is trampled and mixed with soil particles, which makes the available manure unusable. Some management practices also produce high-moisture manure that requires extensive dewatering prior to the recovery process.

* Research Scientist and Associate Manager, respectively, Waste Control and Process Technology Section, Batelle's Columbus Laboratories, Columbus, Ohio.

Because of recent advances in agricultural sciences and technology, however, large numbers of modern livestock operations can meet some of the characteristics necessary for economical fuel recovery from livestock manures. The sizes or weights of animals in farms are increasing. Larger numbers of livestock are being raised in confinement lots. It is conceivable, therefore, that a well-engineered energy recovery process that is tied into a modern livestock operation will not only provide a valuable source of supplemental low-sulfur fuel but also help to solve some of the waste management and pollution problems.

In this paper, thermochemical and biological conversion processes are discussed with specific analyses presented for anaerobic digestion and pyrolysis. For anaerobic digestion, technical and economic considerations are presented for manure generation rates of 0.1, 1, and 10, and 100 tons/day dry basis. For pyrolysis, the technical and economic considerations are reviewed and technical and economic feasibilities are extrapolated to the 0.1 to 100-ton/day range. The lower ranges reflect the small-to-medium livestock operations, while the higher ranges are the large-scale operations. Recovery processes were evaluated for the practical farmer in dairy, beef, and hog production.

U.S. LIVESTOCK-MANURE GENERATION

An analysis was made of manure generated nationwide from beef, dairy, and hog operations. On the basis of a USDA inventory (1), an average number of animals per farm was calculated for each state. Manure production rates (2) per day at the average farms were then calculated. Figures 1 through 3, respectively, show the number of dairy, beef, and hog farms generating various quantities of manure per day. Although some information is lost in the averaging process, demonstration of the most probable number of farms at various manure-generation rates is useful. A fuel recovery process at a given manure-generation rate that is evaluated to be suitable with the largest number of farms reveals such a process to have a promising market. Figures 1 and 2, showing dairy and beef manure-production rates, respectively, reveal that most farms generate manure at a rate between 0.05 to 0.15 ton/day (dry basis). This value gives about 17 heads per farm for the dairy, assuming 1200 pounds live weight per animal, and 32 heads per farm for beef, assuming 900 pounds live weight per animal.

In Figure 3, the hog-manure-generation rates, as expected, are smaller. The most probable hog-farm sizes generate quantities of manure ranging from 0.02 to 0.06 ton/day (dry basis). None of the

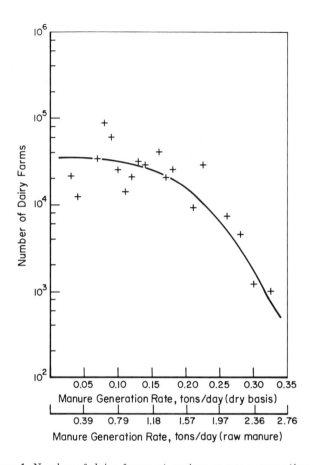

Figure 1. Number of dairy farms at various manure generation rates.

average manure generation was up to 0.1 ton per day, which indicates that the average farm contains less than 238 hogs, assuming 140 pounds live weight per hog.

It is pointed out that with present farming practices, there are practically no dairy and hog farms in the U.S. generating more than 10 tons of manure/day (dry basis). However, in beef production, the generation of more than 10 tons/day is not uncommon. The data presented in Figures 1 through 3 will, of course, be substantially lower because of some operational practices (*e.g.*, open feedlot, unpaved floors) which seem to lower the quality and quantity of collectable manure.

GENERAL REVIEW OF FUEL RECOVERY PROCESSES FROM ANIMAL WASTES

A number of different processes for recovering synthetic fuel from animal manure are in varying stages of development. For ease of discussion, these may be classified into two groups: biological and thermochemical.

Biological Processes

In biological processes, microorganisms consume nutrients present in wastes to increase their own biomass, and consequently release various gases and/or simple carbohydrate materials. Biological processes can utilize a wide spectrum of candidate microorganisms capable of consuming—and thus disposing of—animal

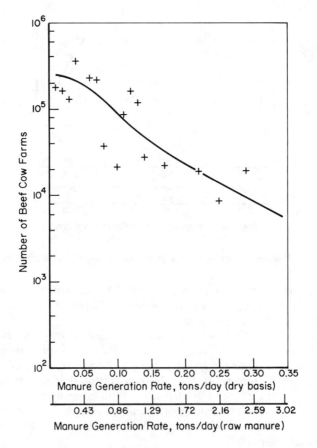

Figure 2. Number of beef cattle farms at various manure generation rates.

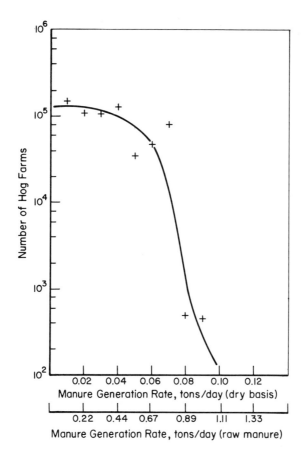

Figure 3. Number of hog farms at various manure generation rates.

waste materials. There are two main classes of biological processes: 1) biogas, or anaerobic digestion (fermentation) and 2) biochemical, or enzymatic hydrolysis. These are shown schematically in Figure 4. While biochemical processes are still at the bench stage of development, the anaerobic digestion process has been studied at pilot levels. The biochemical processes which produce protein and alcohol are not considered in this study. Rather, it is planned to focus on the biological alternative—anaerobic digestion (fermentation).

Methane is produced by anaerobic decomposition (fermentation)— a process which takes place under oxygen-deficient or chemically reducing environment. Although the exact chemistry of anaerobic digestion of organic matter is not well understood, consecutive but simultaneous phases of the process shown in Figure 5 can be defined: solubilization, acidification, and methanation.

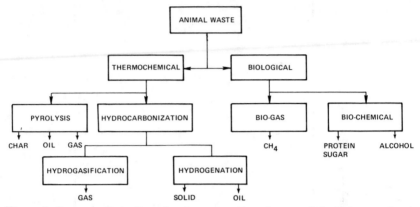

Figure 4. Process alternatives for the generation of renewal fuels from animal manures.

Solubilization is generally believed to be the initial step in the process which may or may not involve bacterial action. At this stage, the complex organic materials consisting of proteins, carbohydrates, and fats are solubilized. Acidification is the second step in which bacteria reduce the soluble organic materials to soluble simple organic acids. In the final step, methane bacteria reduce the organic acids and certain other oxidized compounds primarily to methane and carbon dioxide. The production and loss of methane and carbon dioxide cause the stabilization of the animal waste. It has been estimated that each cubic meter of methane produced at standard temperature and pressure (STP) removes 2.9 kg of COD or BOD (4) (1000 cubic feet of CH_4 at STP removes 178 lb COD or BOD). Therefore, besides the production of methane for fuel, the anaerobic digestion process can also be used to stabilize livestock manures. While anaerobic digestion has long been used by municipalities in the sewage-sludge disposal, its application to the disposal of animal waste has not been widely practiced. In both bench and small pilot studies, methane production from animal wastes has been successfully demonstrated. A logical extension of this present study is an evaluation of the process for practical farm use.

Thermochemical Processes

The thermochemical processes may also be subdivided into two classes: hydrolysis and hydrocarbonation. Depending on the operating conditions, pyrolysis may yield solid (usually called char), oil, and gaseous fuels. Hydrocarbonation may be further sub-classified into hydrogasification and hydrogenation. Hydrogasification yields gaseous fuels, while hydrogenation yields oils and solid fuels.

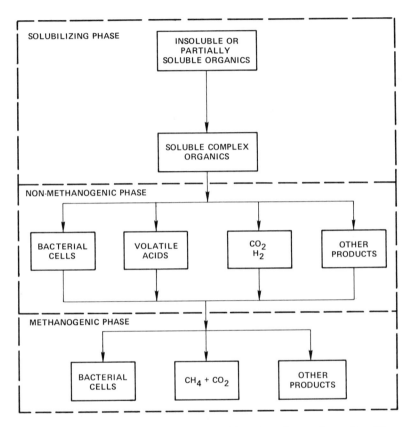

Figure 5. Overall scheme for anaerobic digestion of organic matter (3).

These thermochemical processes have successfully been demonstrated in the laboratory for livestock manure, as briefly described below. The greatest handicap in the application of these processes is the large amounts of moisture which must be removed for the processes to be feasible. Consequently, they hold greatest promise in the treatment of manure initially low in moisture content. Such manure may be found in open beef-cattle feedlots in the southwestern United States where the solar energy substantially reduces the moisture content of the manure. The quantity of manure available for processing is another important factor. Review of the published work shows that while small pyrolysis plants may be feasible at under 100 tons/day (4), hydrogasification and hydrogenation processes require well above 100 tons/day to be promising (5, 6).

In the pyrolysis method, the waste is chemically decomposed in a closed system by means of heat. The waste is thereby converted to fuel gas, oil, char, and water containing some dissolved organic compounds. Pyrolysis processes were principally designed for solid-waste disposal and have not been optimized for synthetic fuel recovery. Costs of the processes vary with plant size, the type of waste processed, and the credits for recovered fuels or salable materials. Changes in operating temperatures and pressures, the use of steam or oxygen, and a shift from gas to liquid fuels to generate process heat may change the economics of the process and increase the synthetic-fuel return.

In the hydrogenation process, the manure is heated under pressure in the presence of carbon monoxide, steam, and a catalyst, also in a closed system. The hydrogenation of cattle manure produced a heavy, largely paraffinic oil with a heating value of 14,000 to 16,000 Btu/lb. The oil has a sulfur content of less than 0.4%, which is much less than that of most commercial fuels. The yield in the hydrogenation process is about three barrels of liquid fuel per ton of dry organic waste (7).

The hydrogasification process for manure is patterned after the Hydrane Process (8) developed by the Bureau of Mines for transforming unpretreated coal to pipeline gas. The Hydrane Process is a direct hydrogasification process generating most of the methane by the direct reaction of carbon in the feedstock with hydrogen, rather than by the well-known catalytic methanation reaction, $CO + 3H_2 \rightarrow CH_4 + H_2O$. For conversion of animal waste, manure can be reacted with hydrogen under a variety of conditions to produce a variety of hydrogasification products. The hydrogasification plants can consist simply of the hydrogasification reactor into which manure and hydrogen are fed. The study (5) pointed out that the hydrogasification reactor will operate at about 1000 psig because high pressure favors methane formation, lowers investment in the gas-purification equipment, and reduces compression costs for introducing the final product gas into the natural-gas transmission line. The hydrogasification reaction is exothermic, and the heat in excess of that required to bring the manure and hydrogen to reaction temperature and that which is lost can be used to generate steam. Also, energy for the generation of plant steam requirements and for drying the raw manure can be obtained at least in part from the combustion of char from the hydrogasifier.

TECHNICAL, ECONOMIC, AND ENVIRONMENTAL ASSESSMENTS

On the basis of a preliminary analysis, two processes, anaerobic digestion and pyrolysis, appear to be the most inviting energy-recovery processes within the manure quantities, 0.1 to 100 tons/ day, selected for this study. The technical, economic, and environmental aspects of these two processes are explored below. It must be recognized that because of the present status of development of these processes, some of the discussions need refinement through pilot field studies. Furthermore, while a detailed economic analysis of anaerobic digestion is provided, only brief descriptions of factors are given for pyrolysis, the detailed analysis being left for future studies.

Technical Evaluation

Methane Production Process. Figure 6 is a schematic of a conceptual single-phase methane-generation system. It consists of raw-manure handling and feed preparation, the digester, gas treatment and storage, and effluent disposal.

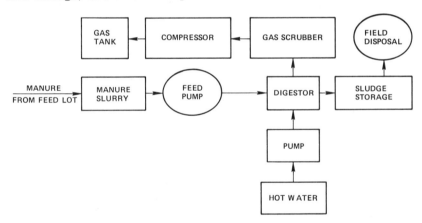

Figure 6. Single-phase anaerobic digestion for methane production.

Manure and Feed Preparation. Raw manure is usually high in solid content and, therefore, will require dilution to a pumpable slurry before being fed into the digester. Various slurry solid concentrations ranging from 4 to 10% solids content have been digested (9-12). The feed rate or loading rate is usually based on the volatile solids or organic-matter content of the manure. Since volatile-solids concentration varies from one type of livestock to another, so will the loading rates of the manure to the digester. Furthermore, for a particular livestock manure, the loading rate varies with process systems used and the operating condition, such as temperature and retention time. The two-phase digester has the potential benefit of smaller digester volume requirement than that for the single-phase digester because of higher loading rates (13). Also, the thermophilic digestion has higher loading rates, shorter retention times, and greater digestion of the organic matter, hence higher production of methane than the mesophilic digestion (12).

The analysis in this study is based on the less favorable condition of single phase of a conventional high-rate design under a mesophilic digestion.

Digester. The digester or reactor for the methane production may be a specially fabricated steel or concrete vessel. The loading rates, which have been successfully demonstrated for hog, dairy, and beef cattle, vary from 0.1 to 0.3 lb of volatile solids/ft^3/day at 95°F (1.6 to 4.8 g/l/day at 35°C(9-11). The retention time varied from 10 to as much as 30 days.

Although heating may not be necessary in the warmer climates, the mesophilic digester may require external heat during cold weather. The optimum pH range of methane bacteria activity is 6.6 to 7.6, and below pH 6.2, the acid conditions may be toxic to methane bacteria. Because manures are rich in nitrogen, phosphorus, and other chemicals, nutrient supplements are generally not required for the digestion of livestock manure.

Other variables known to affect the retention time include mixing and seeding. Mixing is observed to prevent formation of an undesirable scum on the surface of the digester contents; also it tends to bring bacteria into more immediate contact with the organic matter while at the same time, keeping the temperature in the digester more uniform (9). Seeding an anaerobic digester with sludge from an active digester significantly reduces the time it will otherwise take the microbes to multiply into a large, efficient methane-producing population.

Since methane bacteria are strict anaerobes, the tank must be free from oxygen. The presence of small amounts of oxygen will change the system from a reducing to an oxidizing environment, which will destroy the methane formers. The tank is usually equipped with a floating cover to maintain the system at constant pressure. Usually, the digestion process is semicontinuous. A fraction is removed and added daily, the amount of which is based on the retention time. Continuous, integrated operation is, of course, feasible.

Methane Production, Treatment, and Storage. The methane produced varies in quantity and quality. From the foregoing it is clear that the total quantity of methane production will vary, depending on the organic material (type of animal manure) being digested, the rate of loading, and the environmental conditions in the digester. Table 1 shows the methane production rate for a digester temperature of 95°F (35°C) and a retention time of 30 days with loading rates between 0.1 to 0.3 lb of volatiles/ft³/digester volume/day Hog manure gives the highest methane production rate, almost twice that given by beef manure. The next highest production comes from dairy manure, and the lowest from beef manure.

Table 1. Methane production rates.

Livestock, (1000 Pounds Live Body Weight)	Methane Production Range, ft³/day (at 95°F)	Methane, percent	Reference
Dairy	42 - 60	60 - 80	11, 14, 15
Beef	30 - 36	60 - 80	15, 16
Hog	29 - 100	55 - 75	8, 10, 15

The gas produced contains other gas components besides methane. Major impurities include CO_2 and small quantities of H_2S and H_2O. Methane content varies from 55 to 80% and CO_2 from 20 to 45%. As shown in Table 1, methane produced from hog manure seems to have a lower methane fraction. Methane in a mixture of CO_2, H_2O, and other impurities is usually regarded as a low-Btu with a heating value of about 600 Btu/ft³. Therefore, to increase the heating value in addition to controlling potential pollution problems created by the presence of H_2O while burning the gas, the methane must be purified. Hydrogen sulfide can be removed by passing the gas through iron-impregnated wood chips, moisture can be removed by dry glycol, and CO_2 can be removed by an alkali. Alternatively, certain acid-gas removal processes will remove both CO_2 and H_2S.

After purification, the methane must be stored under pressure and ready for use. Carbon-steel pressure vessels or vessels made from other materials may be used.

Effluent Disposal. During the operation of the digester system, water and nondigestible and undigested matter are removed at a rate equal to the feed rate. As much as 50% or more volatile-solids reduction may be achieved by the digestion process. Part of the effluent removed daily may be recycled and part may be disposed of directly or treated before final disposal. Under the farm situation, the effluent for disposal may be collected in lagoons from where it may be spread on the field.

Pyrolysis. Figure 7 shows the several types of pyrolysis reactor types. Because the technology of manure pyrolysis is undeveloped, it is difficult to make comparative statements of the various types of reactors. At least five commercial plants for pyrolysis of refuse are currently under construction; it can be expected that data emerging from these operations will suggest which processes and

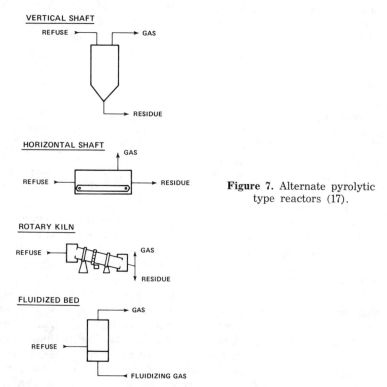

Figure 7. Alternate pyrolytic type reactors (17).

operating conditions are best suited for economical and technically feasible operation, at least for refuse-type waste. Among all the reactors, the vertical shaft reactors are conceptually the simplest and lowest in capital cost. In the vertical type, manure is fed into the top of the reactor and settles into the reactor under its own weight. Generated pyrolysis gases pass upward through the shaft and are removed from the top. Typical feed mechanisms include screw containers, rotary devices, and rams. Also, a residue discharge device and gas take-off manifold must be provided.

Usually, two heating methods can be employed: direct and indirect. Direct refers to a method whereby heat is supplied to the

reaction mixture by partial combustion of the waste and/or supplementary fuel within the reactor. Oxygen must be supplied in
the direct case and the reactor product gas contains significant
amounts of CO_2 and H_2O with a penalty in heating value. Indirect
heating methods are provided by a heating zone separated from
the reactor insides. Manure heating is then provided by heat con-
ducting through a vessel wall.

Process variables associated with the pyrolysis reactor include
temperature, residence time/heating rate, and residue feed conditions. Decomposition of manure commences at approximately 360°F.
At the lower temperatures, primary reactor products are a gas
phase rich in higher-molecular-weight hydrocarbons with a liquid
and solid phase also present. At the higher temperature ranges (2000
to 3000), the reactor products consist primarily of gas and slag
phases.

Residence times can usually range from less than 1 hour to up to
about 6 hours. These times are usually defined in terms of the time
required to bring the residue particles to a uniform described temperature. Feed conditions for processing animal manures usually
favor a dried, finely shredded feed.

The generalized pyrolytic reactor process is depicted in Figure 8.
Properly dried manure is stored in a covered weatherproof storage
bin with an output connection to a feed auger. The auger feeds as-
received manure into a shredder which shreds and feeds the reactor. In addition to the manure feed, the reactor burners receive
the proper mixture of fuel and air during reactor operation. An
exit augur provides a path for the solid products and gases to a
separator. Separated solids are cooled and placed in char storage,
and the gases and oil are passed into a gas scrubber and pump to
oil storage. Scrubbed gases receive further purification and are
recycled to assist in primary reactor heating.

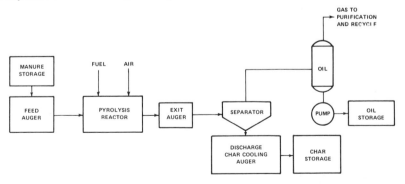

Figure 8. Schematic of pyrolysis process for conversion of manure to synthetic
fuels.

Economic Evaluation of Methane Production by Anaerobic Digestion

Where available, the economic evaluation of the process is based on data obtained by the developers of the process. In most cases, however, this approach was not possible because the equipment used was not within the capacity ranges of this study, which invalidated the use of cost-capacity concept of the six-tenths factor rule (18). When this was the case, capital costs of the major equipment making up the larger fraction of the capital costs (capital-intensive items) are developed on the basis of current equipment prices using process engineering principles. This is a process of accounting for the major equipment and other indirect costs for each unit process of the system together with the system operating costs.

Detailed analyses of the cost-intensive items are made on the basis of the design data shown in Table 2.

Table 2. Anaerobic digestion design parameters for livestock manures.

	Dairy Cow	Beef Feeder	Swine Feeder
Waste Characterisitcs			
Manure Production (lb/day/1000 lb live weight)	82	60	65
Total Solids (TS), percent manure produced	12.7	11.6	9.2
Volatile Solids (VS), percent TS	82.5	85	80
Weighted Live Weight, average lb/animal	1,200	900	140
Farm Sizes			
(1) 0.1 Ton TS/Day			
Total Number of Animals on the Farm	17	32	238
Total Farm Volatile Solids, lb/day	165	170	160
Loading Rate, lb VS/cu ft	0.155	0.16	0.15
Digester Capacity, cu ft	1,065	1,063	1,066
(2) 1.0 Ton TS/Day			
Total Number of Animals on the Farm	170	320	2,380
Total Farm Volatile Solids, lb/day	1,650	1,700	1,600
Loading Rate, lb VS/cu ft/day	0.155	0.16	0.15
Digester Capacity, cut ft/farm	10,645	10,625	10,667
(3) 10 Ton TS/Day			
Total Number of Animals on the Farm	1,700	3,200	23,800
Total Farm Volatile Solids, lb/day	16,500	17,000	16,000
Loading Rate, lb VS/cu ft/day	0.155	0.16	0.15
Digester Capacity, cu ft/farm	106,452	106,250	106,666
(4) 100 Ton TS/Day			
Total Number of Animals on the Farm	17,000	32,000	238,000
Total Farm Volatile Solids, lb/day	165,000	170,000	160,000
Loading Rate, lb VS/cu ft/day	0.155	0.16	0.15
Digester Capacity, cut ft/farm	1,064,516	1,062,500	1,666,666

Capital Costs of a Single-Phase Anaerobic Digestion. The equipment costs for each unit process shown in Figure 6 were estimated on the basis of published data, and then updated using the latest cost index factor (February, 1975). It must be emphasized that equipment costs provided are best engineering estimates, which may vary by 25% or more.

The total capital investment was obtained by summing the equipment costs and the following indirect capital cost items (18).

Indirect Cost Item	Charges
Installation	40% of equipment cost
Piping and Insulation	15% of equipment cost
Engineering & supervision	10% of equipment cost
Electrical services	10% of equipment cost
Contractors fee	5% of equipment cost
Contingency	5% of equipment cost

The total capital costs for the four plant capacities are given in Table 3.

Two major cost-incentive items are evident in Table 3. They are the digester or reactor tank and the methane storage tank. For example, the digester cost varies from about 52% of the total equipment costs for the 0.1 ton/day capacity to as much as 73%

Table 3. Equipment costs (dollars) for various plant capacities, ton/day (dry basis).

Process System	0.1	1	10	100
(1) Animal waste slurry carbon steel, mixer steel, including cost of driving units[a]	500	1,800	5,000	16,000
(2) Feeding[b]	—	—	—	—
(3) Digester with Mixer[c]	5,100	20,000	80,000	440,000
(4) Gas Cleaner[d] Iron Impregnated with Wood Chips	500	1,000	3,000	10,000
(5) Compressor[e]	1,000	1,500	2,000	5,500
(6) Methane Storage Tank[f]	1,000	1,500	10,000	100,000
(7) Hot Water Tank	1,000	3,000	12,000	25,000
(8) Hot Water Pump	750	1,300	2,600	5,000
Total Equipment Cost (Direct Cost)	9,850	30,100	114,600	601,500
Total Fixed Capital Cost (Direct + Indirect Costs)	18,200	55,700	212,000	1,113,000

(a) 5% solid content of slurry.
(b) Gravity feeding at 0.155 lb VS/cu ft/day.
(c) Capacity for waste material for 1 day — pressure tank at 15 psi.
(d) Cartridge-type filter at STP with wood chips impregnated with iron. Assuming 1% of sulfur content appears in the biogas.
(e) Single-stage reciprocating compressors.
(f) Storage tank at 15 psi.

for the 100 tons/day capacity. The effect of gas storage is not so severe but it is still significant. It is clear that cost reduction in capital equipment for the anaerobic digestion must concentrate on reducing the cost of the digester and storage tank, probably through the selection of cheaper materials of construction.

Operating Costs of Single-Phase Anaerobic Digestion

The total operating costs are estimated by summing labor requirements and the following items (18):

Operating Cost Item	Charges
Maintenance and repair	5% of capital invested
Depreciation	5% of capital invested
Tax and insurance	0.5% of capital invested
Interest	4% of capital invested

The cost estimates were chosen to reflect the type of process and complexities of instumentation adaptable to the farm situation. The total operating costs are shown in Table 4. The indirect capital

Table 4. Operating costs (dollars) for various plant sizes.

	Plant Capacities, tons/day (dry basis)			
Items	0.1	1	10	100
Raw Materials[a]	—	—	—	—
Operating Labor[b]	7,300	10,850	14,600	30,000
Maintenance and Repair	910	2,785	10,600	55,650
Depreciation	910	2,785	10,600	55,650
Interest	728	2,228	8,480	44,520
Tax and Insurance	91	279	1,060	5,565
Utility[c]	197	1,970	19,700	197,000
Total Operating Costs/Year	10,136	20,897	65,040	388,385

(a) Animal manure has no cost to the processor.
(b) 4 hours/day at $5/hour.
(c) Includes steam for heating the digester. Assuming 20% of total methane sale.

cost factors and those of the operating costs vary, depending on the plant location and other factors. As will be expected, the operating labor and maintenance and repair are the major cost-intensive items, and consequently will demand attention in any cost-reduction efforts.

Analysis of Costs and Credits Due to Anaerobic Digestion

Credits may be taken for the sales or use of methane produced and for handling and disposal of the digester portion of the raw manure which the farmer should have paid. On the other hand, cost of treatment and disposal of the undigested residue must be added to the capital and operating costs of methane production.

Table 5 shows that at the delivery price of $2/mm on Btu, the anaerobic digestion of dairy manure, for example, was profitable at 10 and 100 tons/day plant capacity while, at the pipeline gas price of $0.8/mm Btu, it was not profitable even at 100 tons/day plant capacity. The above calculation did not include the effects of manure reduction due to methane production or the disposal of the undigested residue.

Table 5. Summary of costs and credits of methane production by anaerobic digestion.

	Plant Capacities, tons/day			
	0.1	1	10	100
Total Operating Costs/Year	10,140	20,900	65,000	388,000
Methane Sales/Year at $0.8/mm Btu[a]	340	3,400	33,800	338,000
Difference[b]	9,800[c]	17,500[c]	31,200[c]	50,000[c]
Methane Sales/Year at $2/mm Btu[a]	840	8,500	85,000	845,000
Difference[b]	9,300[c]	12,400[c]	20,000[d]	567,000[d]

(a) Low-Btu gas, 600 Btu/cu ft.
(b) Operation cost minus methane sales.
(c) Loss.
(d) Profit.

However, with the consideration of manure disposed by landspreading and a 50% volatile solids reduction by anaerobic digestion, the economics of the system were a little different. The manure volume had been increased from 80 to 95% because of dilution for the anaerobic digestion process. Therefore, the volume reduction of the manure for a particular plant due to anaerobic digestion is only 35%, with 65% of the volume left to be disposed of. The cost of landspreading of the manure varies from one livestock production system to another, and from one region to another. For the dairy production system, where it is assumed that the manure disposal is by mechanical scraper and pumped to an outside storage from where it is spread on the field, the cost of landspreading the raw manure was estimated at $37.15/cow/year (19). After digestoion, the cost of disposal will be about 65% of this cost or $24.15/cow/year. The credit derived from digestion is the difference or $13/cow/year. Total revenue from digestion is the sum of the methane sales plus the disposal-cost-benefit differential. Accounting for the effect of effluent reduction due to digestion has improved the economics. The cost variations with plant capacity for the digestion and landspreading of the residue are shown in Figure 9. It is clear that the cost of methane is a major factor in the economics of the anaerobic digestion. At \$2/mm Btu, the breakeven is a little less than 10 tons/day plant capacity, while at \$0.8/mm Btu, the break-even is at about 30 tons/day. It will be expected that with increase in gas prices, lower plant capacities will be increasingly more attractive.

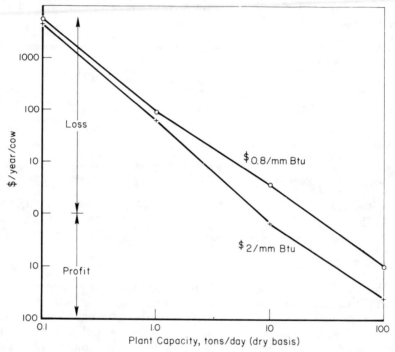

Figure 9. Digestion of dairy manure to produce methane—profit and loss investment at various plant capacities.

It must be emphasized that in the U.S. there are hardly any dairy and hog farms generating 10 tons or more of manure per day. Table 2 showed that at 10 tons of manure per day, the dairy and hog farms will have about 1700 and 24,000 heads, respectively. Farm capacities in this range are outside the average farm sizes in the U.S., as shown in Figures 1 through 3. Before methane production will be attractive to the dairy and hog operations, on a commercial scale, the cost of gas must rise substantially above the present gas rates. For beef-cattle production, the situation is relatively different. Although the average beef-cattle-farm capacity is also outside this range, there are few farms, especially in the southwest United States, that have feedlots generating 10 tons or more of manure per day. The same assumptions and calculation procedures used for determining costs of the digestion and disposal of dairy manure were used for the beef-cattle manure. At $0.8/mm Btu and for 100 tons/day, a profit of $3/beef animal year was obtained for the digestion and disposal of residue compared with $3/cow/day for dairy farms. However, at $2/mm Btu and 10 tons/day, there is a deficit of $7/beef animal/year, compared with a profit of $2/cow/year for dairy farms. Also at $2/mm Btu and 100 tons/day, a profit of $23/beef animal/year was obtained, compared with $40/cow/year profit for dairy farms.These values show conclusively that the quantity of manure, in other words the farm capacity, and gas market prices have marked impact on the economic feasibility of anaerobic digestion.

Pyrolysis. An economic evaluation of pyrolysis processes that could be used to recover fuel from animal manure is difficult because there are many types of designs. About 24 types of pyrolysis designs have been identified (4). Most are in bench and pilot stages and none is as yet in full production for the solid-waste or animal-manure treatment. Functional parameters of pyrolysis such as maximum temperature, material composition, particle size, heating rates, and pressure are specific with the material being pyrolyzed.

One manure property that is not conducive to good operation of a pyrolysis process is the high moisture content. However, a dry animal manure is well suited for a pyrolysis process. Therefore, an additional cost of dewatering and drying wet manure must be incurred before the manure is pyrolyzed. Hence, manure stored in pits filled with water, as is the case in many hog confinements, will incur considerable expenses of dewatering and drying if pyrolysis is chosen as the process for recovering fuel from the manure. Hence, it is more economical to start with manure with low moisture content. Feedlots in semiarid regions southwest U.S. where farmers enjoy the full solar energy have their manure substantially dried in the field.

Another economic consideration important to the pyrolysis of animal manure is the quantity of collectible manure. Large beef feedlots of the southwestern part of the United States will be most adaptable to fuel recovery by pyrolysis from the animal manure. The plant capacities considered in this study, which are between 0.1 and 10 tons/day, are too low for a viable pyrolysis process at the present state of the art. More cost analysis is needed to determine whether inputs from 10 to 100 tons/day will be viable. Table 6 shows classification of cited pyrolysis of animal manure. There are insufficient experimental data and process economics to show which process or procedure is most economical.

The following pyrolysis-process variables must be adequately considered in order to reduce the cost of the pyrolysis of animal manure.

- Reactor type — should be simple and low in initial cost. Shaft reactors — horizontal and vertical — satisfy these conditions.

- Heating method—indirect method is generally regarded as being less efficient than direct methods. However, it avoids the problem of excessive CO_2 and H_2O formation which reduce heating value and high NO_x and N_2 content (4).

- Feed conditions — feed materials that are dry and finely shredded with solid inorganics removed are desirable. Fortunately, the animal manure does not contain inorganics as does municipal refuse, but it has high water content which must be reduced.

- Products output — pyrolysis output consists of liquid (oil), solid (char), and gases (fuel gases) phases. The compositions and amounts of each phase during the pyrolysis of animal manure are dictated by a) preprocessing for water removal and sizing; b) reactor temperature and residence time; and c) heating method (direct or indirect).

Environmental Considerations

The environmental-impact evaluation of a plant involves the identification of kinds, quality, and quantity of pollutant emissions from the unit processes of the plant and then assessment of the impact of these emissions on the three media—land, air, and water. At the present status of process development in fuel recovery from animal manure, no emissions data are available. Consequently, most of the discussion in this section is qualitative.

Methane Production. Main sources of pollutant emissions during methane production by anaerobic digestion are raw-manure handling, the digester, methane storage, and effluent disposal. Odor is the greatest problem of raw-manure handling, and the odor intensity may be serious if the raw manure is stored before it is digested. The odorous compounds, such as H_2S, mercaptan, ammonia, and sulfides, are products of anaerobic decomposition emitted at various stages of the digestion. The logical control technique is to handle the manure in enclosed systems so that the emissions can be collected and treated, *e.g.,* by passing through an activated carbon filter or by means of other odor-control systems.

Table 6. Classification of cited pyrolysis of animal manure.

Reactor Type	Process	Heating Method	Products Gas	Products Oil	Products Solid Char	Feed Rate/ Animal Type	Status	Cost, $/ton	Comment	Reference
Shaft	Batch	—		X	X	Beef	Bench	5.60	80% moisture	(20)
	Batch	—			X	Various Animals	Bench	—		(21)
	Batch	—	X	X	X			—	3.6% moisture; 0.3 barrel/ton processed	(22)
Vertical retort	Continuous	—	X	X	X	Beef		—	29.1% moisture; Oil yield 1 barrel/ton of manure processed	(23)
Retort	Continuous	Indirect				200 tons/day	Pilot	3.71	30% moisture; No credit taken for oil and char	(24)

Under normal operation, the digester will not produce air-pollution problems. Problems arise when it turns sour because of changes in process variables, such as an increase in volatile acids in the digester. Again the emissions are H_2S, mercaptans, sulfides, volatile acids, etc.

The environmental concern of the product (methane) is that it is explosive at air concentrations between 6 and 15%. Safety procedures should be taken during storage and gas compression. The effluent is largely odorless, although it contains most of the nutrients which are thought to be in more available form for the plants. Furthermore, it can be better handled than the raw manure.

Pyrolysis. The sources of pollution during pyrolysis are manure pretreatment, consisting of drying and size reduction, producing odor and particulate emissions. At the reactor, the emissions may consist of NO_x, particulates, and possibly odorous gases. Under normal operation, however, it will be expected that these emissions will be reduced because of the high temperature of operation.

CONCLUSIONS AND RECOMMENDATIONS

1. Presently, the quantity of manure produced from average dairy, beef, and hog production facilities in the U.S. is too small for an economic recovery of synthetic fuel from the manures. The average U.S. dairy and beef facility generates about 0.1 ton of manure per day (dry basis) and for a hog facility it is about 0.04 ton of manure per day (dry basis). Efforts to recover fuel economically from animal manure will be directed to large animal facilities. For better manure accountability, it is recommended that U.S. livestock production inventory be published in terms of number of farms at various animal populations.

2. Alternative processes for recovering fuel from animal manure include biological and thermochemical processes. The biological processes include methane production and biochemical processes for protein and alcohol production. The thermochemical processes are pyrolysis, hydrogenation, and hydrogasification. The high moisture content of the livestock manure is a major disadvantage in the application of the thermochemical processes. For fuel synthesis, only methane production and pyrolysis may hold promise at above 10 tons/day plant capacity.

3. Capital-intensive items in the biological system are the anaerobic digester and the storage tank, while the operating cost-intensive items are labor and maintenance. The process-cost-reduction effort must, therfore, focus on these items. The analysis of the anaerobic digestion of dairy manure and disposal of the residue gave a net profit only at 100 tons/day plant capacity when the price

of methane was \$0.8/mm Btu at 600 Btu/cu ft. The value obtained was \$10/cow/year. At the methane price of \$2/mm Btu, net profits for 10 and 100 tons/day plant capacities were \$14 and \$40 per cow per year, respectively. Fuel selling prices and the quanity of manure to be processed, therefore, are important factors in the economic feasibility of the process.

4. Refinement in the engineering-cost analysis made for methane production is recommended. This can be accomplished through thorough cost inventory for a pilot anaerobic digestion plant. Although an energy conversion system may be too complex and time-consuming for an average farmer, a package plant system for fuel recovery by a private investor may be economically built adjacent to large livestock-production facilities.

REFERENCES

1. United States Department of Agriculture Statistical Reporting Service, Crop Reporting Board, Da 1-1. Washington, D.C. (January, 1975).
2. American Society of Agricultural Engineers. Data adapted from Structures and Environment Committee 412 Report AW-D-1. Revised (June 14, 1973).
3. Ghosh, S., and D. Klass. Biogasification of Solid Wastes. Research proposal to NSF (January, 1973).
4. Antunes, G. E., *et al.* Energy Recovery from Solid Waste. NASA-ASEE System Design Institute, University of Houston, Johnson Space Center, Rice University, Volume 2 (1974).
5. Feldmann, H. F., K. Kiang, C. Y. Wen, and P. M. Yavorski. Cattle Manure— A Process Study. Mechanical Engineering (October, 1973).
6. Friedman, S., H. H. Ginsberg, I. Wender, and P. M. Yavorski. Continuous Processing of Urban Refuse to Oil Using Carbon Monoxide. Bureau of Mines paper presented at Third Mineral Waste Utilization Symposium, Chicago (March, 1972).
7. Anonymous. Process Converts Animal Wastes to Oil. Chemical & Engineering News (August 16, 1971).
8. Feldmann, H. F., C. Y. Wen, W. H. Simons and P. M. Yavorsky. Supplemental Pipeline Gas from Coal by the Hydrane Process. Presented at the 71st National AIChE Meeting, Dallas, Texas (February 20-23, 1972).
9. Lapp, H. M., D. D. Schultc, and L. C. Buchanan. Methane Gas Production from Animal Wastes. Canada Department of Agriculture, Ottawa, Publication 1528 (1974).
10. Gramms, L. C., L. B. Polkowski, and S. A. Witzel. Presented at 1967 Annual Meeting American Society of Agricultural Engineers, Purdue University, Lafayette, Indiana, Paper No. 69-462 (June 22-25, 1969).
11. Taiganides, E. P., E. R. Baumann, and T. E. Hazen. Sludge Digestion of Farm Animal Wastes. Compost Science (Summer, 1963).
12. Varel, V. H., H. R. Isaacson, and M. P. Bryant. Thermophilic Methane Production from Cattle Waste. University of Illinois Urbana-Champaign, Illinois (April, 1975).
13. Ghosh, S., J. R. Conrad, and D. L. Klass. Anaerobic Acidogenesis of Wastewater Sludges. Journal of Water Pollution Control Federation 47 (1) (January, 1975).
14. Grout, A. R. Methane Gas Generation from Manure. The Pennsylvania State University, University Park, Pennsylvania (1974).

15. Meenaghan, G. F., D. M. Wells, R. C. Albin, and W. Grub. Gas Production from Beef Cattle Waste. Presented at Winter Meeting of American Society of Agricultural Engineers, Chicago, Illinois (December 8-11, 1970).
16. Converse, J. C., and R. E. Graves. Facts on Methane Production from Animal Manure. University of Wisconsin, Madison, Wisconsin (July, 1974).
17. Kuester, J. L., and L. Lutes. Pyrolysis of Solid Waste. Battelle's Columbus Laboratories (October 22-24, 1974).
18. Peters, M. S., and K. D. Timmerhaus. Plant Design and Economics for Chemical Engineers. 2nd Edition, McGraw-Hill Book Company, New York (1968).
19. Johnson, J. B., C. R. Hoglund, and B. Buxton. An Economic Appraisal of Alternative Dairy Waste Management Systems Designed for Pollution Control. Journal of Dairy Science 56, 1354-1366 (October, 1973).
20. Garner, W., and I. C. Smith. The Disposal of Cattle Feedlot Wastes by Pyrolysis. Environ. Protection Tech. Series, EPA R2-73-096 (1973).
21. White, R. K., and E. P. Taiganides. Pyrolysis of Livestock Wastes. Proceedings of International Symposium on Livestock Wastes (April 19-22, 1971).
22. Schlesinger, M. D., W. S. Sanner, and D. E. Wolfson. Pyrolysis of Waste Materials from Urban and Rural Sources. Proceedings of 3rd Mineral Waste Utilization Symposium, pp 423-428 (1973).
23. Massie, J. R., Jr., and H. W. Parker. Continuous Solid Waste Retort Feasibility Study. Presented at AIChE 74th National Meeting, Paper 43a, p 31 (1973).
24. Parker, H. W., C. J. Albus, Jr., and G. L. Smith. Cost for Large Scale Continuous Pyrolysis of Solid Wastes. Presented at AIChE 74th National Meeting, Paper 43b, p 25 (1973).

28.

Methane-Carbon Dioxide Mixtures in an Internal Combustion Engine

S. Neyeloff and W. W. Gunkel*

In the last few years the "energy crisis" has triggered a great deal of research in the field of energy conservation and utilization. Most investigators agree that the world's supply of energy has been so heavily taxed that, unless measures are taken immediately to 1) decrease the uses, 2) augment the efficiency with which energy is used, and 3) find new sources of energy, in the next few decades the world is going to face a grave problem with potential critical consequences.

Food production, particularly in the U.S., is highly energy-dependent. In the last 50 years, agriculture has changed from a net producer to a net user of energy (1). For example, corn yield per acre in the last 30 years has more than doubled; however, the mean energy inputs to achieve this level in production have more than tripled (2). This is due to the fact that the increase in production has been largely due to an increase in the use of machinery, inorganic fertilizers and pesticides, all of which demand a large input of fuel.

Agriculture also produces a considerable amount of organic waste with its consequent impact on the environment. Suitable methods for utilizing agricultural waste profitably from an en-

* Graduate student and Professor, respectively, Department of Agricultural Engineering, College of Agriculture and Life Sciences, Cornell University, Ithaca, New York.

vironmental and economic standpoint are still in the embryonic stage (3).

Extensive research is being conducted in the area of anaerobic digestion of farm wastes. Although the process is not new, its application in U.S. agriculture is certainly not common. Preliminary calculations show that organic wastes from a 60-head dairy farm, if properly handled and treated, contain enough energy to satisfy most energy requirements of the farm (4). This is, of course, theoretical, and current research to determine the feasibility of incorporating anaerobic digestion systems to farm operations indicates that under certain conditions this may be likely, although the technology required for the process has to be developed further (5).

In the future, anaerobic digestion of farm manures and refuses may become common, and the methane gas thus generated may be used in place of other fuels for heating, cooking, or as a fuel for internal combustion (IC) engines.

Many gases make excellent fuels for IC engines and some have been and are being used for this purpose. Products and by-products of the coal industry have been used since raw coal contains considerable amounts of volatile hydrocarbons (6). Liquefied petroleum (LP) gas is extensively used in agriculture, while biogas extracted from domestic waste is used in some sewage treatment plants to operate both IC engines and heating equipment.

It is known that gaseous fuels have many advantages over other kinds of fuels. However, storage creates a problem, especially with methane since very high pressure is required for it to be liquefied. Gaseous fuels, particularly methane, have a high octane rating. The combustion efficiency of an engine, then, can be increased without producing more heat by increasing the compression ratio. Gaseous fuels generally build up minimum carbon deposits; they also mix more thoroughly in air to produce a mixture which burns more completely than gasoline mixtures. Other advantages (7) of gaseous fuel operation are

1) excellent anti-knock qualities,
2) small amount of contaminating residues,
3) less sludge in oil,
4) no wash down of cylinder wall lubrication during engine starting,
5) no tetra-ethyl lead to foul spark plugs and other engine parts,
6) a nearly homogenous mixture in cylinder,
7) less valve burning.

The gaseous fuels mentioned above present the advantage of being consistently made up of the same proportion of constituents. However, in the case of gas generated from manures and refuse this might not be the case due to the wide variety of wastes that may be introduced into the digester. This heterogeneity of the waste to be digested added to operational conditions such as care, maintenance, temperature, and other factors, results in altering the composition of the gas generated.

In experiments run at Cornell, for example, when organic wastes such as dairy manure, poultry manure, potato refuse, blood, leaves, oats, and other organic wastes, were anaerobically digested, it was found that, although the composition of the gas was consistently methane (CH_4) and carbon dioxide (CO_2), the amount of CH_4 in the mixture varied from 13 to 61%, the rest being mostly CO_2 with traces of N_2 and other gases (8).

It is found in the literature that when conditions are best, organic wastes in a digester should generate a gas consisting of up to 80% CH_4, 67% CH_4 on the average (3). It may be expected that in a field situation, the composition of the gas will vary from practically no methane to almost 80% methane. Due to the nature and kinetics of anaerobic digestion this variation may occur in a short time if input conditions change. This change in the fuel mixture will have a direct effect on an IC engine run on the gas.

Based on the assumption that gas from anaerobic digestion of agricultural refuse may become available for use in IC engines, it was decided to investigate how an engine would perform using this gas as the fuel. Although the biogas may contain traces of sulfides and other gases, methane and carbon dioxide make up more than 98% of the gas. Consequently, the investigation was limited to the effects of carbon dioxide alone on the characteristic properties of methane as a fuel.

In analyzing the problem, two questions became readily apparent. First, how will carbon dioxide dilution affect the performance of methane as a fuel, and second, how will the limits of inflammability of methane be changed, if at all? An extensive review of literature produced very little specific information on this problem and correspondence with personnel at manufacturing firms indicated that they did not have any information on the topic either. One of the firms contacted indicated that they could find no one who had this kind of data, and although some work was done in the early nineteen fifties, that data was no longer available (9).

Under standard conditions a methane-air mixture will ignite if it is composed of 5.3% to 14% methane (10), the ideal stoichiometric mixture being abut 9.4% methane (11). These figures represent fuel to air (methane/air) ratios by volume of 0.056:1, 0.163:1 and 0.104:1, respectively. The ignition limits of methane in air are affected by carbon dioxide dilutions and under atmospheric conditions the mixture will not combust if the amount of carbon dioxide is greater than three times the amount of methane (12). Also, ignition limits of methane should be little affected by the temperatures and pressures developed in an IC engine (13).

EQUIPMENT

A diagram of the equipment used is illustrated in Figure 1. Basically it consists of sources of CH_4 and CO_2, regulating and metering devices, an engine and a dynamometer. CH_4 and CO_2 come out of the pressurized cylinders at high pressure and expand through two-stage pressure regulators and an expansion coil to approximately 25 psi; a second regulator drops the pressure to metering pressure (20 psi). After passing a pressure gauge the gas is metered through rotometers (manufactured by Brooks Instrument Division, Emerson Electric Co., Hatfield, Pa.) and the flow is controlled by flow controllers. From these controllers the gases mix in a gas carburetor (manufactured by Marvel-Schebler/Tillotson, Division of Borg Warner) and flow into the engine. The gas carburetor acts only as a mixing valve since the mixture is metered before going into it.

Air is naturally aspirated through a 3-in. diameter pipe, metered through a sharp edge orifice with pipe taps connected to an inclined

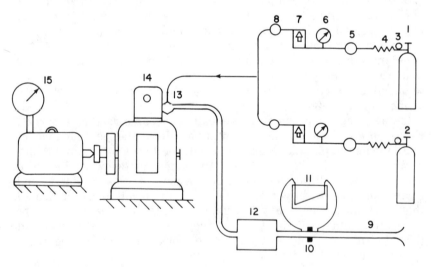

1. Methane cylinder	6. Pressure gauge	11. Inclined manometer
2. Carbon dioxide cylinder	7. Rotometer	12. Surge chamber
3. High pressure regulator	8. Flow controller	13. Carburetor
4. Expansion coil	9. Air intake	14. CFR Engine
5. Low pressure regulator	10. Sharp edge orifice	15. G.E. Dynamometer

Figure 1. Schematic diagram of equipment.

Figure 2. Equipment.

manometer. Air then goes into a 5 cu ft surge chamber and then to the carburetor.

The engine is a high speed CFR (Cooperative Fuel Research) manufactured by Waukesha Motor Company under SAE (Society of Automotive Engineers) specifications (see Engine Specifications, Appendix A). This is a spark ignited variable compression engine. Its output was absorbed by a General Electric DC cradled dynamometer, type TLC-2000.

PROCEDURE

Studying the characteristics of a fuel and being interested in finding out how it will perform in an IC engine is more complicated than one would think and we found that it was a very difficult task to determine how it would perform under all possible conditions. Therefore during these tests some of the engine parameters were fixed and some were changed. Valve timing was kept constant at settings recommended by the SAE specifications for this particular engine (see Appendix A) and all tests were run at 900 ± 30 rpm.

During some preliminary tests it was found that the efficiency of the CFR engine was highest when ignition timing was set at 30 degrees before top dead center (BTDC). Consequently this timing was kept constant throughout the tests. Occasionally during the test runs ignition timing was varied slightly from 30° BTDC and it was found again that the change in output, if any, was not significant.

Operating coolant temperature was constant (at 195°F ± 5°F) in all test runs. The engine was installed inside a building where ambient temperature was 70° ± 3°F, relative humidity 35% ± 10% and the pressure atmospheric.

Methane is a gas at standard conditions (70°F and atm P). It is lighter than air, and ignites in air at mixtures of 5.6% to 16.3% methane/air by volume. Carbon dioxide is a gas at standard conditions, is heavier than air, and does not burn. Properties of these gases are given in Appendix B.

The test procedure used was the same throughout all the test runs. The engine was cranked by the dynamometer and methane gas and air were allowed to flow into the engine. Once the engine was started it was allowed to operate for approximately 5 minutes until the coolant water temperature reached operating conditions at 195°F. When the coolant temperature had reached 195°F the engine was shut off and the gap in the valves and ignition points were checked and reset if necessary. Then the engine was started again and the ignition timing was set at 30° BTDC.

A few preliminary tests were run to develop a curve of specific power output vs. compression ratio using methane alone (no CO_2). This was done at two different fuel to air (F/A) ratios and the curve was used to establish a compression ratio desirable for use throughout the experiment.

To obtain curves for specific power output, horsepower per liter of methane per minute (HP/lt CH_4/min), versus fuel to air ratio at several dilution conditions the following procedures were used. With methane alone as a fuel the flow was adjusted so that the fuel to air ratio varied from 0.50:1 to 0.17:1. Several trials were made to establish a 100% methane curve. Some runs were started at low F/A ratios while others at high ratios in order to obtain a certain degree of randomness in the data.

When carbon dioxide was added to the fuel mixture the test procedure was essentially the same except that flows were calculated so that a fixed percent of CO_2 in the mixture could be maintained for the range of F/A ratios. The meters were then set to the calculated flows so that curves for methane-carbon dioxide mixtures could be developed.

Measurements of methane, carbon dioxide, and air flows, revolutions per minute of the engine, and torque were recorded at about 10 or 12 increments through the range of F/A ratios indicated above.

RESULTS

Horsepower output of the engine was obtained from the measurements of torque and rpm's. The efficiency of the engine in converting the fuel's energy to mechanical power output is expressed as specific power output (SPO); the richness of the fuel mixture is expressed as fuel to air ratio in percent by volume (CH_4/air x 100) and the dilution of the fuel is expressed as the percent of CO_2 relative to methane gas (CO_2/CH_4 x 100) by volume.

The results obtained from the test are shown in Figures 3-6. Figure 3 shows SPO versus compression ratios for two F/A ratios. The dotted lines enclose the region of best efficiencies obtained. Percent enthalpy efficiency for the engine can be calculated by multiplying SPO by a factor of 120, *i.e.*, if SPO is .18 then the enthalpy efficiency is 21.6%.

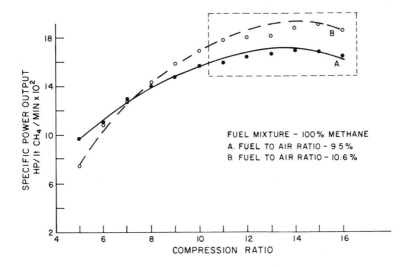

Figure 3. Compression ratio vs. specific power output.

Figure 4 shows SPO versus F/A ratio for straight methane air mixtures at a CR of 15:1. Figure 5 shows SPO versus F/A for six different fuel mixtures, ranging from no CO_2 dilution of methane to 100% dilution (equal parts CH_4 and CO_2). Figure 6 shows SPO versus F/A under two conditions. Curve A shows the SPO obtained from the test engine using pure methane as a fuel, and a CR of 15:1. Curve B shows SPO obtained at a CR of 7.5:1 and a methane dilution of 50%.

DISCUSSION AND CONCLUSIONS

Since this project was conducted to study some of the characteristics of gas generated from anaerobic digestion as a fuel for IC engines, it should first be stated that the gas can, in fact, be used successfully for this purpose.

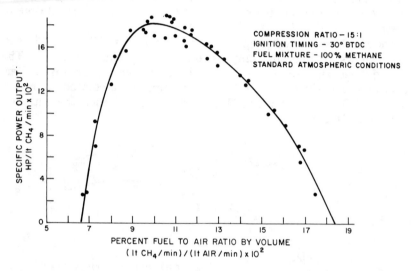

Figure 4. Specific power vs. air to fuel ratio.

From the data obtained it is apparent that the best results in terms of specific power output of the engine, are obtained at compression ratios ranging from 11:1 to 16:1. Specific power output peaked at CR or 15:1 although at 15:1 and above audible knocking occurred. On the low end of the scale one can see that below CR of 10:1 SPO dropped rapidly indicating a loss in efficiency (Figure 3).

From Figure 4 it can be observed that pure methane ignited at fuel to air ratios ranging from 0.065 to 0.185 by volume. These limits are both higher than those reported for atmospheric conditions. The highest specific power output was obtained at a fuel to air ratio of 0.10. Figure 4 also shows the kind of data that was obtained with the equipment used and the variability that was present in the readings. In order to compensate for this sensitivity and operational problems, several sets of data were taken on different days in order to develop each of the curves shown in the figures.

From Figure 5, one can see that the range of F/A mixtures at which the diluted methane ignites decreased with increasing dilution. The SPO's obtained under diluted conditions also dropped proportionately with dilution. The SPO was calculated per unit of methane, so that the decrease in performance is solely due to the effect of CO_2 dilution, rather than a decrease in heating value of the fuel mixture.

It can also be observed that the F/A ratio for best performance increased slightly with increased dilution, but that in general, best performances were obtained between F/A ratios of 0.085 and 0.130.

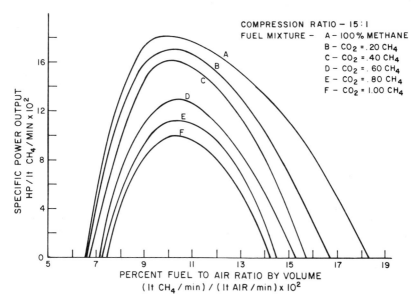

Figure 5. Specific power vs. air to fuel ratio — six settings.

Figure 6 shows a comparison of SPO obtained using undiluted methane at a CR of 15:1 and SPO using a 50% dilution at a CR of 7.5:1. This figure then represents the engine's output under what could be called ideal conditions, as compared to what would be obtained if a typical gasoline engine were run on straight AD gas. The loss in output is substantial, about 40%. If one assigns a value of 100 to the best SPO obtained, a comparison of outputs would yield the following table for several different operating conditions.

Compression Ratio	Percent CO_2 Dilution $CO_2/CH_4 \times 100$	Highest SPO HP/lt CH_4/min	Comparison Factor
15:1	0%	18.3	100
15:1	50%	14.4	79
7.5:1	0%	13.3	73
7.5:1	50%	11.0	60

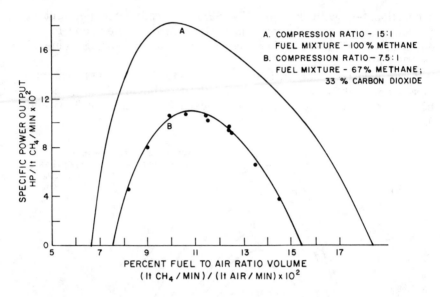

Figure 6. Specific power vs. air to fuel ratio — two settings.

One can see then that while the use of gas from a digester to operate a spark ignited IC engine is perfectly feasible, the output obtained from the engine per unit of methane greatly depends on the engine design as well as on the degree of CO_2 dilution of the methane gas.

Finally, it should be noted that in a typical spark ignited engine, of the heating value of the fuel, only about 20% results in mechanical output. This 20% is what SPO refers to throughout this study and it should be kept in mind that the results obtained by using the CFR engine may not exactly correspond to what would be obtained using a different type of engine.

REFERENCES

1. Price, D. R. and W. W. Gunkel. The Energy Crisis—Its Affect on Agriculture. New York State Horticultural Society Proceedings, Volume 119, 1974.
2. Pimentel, D., L. E. Hurd, A. C. Bellotti, M. J. Forster, I. N. Oka, O. D. Sholes, and R. J. Whitman. Food Production and the Energy Crisis, Science 182 (1973).
3. Loehr, R. C. *Agricultural Waste Management*, Academic Press, New York (1974).
4. Jewell, W. J. Energy from Agricultural Waste—Methane Generation. Agricultural Engineering Extension Bulletin 397, New York State College of Agriculture and Life Sciences, Cornell University, Ithaca, New York (1974).
5. Williams, D. W., Personal communication, Unpublished findings on a Feasibility Study of the Bio-Conversion of Agricultural Waste for Energy Conservation and Pollution Control (1975).

6. Moss, S. A. *Elements of Gas Engine Design.* D. Van Nostrand Company, New York 1906.
7. ONAN Technical Bulletin T-015, Use of Gaseous Fuel with Onan Electric Generating Sets, Minneapolis, Minnesota, 1974.
8. Group project for Treatment and Disposal of Agricultural Waste, AE 677, Cornell University, Ithaca, New York (1974).
9. International Harvester, Personal correspondence with R. E. Wallace, Manager Engineering Sciences, August, 1974.
10. *Matheson Gas Data Book,* The Matheson Company, Inc., Ontario, 1966.
11. Litchy, L. C. *Internal Combustion Engines,* 6th Edition McGraw Hill Co., Inc., 1951.
12. Coward, H. F. and G. W. Jones. Limits of Flammability of Gases and Vapors. U.S. Bureau of Mines, Bulletin No. 503, 1952.
13. Khitrin, L. N. *The Physics of Combustion and Explosion,* translated from Russian *Fizika goreniya i vzkyva.* Published for the National Science Foundation, Washington 1962.

APPENDIX A

Engine Specifications

Engine—CFR high speed crank case
Manufactured by Waukesha Motor Company, Waukesha, Wisconsin

Compression Ratio	4:1 to 16:1
Standard Bore	3.25 in.
Stroke	4.50 in.
Displacement	37.33 cu in

Operating Conditions

Spark Plug Gap	0.020 ± 0.005 in.
Breaker Point Gap	0.020 in.
Valve clearance (hot)	0.010 ± 0.001 in.

Valve Timing with 0.010 Clearance

Inlet opens	$15°$ ATDC$\pm 2.5°$
Inlet closes	$50°$ ABDC$\pm 2.5°$
Exhaust opens	$50°$ BBDC$\pm 3.0°$
Exhaust closes	$15°$ BTDC$\pm 3.0°$
Cooling Water Temperature	$195°$ F
Revolution per Minute	900 ± 30

APPENDIX B

Properties of Methane and Carbon Dioxide (10)

METHANE   (Synonyms: Marsh Gas; Methyl Hydride) CH_4

Physical Constants

Molecular Weight	16.04
Specific Volume at 70°F, 1 atm	23.7 cu ft/lb
Boiling Point at 1 atm	−258.9°F (−161.61°C)
Freezing Point at 1 atm	−296.5°F (−182.5°C)
Specific Gravity, Gas 60°F, 1 atm, Air=1)	0.5549
Density, Gas at 32°F, 1 atm	0.04475 lb/cu ft
Critical Temperature	−115.8°F (−82.1°C)
Critical Pressure	673.3 psia (45.8 atm)
Critical Density	0.162 g/cc
Latent Heat of Vaporization at bp	121.87 cal/g
Latent Heat of Fusion at mp	14.03 cal/g
Specific Heat, Gas at 60°F	
Cp	0.5271 Btu/(lb) (°F)
Cv	0.4032 Btu/(lb) (°F)
Ratio, Cp/Cv	1.307
Flammable Limits in Air	5.3-14% (by volume)
Heat of Combustion at 25°C	97.8 Btu/cu ft
Gross Heat of Combustion, 60°F., 1 atm	1011.6 Btu/cu ft
Viscosity at 60°F	0.012 centipoise
Viscosity, Gas, 32°F, 1 atm	0.0109 centipoise

CARBON DIOXIDE   (Synonym: Carbonic Anhydride) CO_2

Physical Constants

Molecular Weight	44.01
Vapor Pressure at 70°F	830 psig
Specific Volume at 70°F, 1 atm	8.76 cu ft/lb
Sublimation Point at 1 atm	−109.3°F (−78.5°C)
Triple Point at 5.11 atm	−69.9°F (−56.6°C)
Density, Gas at 70°F, 1 atm	0.1146 lb/cu ft
Specific Gravity, Gas (Air=1)	1.5239
Critical Temperature	87.8°F (31.0°C)
Critical Pressure	1071.6 psia (72.9 atm)
Critical Density	0.468 g/ml
Latent Heat of Vaporization	
at tp	149.6 Btu/lb
at 0°C	101.03 Btu/lb
Specific Heat at 60°F	
Gas Cp	0.1988 Btu/(lb) (°F)
Gas, Cv	0.1525 Btu/(lb) (°F)
Ratio Cp/Cv	1.303
Thermal Conductivity	
at 32°F	0.0085 Btu/(hr) (sq ft) (°F/ft)
at 212°F	0.0133 Btu/(hr) (sq ft) (°F/ft)
Viscosity of Gas at 70°F, 1 atm	0.0148 centipoise

29.

Limitations of Animal Waste Replacement for Inorganic Fertilizers

D. A. Lauer*

Animal manure has been used for centuries as a source of nutrients for growing crops. Only recently has animal manure become regarded principally as a waste product. One definition of the word "manure" is "any natural or artificial substance for fertilizing the soil." Another definition states the more commonly associated meaning as "dung or refuse of the stable or barnyards." The derivation of the word, manure, obviously predates the extensive use of manufactured inorganic fertilizers. The meaning also reflects an era when fertilization of crops was synonymous with spreading of animal manure on the soil. This does not imply that handling and use of manure was necessarily more efficient in that era but that the general attitude was one of utilization as a fertilizer material rather than disposal as a waste material.

In the United States use of manufactured fertilizer has increased dramatically over the past 25 to 30 years. From 1950 through 1974 use of plant nutrients from inorganic fertilizer materials has increased from 3.7 to 17.5 million metric tons including N, P_2O_5 and K_2O (1). The most dramatic increase has been with N which has increased over 9-fold over this period. Combining technology and inexpensive energy these large quantities of plant nutrients have been manufactured into a product with much lower nonnutritive

* Department of Agronomy; Cornell University, Ithaca, N. Y., presently affiliated with Department of Agronomy, University of Georgia; Coastal Plain Experiment Station, Tifton, Georgia.

material than manure. The physical characteristics of the inorganic fertilizers promote an ease of transport and handling not possible with manure. With all factors combined nutrients supplied by manufactured fertilizer have been very inexpensive. The large non-nutritive content of animal manure, largely water and carbonaceous organic matter, has caused the manure to become regarded largely as a nuisance waste material. This attitude has been reinforced by massive accumulations of animal manure near large fed-beef and dairy facilities.

Recent happenings have caused some reevaluation of the importance of manure. Concentrations of animal population in small geographic areas have produced large accumulations of manure. This is particularly true where the animal production facilities are located remotely from major feed-grain production areas and where feed is imported for the animals. A general increase in awareness of actual and potential problems such as odor, esthetics, runoff, and leaching has focused attention of proper handling and disposal of massive quantities of animal manure. This same pollution awareness has also raised questions about more efficient use of fertilizers on cropland to prevent degradation of water quality from nutrients lost by runoff and leaching from fertilized fields. These are concerns that must be considered in utilization of animal manure in soil fertility and crop production. Precautions against pollution hazards must be an integral part of using animal manure as a fertilizer.

Recent increases in energy costs and availability have also triggered reevaluation of manure as a source of nutrients and as an energy source itself via methane generation. Parallel to the concern over energy supply there has been a broader realization that resources such as fossil fuels and mined fertilizer materials are finite and nonrenewable. Manufacture of fertilizer is an energy requiring process which with increasing cost of energy has increased the cost of nutrients in the fertilizer. These factors combined have turned attention to more efficient use of fertilizer and possibilities for recycling nutrients in manure, sewage sludge and other biological waste materials. The value of the nutrients in these waste materials increases as the cost of nutrients from commercial fertilizer increases.

INTERRELATIONSHIPS OF FERTILIZER, CROP AND ANIMAL PRODUCTION

The various processes associated with the production of fertilizer, crops and animals are outlined in Table 1. Production of fertilizer consists of obtaining raw materials, processing these materials, transport and application of the soil. The link between fertilizer

Table 1. Processes of fertilizer, crop and animal production.

Fertilizer Production	Crop Production	Animal Production
Raw Material Procural	Nutrient Application	Animal Ingestion and Digestion
Various Manufacturing Processes	Return of Plant Residue to Soil	Animal Product Processing and Transport
	Soil Chemical Reactions	
Product Transport	Leaching and Runoff	Manure Deposition, Leaching and Runoff
Application to Land	Volatilization[a]	Volatilization[b]
	Plant Uptake and Growth	(Manure Treatment)
		Manure Transport
	Harvest, Processing and Product Transport	Land Application of Manure

a. Volatilization of N_2, N_2O as products of denitrification and NH_3 from manure.
b. Volatilization of NH_3 from manure.

and crop production is through the application of fertilizer material to the soil to supply plant nutrients. The nonrenewable nature of fertilizer raw materials is revealed by the lack of obvious feed-back from either soil-plant or animal systems. An exception is denitrification which produces N_2 which in turn renews atmospheric N_2 a basic raw material for N fertilizer production. Mined materials such as phosphorus (P) and potassium (K) deposits are being extracted much faster than geologic deposition is occurring.

Crop and animal production exhibit a more integrated set of processes with each other than with fertilizer production. A major objective in crop production in the United States is to produce feed for animals. The feed is ingested and digested by the animals. The animals in turn produce meat, milk, eggs, etc. and a quantity of manure. The manure and the nutrients contained in the manure then link-up with the soil-plant system through land application of the manure.

Between defecation and land application the manure is subject to a variety of processes. Exposed manure is subject to leaching and runoff caused by water moving over and through the manure. Chemical reactions within the manure produce ammonia which is then subject to volatilization. Manure treatment pertains to engineered processes which alter the chemical and physical characteristics of manure prior to ultimate disposal or utilization. Treatment of manure is meaningful only with respect to manure produced by animals in confinement. The final processes in manure production consist of transport of the manure or its treated residue for application to the land.

Manure production by animals on pasture or range is basically the same as by animals in confinement with some important variations. Ingestion consists of grazing on the forage growing directly on soil. Manure deposition is different in that urination and fecal drop may not occur together. The physical and chemical interactions of the urine and feces with the soil are quite different than the mixture of urine and feces hauled and spread from confinement facilities. The urine deposited by ranging animals will infiltrate much like rainfall. Fecal material probably contacts the soil via transport by insects, burrowing animals and somewhat by direct contact with the soil surface. Materials in the feces and infiltrated urine are subject to leaching. Runoff water will act on the feces residing on the soil surface. Volatile losses of ammonia from the urine and feces deposited by ranging animals probably occurs but as separate processes. Manure treatment has no meaning in this context. Transport and land application in the range and pasture setting are inherent in the deposition and various interactions of the manure with soil just discussed. The soil-plant relationships in this situation are largely the same with the exception of harvesting, etc. which is substituted for by the grazing of the animals.

Overall the relationships between fertilizer, crop and animal production are linked through the flow of nutrients required for growth of plants and animals. Fertilizer nutrients are an external input of nutrients into the soil-plant system. Cycling of nutrients is possible between soil-plant and animal systems via exchange of nutrients in feed supplied by crops and nutrients in the manure deposited back on the soil. A logical next question is what is the substitutability of nutrients from manure for nutrients supplied currently by inorganic fertilizer. To evaluate the potential replacement of fertilizer by manure the fates of nutrients during the manure production processes described above are important. Also important are the quantity of manure and the contained nutrients, quantities and types of cropland and crops requiring fertilization and the quantities of fertilizer nutrients to be replaced. In the following sections some of these factors are discussed as well as the distributional characteristics of manure, cropland and fertilizer.

DISTRIBUTIONAL CHARACTERISTICS OF MANURE, CROPLAND AND FERTILIZER

Manure Production and Land Resources

Total production of manure in the United States is shown in Table 2 according to classes of animals and their relative contributions to

Table 2. Manure production in the United States.

Animal[a]	Dry Measure $\times 10^6$ metric tons	Percent of Total
Cattle	195	82
Swine	22.4	9.4
Horses	12.6	5.3
Poultry	4.59	1.9
Sheep	2.89	1.2
All	237[b]	100

a. Animal population numbers were obtained from reference (2) and quantity of manure produced per animal was obtained from reference (3).
b. Wet weight $= 1.5 \times 10^9$ metric tons at 16.1% dry matter.

the total. Clearly cattle and swine as an aggregate produce most of the animal manure in the nation with cattle the largest single source of manure. The 237 million metric tons of dry manure exceeds the total quantity of fertilizer material used in recent years by approximately 6-fold (1). The N, P and K contents of manure and fertilizer will be compared later but it can be inferred that in general the distribution of nutrients in manure parallels the distribution of the manure itself. The different contents of nutrients in manure as a reflection of species differences, however, may modify the latter statement somewhat.

Since land is the target of nutrients from either fertilizer or manure, a brief examination of total land resources will give some prospective on where these nutrients are required. A breakdown of total land resources is given in Table 3. About 18% of the total land area is actively used for cropland or 138 million hectares. This percentage varies from year to year depending upon the markets for crops produced and the plans made by farmers each year in response to the markets. The category labelled "pasture" includes idled cropland and therefore this category fluctuates inversely with cropland as land is idled or brought back into production. The other major feature of total land utilization is that 42% of the

Table 3. Land resource utilization in the United States[a].

Use Category	Area $\times 10^6$ hectares	Percent of Total
Active cropland	138	18
Pasture and idle cropland	54	7
Grazing land	323	42
Nongrazed forestland	169	22
Urban	9.2	1.2
Other	77	10
Total	770	100

a. In 48 conterminous states from census year 1969; reference (2).

total is used to graze animals. This large area for grazing indicates that a relatively large portion of the animal population with their associated manure production are on pasture or range. More will be discussed in connection with this later.

Cropland Utilization and Fertilizer Consumption by Major Crops

In Table 4 the total cropland is divided into individual crop categories. Like the total cropland figure the areas of a given crop category are dynamic in response to changes in market situations. A general picture, however, can be gained from these data. Some insight can be also obtained regarding nutrient use on crops from Table 4. Feed grains occupy the largest single area of cropland, which is consistent with the animal oriented agriculture in the United States. Corn is the major component of the feed grains.

Table 4. Cropland utilization in the United States[a].

Crop	Area × 10^6 hectares	Percent of Total
Feed grans[b]	46.6	34.5
Food grains[c]	24.4	18.1
Hay	24.2	17.9
Soybeans	19.0	14.1
Cotton	5.7	4.2
Vegetables and melons	1.3	0.96
Fruits and nuts	1.3	0.96
Other	15.3	9.1
Total	138	100

a. Reference (2).
b. Corn, oats, barley, and sorghum.
c. Wheat, rye and rice.

Feed grains, corn in particular, then would be expected to have the single largest requirement for plant nutrients. Food grains would likely have the second highest requirement as an aggregate with wheat as the major component of food grains. Alfalfa hay is the single largest crop in the hay category and like soybeans is leguminous. Alfalfa and soybeans, therefore, require fertilizer but not N fertilizer. Finally, cotton makes up a significant portion of the cropland and requires complete N, P and K fertilization.

Fertilizer use by the four major crops grown in the United States is summarized in Table 5. Corn, soybeans, wheat, and cotton use a large fraction of the total fertilizer accounting for about two-thirds of the total N, P and K use in the United States. The soybean as a leguminous crop requires no fertilizer N but a small quantity of N fertilizer is sometimes used as a starter application at planting. Corn and soybeans are grown principally in the North Central or Midwestern states. Wheat is produced principally in the Great Plains extending into western portions of the North and South Central regions and into eastern portions of the Mountain region. Cotton production is centered in the southwestern United States from Texas west to the Pacific coast. If extensive substitution of manure nutrients for fertilizer nutrients is undertaken these regions and these four major crops would be the principal locales and crops involved.

Table 5. Consumption of fertilizer N, P & K by four major crops[a].

Crop	N	Fertilizer P Percent of Total	K
Corn	38.9	35.4	42.6
Soybeans	2.2	10.6	15.6
Wheat	13.3	15.8	13.7
Cotton	3.4	4.4	3.4
Total	57.7	66.2	75.3

a. Reference (1).

Overall it appears that feed grains and food grains are the major crops that have the greatest potential for utilization of nutrients from manure. This is consistent with the objective of recycling as much of the nutrient content of manure as possible since feed and food grains make up 50% of the total cropped acreage and contribute a major fraction of the nutrient input to animal production. Corn and wheat together constitute the major part of feed and food grains and these two crops utilize over 50% of the fertilizer N, P and K in the United States (Table 5).

Having examined, in general, manure production by animal species and cropland by type of crop grown, a prospective of the source of manure nutrients and the crops that could use these nutrients is revealed. Next an examination of the relative distributions of manure, cropland and fertilizer consumption will reveal the location of the manure nutrients relative to location of major cropland and fertilizer utilization.

Regional Distribution of Cropland, Fertilizer Consumption and Manure

Shown in Table 6 are the regional distributions of the cropland for which the manure is intended and the fertilizer to be replaced. The regions are arranged in order of increasing percentage of cropland. Consumption of N, P and K fertilizer roughly parallels available cropland with a few exceptions that reflect soil differences

Table 6. Regional distribution of cropland and fertilizer use[a].

Region[b]	Cropland	N	Fertilizer Consumption P Percent of Total	K
Northeast	3.9	4.0	6.5	6.7
Pacific	5.7	9.5	5.7	2.3
Southeast	9.2	10.7	11.2	18.4
Mountain	10.5	5.5	5.7	0.5
South Central	13.5	21.2	18.8	15.8
North Central	57.3	48.9	51.7	55.3

a. References (1, 2).
b. States in each region
   Northeast: ME, NH, VT, MASS, RI, CONN, NY, NJ, PENN, DEL, MD, WV
   Pacific: WASH, OREG, CALIF
   Southeast: VA, NC, SC, GA, FLA
   Mountain: MONT, IDAHO, WYO, COLO, NM, ARIZ, UTAH, NEV
   South Central: KY, TENN, ALA, MISS, ARK, LA, OKLA, TEXAS
   North Central: OHIO, IND, ILL, MICH, WISC, MINN, IOWA, MO, ND, SD,
       NEBR, KAN

and different types of cropping patterns. Nitrogen and phosphorus fertilizers most closely match the cropland since these elements are the soil borne nutrients most universally required by plants. Potassium deviates most drastically from the cropland distribution. Potassium consumption is lower in proportion to cropland in regions where native soil K is high which is particularly apparent in the two for western regions. Conversely on soils in the Southeastern and South Central regions fertilizer inputs of K exceed proportionally the amount of cropland indicating high leaching and low native levels of soil K. These variations in requirements from region to region indicate an important limitation on substitution of manure for inorganic fertilizers. Inorganic fertilizers can be formulated and sold according to specific nutrient requirements within a region. Manure, however, contains all the plant essential nutrients that must be applied together. Therefore, use of manure to correct a given nutrient deficiency will supply other nutrients that are not required. Excessive application rates of manure may cause nutrient imbalances to occur in plants grown on the soil.

As referred to earlier there is a large area of grazing land utilized in the United States (Table 3). Since grazing animals deposit manure and its nutrients on range or pasture land these nutrients are unavailable for direct substitution for fertilizer. Some benefit would be derived from residual fertility built up in the soil, principally organic N, in situations where temporary pasture is a part of crop rotation sequence. Of the animal groups depicted in Table 2 the grazing animals are cattle breeding stock and calves, sheep and horses. Poultry are raised and live under almost total confinement for the most part. Swine, with the exception of breeding stock, are principally in total confinement or dry lot not used primarily for crop growth. It was assumed then that sheep, beef cows, dairy and beef heifers and all cattle less than 225 kg (500 lb) in weight spend all their time unconfined. In addition dairy cows were assigned 30% of their time on pasture. Horses were not included because state by state data on horse numbers was unavailable. Although these assumptions are difficult to validate, the exclusion of the horse data would tend to offset overestimates in the other animal categories. A state by state accounting of unconfined animals and their manure production, using the same basis as for Table 2, results in an estimated fraction of 57% of the total manure production in an unconfined situation. The nutrients in this manure are therefore dispersed and unavailable for the direct management control for intensive soil fertility programs and crop production programs. Conversely, only about 43% of the manure is produced in confinement and has potential for use to replace inorganic fertilizer.

The regional distribution of all manure closely parallels the cropland distribution as shown in Table 7. This manure, however, includes all that produced by animals on pasture and range as well as animals in confinement. Also in Table 7 is the percentage of manure produced in confinement derived from the assumptions stated earlier. It is easily seen that in the Southeast, Mountain and South Central regions that grazing is relied on heavily for the majority of the animal population. Apparent surpluses of manure over available cropland turn out to be large deficits in these three regions. The amount of manure produced in confinement is not adequate in any region to match the cropland. The relative per-

Table 7. Regional distribution of cropland, all manure and manure produced in confinement.

Region	Cropland	Manure	"Confined" Manure[a]	Relative % of "Confined" Manure and Cropland[b]
			%	
Northeast	3.9	5.1	54.9	72.5
Pacific	5.7	6.1	46.8	49.7
Southeast	9.2	14.7	34.8	55.7
Mountain	10.5	10.9	30.1	31.2
South Central	13.5	17.8	28.3	37.3
North Central	57.3	45.4	54.6	41.4

a. Manure produced in confinement and not on pasture or range.
b. Ratio of percentage of total cropland within a region to percentage of total manure produced in confinement within a region expressed as a percent.

centage of manure produced in confinement and cropland shows the Northeastern region has a most favorable situation because of large dairy and poultry populations and a relatively small percentage of the cropland. As expected the Mountain and South Central regions have the least desirable relationship of cropland to manure production due to large grazing animal populations. The North Central region exhibits a less favorable situation largely because of the extreme concentration of cropland relative to the other regions. Overall, it is significant that on a national scale the maximum potential for use of manure under direct management control is 43% of the total manure. In individual situations and local areas considerable variation from this overall figure will be exhibited. Some farmers may have sufficient animal manure to economically supply all nutrient needs for their crops. Other farmers would need to haul manure over a long distance which may not be economical because of the large non-nutritive bulk of manure.

NUTRIENT CONTENT OF MANURE

The nutrient content of manure from a qualitative standpoint is very favorable. Since manure is of biological origin it contains all plant essential nutrients that are supplied via uptake from soil. Quantitatively, however, considerable variation of manure nutrient content is observed because of differences in animal species, feed consumed and animal condition. After excretion, additional variation in nutrient content of manure is introduced depending on methods of handling, storage, treatment and application to land. Total nutrient content depends on the quantity of manure produced per animal, number of animals and concentration of nutrients in the manure. Individual farmers know the number of animals they have but do not have a satisfactory way of knowing either the quantity of manure produced nor the nutrient concentration. This gap in knowledge from a management standpoint is a major disadvantage. The unknown variations in manure are in contrast to fertilizer in which both nutrient concentration and total quantity of fertilizer material are known precisely. The reader should keep in mind that there is extreme variation in estimates of manure production and nutrition content of manure. Even the variation itself is difficult to estimate.

Comparison of Nutrient Contents of Fertilizer and Manure

In Table 8, a comparison of total N, P and K contents and concentrations is made between fertilizer and manure. The total N contents are quite comparable and, considering the variation, the numbers are identical. Manure exhibits a deficiency of P content

Table 8. Comparison of nutrient contents of manure[a] and fertilizer[b].

	Total × 10³ metric tons	Nutrient Content kg per dry metric ton	lb per wet English ton
Manure N	7367	31	10
Fertilizer N	7540	191	—
Manure P	1463	6.2	2
Fertilizer P	2222	56	—
Manure K	5342	22	7
Fertilizer K	3836	97	—

a. References (3, 4)
b. Reference (1)

relative to fertilizer but manure K actually exceeds the total content of K in fertilizer. The latter is a reflection of K intake by animals grazing or being fed crops grown on soils with adequate available K and therefore not requiring K fertilizer. Also shown in Table 8 are concentrations of N, P and K in manure and fertilizer. The lower concentrations of N, P and K in the manure indicate the large non-nutritive content of manure. This condition is even more drastic when viewed on the basis of manure in its wet condition as in the last column of Table 8. The deficiency of P in manure is further emphasized in these concentration figures as manure contains nearly 9-fold less P per dry metric ton than the average in all fertilizer materials. It should be noted that some fertilizer materials used have considerably higher nutrient concentrations than indicated by this average. Examples of highly concentrated individual fertilizer materials are: anhydrous ammonia with 820 kg N per metric ton, triple superphosphate with 200 kg P per metric ton and raw K ore (sylvinite) frequently containing as much as 200 kg K per metric ton (5).

Regional Distribution of Nutrients in Fertilizer and Manure

Comparisons have been made between nutrient contents of fertilizer and manure to this point without regard for geographical distribution. Comparisons are made in Table 9 of manure and fertilizer on the basis of a nominal rate of application on all available cropland within a region. Comparisons on a national scale reflect directly the total nutrient contents of manure and fertilizer from Table 8. The category labelled "confined manure" includes only the fraction of the manure that is produced in confinement (Table 7) as being available for spreading on cropland. These latter figures are somewhat different than would be derived if the distribution and nutrient variations of each individual animal species were used in the calculation. Considering the variations within species, however, little is gained from this refinement. Examination of Table 9 in more detail reveals considerable variation from region to region. The Pacific and South Central states in particular exhibit deficiencies in manure N relative to fertilizer. The Pacific region has the highest fertilizer consumption rate per hectare of cropland which is the major cause of the deficit there. In the South Central region a high fertilizer N consumption and only 28.3% of the manure produced in confinement combine to give a large gap between manure and fertilizer N. In all other regions the percentage of fertilizer N that could potentially be replaced by manure N is close to one-half or greater. Even in the North Central region which has almost 60% of the cropland the manure replacement potential for N is about 50%.

Replacement potential of manure P for fertilizer P is uniformly low across all regions reflecting the inherently low P content of

Table 9. Nominal rates of N, P and K from manure and fertilizer on a regional basis.

Nutrient and Region	All Manure	"Confined" Manure	Fertilizer Consumption	Potential Replacement of Fertilizer by "Confined" Manure
	kg/hectare			%
Nitrogen				
Northeast	70	38	56	68
Pacific	57	27	91	30
Southeast	85	30	64	47
Mountain	55	17	29	59
South Central	70	20	86	23
North Central	42	23	47	49
United States	53	23	55	42
Phosphorus				
Northeast	14	7.7	27	29
Pacific	11	5.1	16	32
Southeast	17	5.9	20	30
Mountain	11	3.3	8.7	38
South Central	14	4.0	22	18
North Central	8.4	4.6	15	31
United States	11	4.7	16	29
Potassium				
Northeast	49	27	48	56
Pacific	40	19	11	172
Southeast	60	21	56	38
Mountain	39	12	1.3	923
South Central	50	14	33	42
North Central	30	16	27	59
United States	38	16	28	57

manure. The Mountain region has the most favorable situation relative to manure P mostly because of a lower fertilizer consumption in this region. The South Central region exhibits the lowest replacement potential for the same reasons as stated for its N deficit.

Potassium exhibits the most favorable situation for replacement of fertilizer of the three major plant nutrients on a national scale. Closer examination of regional distribution, however, reveals some serious distributional anomalies. The two western regions contain more than enough K to completely replace K consumed in fertilizer. This is caused largely by the lower requirement for fertilizer K in these regions. Replacement values in other regions vary from about 40 to 60%.

Overall, assuming all the potentially available manure (manure produced in confinement) were spread on cropland, then about 42% of the N, 29% of the P and 57% of the K in fertilizer could be replaced by manure (Table 9). This is a maximum potential that does consider distributional problems on a regional or local basis. Neither do these potential figures reflect what an individual farmer may be able to do in his particular situation. For example, in the North Central region where the largest concentration of cropland (57.3%) is located about one-half of the N, one-third of the P and slightly less than two-thirds of the K in fertilizers could be potentially replaced by N, P and K from manure. In addition, the intraregional or local distributional character of the manure production in the North Central region is most favorable also. In 1972 cattle feedlots with capacity of less than 1000 numbered 152,429 (6) (Livestock and Meat Statistics, 1973). About 94% of these smaller and more widely dispersed lots are in the 12 states of the North Central region. In contrast, 79% of the feedlots with capacities over 1000 are in the western states, including Nebraska and Kansas. These large concentrations of animal population may cause local problems of distribution of the manure particularly where nearby cropland areas are limited.

Nitrogen as a Key Nutrient Element in Manure

Of all the nutrients contained in manure, N has received the most attention. There are several reasons for this emphasis on N. As seen from Table 8, N is the single largest nutrient consumed from fertilizers. This is a reflection of N being required in the largest quantity by plants relative to any other nutrient supplied from the soil with the exception of water. Most recently N has received attention because of nitrate as a potential water contaminant. Losses of nitrate from animal and crop production systems are major concerns in the context of potential pollution of

water for human consumption. Considering animal and crop production together N exhibits a larger mass flow than any other single
nutrient except carbon or oxygen. Nitrogen also exhibits significant
mass flow in volatile forms such as ammonia and end products
of denitrification which has implications in its management for
purposes of controlling potential pollution or utilization as a fertilizer.

AMMONIA VOLATILIZATION FROM ANIMAL MANURE

Ammonia Volatilization in Relation to Nitrogen Flow in Crop and Animal Production

As ammonia, N can be lost by volatilization and widely dispersed.
This N is effectively lost for use as fertilizer, which is a decrease in
the fertilizer value of manure. Animals ingest N that has been
taken up by crops in an inorganic form and converted to organic N
in the plant. During digestion enzymatically labile organic N compounds are formed. Following excretion enzymatic reaction with
these labile compounds releases ammonia which is subject to volatilization. Inherently animal production through volatile losses of
ammonia from manure allows a potentially significant "leak" or
mass flow of N from the agricultural N cycle. Ammonia volatilization decreases the amount of N that could be recycled back to a crop
plant where the N originated. Ammonia loss by volatilization occurs in addition to other avenues of loss including leaching and
runoff from manure in animal facilities. Volatile losses of ammonia
may be more significant in total flow of N than those other pathways because of the rapid dynamics of the volatilization process.

To gain some perspective of N flow in crop and animal systems
in the context of human food production estimates were made of
quantities of N flowing in and out of crop and animal production on a
national scale. The units of N flow were expressed as per capita of
human population to be consistent with per capita N consumption
as protein in the human diet. About 50 kg N/person/year goes into
crop production with about 36 kg N/person/year from fertilizer
(see Table 8) and the balance from symbiotically fixed and soil
derived N. The latter was estimated from N contents of legumes
and unfertilized crops. About 34 kg N/person/year is consumed
by animals in feed (2, 7) with an additional unknown quantity consumed from grazed range and pasture forage. Human consumption of
N from animal protein amounts to 4 kg N/person/year (2). The balance of the N associated with animal production, about 36 kg N/
person/year, is contained in the manure (see Table 8). There is almost 10 times as much N in animal manure than is consumed by

humans from animal products. Even a 10% loss of N as ammonia from manure represents a significant flow of N and estimates have been as high as 50% of N in manure lost by ammonia volatilization.

There is a substantial potential for ammonia volatilization from N flow on a national scale because of the large quantity of N in animal manure in the United States. Volatile movements of ammonia on a national scale are not known because relatively little quantitative data is available about its magnitude. The scientific literature confirms that ammonia volatilization does occur from animal production facilities (8-12). Measurement of ammonia in the air around these facilities has shown elevated levels of ammonia over levels in the surroundings (11, 12). Manure removed from animal production facilities and spread on land for disposal or as a fertilizer is subject to further volatile losses of ammonia. A thin layer of manure on the soil surface is exposed allowing it to dry rapidly. During the drying process ammonia volatilization takes place.

Experimental Measurements of Ammonia Volatilization From Dairy Manure

A series of experiments on volatile losses of ammonia from dairy manure spread on the soil surface was undertaken to measure quantities of N flowing by this process. The experiments were performed under field conditions to duplicate actual situations. Five experiments were carried out over a period of 2 years under winter, spring and summer conditions. Only a summary of the experiments follow with details reported elsewhere (13). Experimental periods, rates of manure application and analysis of the manure for dry matter and N content are given in Table 10.

Table 10. Rates of manure application and manure analysis.

Experimental Period	Rate of Wet Manure Application metric tons/ha	Dry Matter %	Total N	Ammonia
			N Analysis % of dry matter	
April 9-30, 1973	225	17.1	2.85	0.909
August 13-Sept. 7, 1973	225	18.4	2.41	0.457
January 7-March 4, 1974	200	27.5	1.81	0.481
January 10-March 4, 1974	34	21.9	1.86	0.503
April 24-30, 1974	200	21.4	1.78	0.426
April 25-30, 1974	34	21.7	1.37	0.271
May 27-June 12, 1974	200	21.2	2.18	0.653
May 31-June 12, 1974	34	30.0	2.04	0.444

Losses of ammoniacal N were monitored with time after spreading by measuring the total ammoniacal N content in samples of manure collected directly from the soil surface. Rates of loss were determined by sampling at close intervals from the time of spreading. Dry matter content was also measured on each manure sample to ascertain changes in moisture status. Major precipitation or thaw events which caused runoff or leaching were criteria for termination of an experiment, since ammoniacal N losses were obtained by difference between successive sampling dates. Soil samples were taken periodically to monitor movement of ammoniacal N in the soil profile.

Because the total time span of the experiments varied from 5 to 25 days and the loss curves (Table 11) approximated first order rate processes the parameter denoted half-life (time required for 50% loss) was adopted to compare rates of ammonia volatilization. In the non-winter experiments for a period of 5 to 7 days after spreading, rates of ammonia loss are represented by mean half-lives of 1.86 to 3.36 days, respectively, for the 34 and 200 metric ton/hectare rates of manure application. After the initial period of loss, the rate of ammonia volatilization slowed in most cases. The cause of

Table 11. Loss curves for ammoniacal N loss by volatilization[a].

Elapsed Time After Spreading Days	Experimental Period					
			April 74[c]		May-June 74[b]	
			34[d]	200[d]	34	200
	April 73[b]	August 73[b]	metric ton/ha		metric ton/ha	
			percent remaining[e]			
0	100	100	100	100	100	100
1	75.1	73.8	69.5	82.5	90.9	89.5
2		60.3	47.0	54.2	34.0	63.5
3	60.3	55.8	30.0	43.1	25.2	57.7
4		43.5	23.0	34.3	30.7	35.2
5	38.6	51.1	16.0	38.8	16.1	30.0
6		55.9			16.8	26.2
7	31.4				13.0	27.2
8		40.6				
9					15.2	18.5
10	23.4	27.8				
11					8.6	16.7
12		27.2				
13						9.5
14	18.9					
15		12.3				12.4
19		3.1				
21	18.6					
25		0.5				

a. Excludes January 1974 experiment.
b. Experiment terminated by rainfall event.
c. Experiment terminated by plowing down of manure.
d. Rates of manure application.
e. Relative to quantity of ammoniacal N applied.

the more rapid loss at 34 metric ton/ha is more rapid drying of the thinner layer of manure on the soil surface. Therefore, with a fixed quantity of manure applied over a larger area of land, as compared to less land area at high rate of application, the probability of ammonia loss is increased at a lower rate of manure application. This means that higher total losses of ammonia by volatilization are more probable under current practice than under higher rates of application.

The results of the ammonia losses are summarized in Table 12. The mean loss, excluding the January 1974 trial, was 85% of the ammoniacal N content of the manure at spreading. In the winter trial, ammonia volatilization was precluded by a combination of subfreezing temperatures, snow cover and a rapid thaw which leached the ammoniacal N into the soil. The absolute quantities of N loss by ammonia volatilization were large except for the single winter trial. Even the quantities of N lost from the 34 metric ton/ha rate of manure were significant. This rate of manure application is more typical of current farmer practice, and the rate of loss of ammonia from the 34 metric ton/ha rate is more rapid than at the higher manure rates.

Table 12. Quantities of ammonia lost from dairy manure spread on the soil surface.

Experimental Period	Rate of Manure Application metric tons/ha	Total NH₃ lost kg N/ha	NH₃ Lost as % of Applied Ammoniacal N
April 1973	225.	285.	81.
August 1973	225.	189.	99.
January 1974	34.	0	0
	200.	0	0
April 1974	34.	17.	85.
	200.	112.	61.
June 1974	34.	41.	91.
	200.	248.	90.

Estimated Quantities of Ammonia Loss from Animal Manure

The investigation of ammonia losses from dairy manure deals specifically with land-applied manure. There is evidence, however, from comparing the literature on N analysis of manure from feeding trials and from normal feeding situations that significant losses of ammonia occur within a very short time after defecation. This is supported by the elevated ammonia concentrations observed around animal production facilities. Chemical conditions are favorable (pH 8, high concentration of ammoniacal N) in dairy manure within 8-10 hours from defecation (Personal communication with R. J. Cummings, Research Specialist, Cornell University).

In Table 13 analysis of dairy manure is compared under various conditions as a sequence from defecation to several days after land application. The trends are clear. Through the sequence of events there is a general drying which is probably a combination of evaporation and liquid separation. Total N decreases throughout the sequence. This decrease is apparently due largely to losses of ammoniacal N as the fraction of this form of N relative to total N decreases more than 10-fold from defecation to land application. The implications of this data are that ammonia losses are significant throughout the sequence depicted in Table 13. The probability is very high that volatile losses of ammonia is the major pathway of these losses. Confirmation of this hypothesis, however, requires direct experimentation.

Table 13. Decrease in ammoniacal N from defecation through spreading on land.

Manure Condition	Dry Matter %	Total %	Ammoniacal N as % of total N
As defecated[a]	11.4	5.65	60.7
Farm-fresh[b]	15.0	3.07	36.2
Farm-stored[c]	24.0	1.84	25.1
Farm-stored after spreading[d]	46 to 85	1.46	4.52

a. Derived from experimental feeding trials where urine and feces were collected and analyzed separately (Personal communication with Carl E. Coppock, Animal Sci. Dept., Cornell Univ. and reference (14)).
b. Obtained from dairy used as source of manure in this investigation. This manure was about 24 hours old and obtained at daily cleaning time. (Collected February 29, 1972).
c. Obtained from same dairy farm but stored in an unprotected pile for an indefinite period (mean of six samples).
d. This is mean (six samples) condition of the "Farm-stored" manure from the 1974 series of experiments several days after spreading on the soil surface.

Assuming ammonia volatilization from manure removes about 50% of the total N in animal manure as the evidence just presented suggests then very large quantities of N flow are indicated. From data previously presented 43% of the manure N produced in confinement amounts to 3.2 million metric tons of N. Loss of 50% by volatilization of ammonia would be 1.6 million metric tons or 12 kg N/ha of cropland. This also amounts to 7.6 kg N/person/year in the United States which is more than the combined plant and animal protein dietary consumption of 5.8 kg N/person/year. These estimates may be conservative since the estimate of total N in manure is based on analyses of manure from which significant losses of N could have occurred previous to and during the analysis.

**Implications of Ammonia Volatilization from
Animal Manure to Management for Pollution Control
and Soil Fertility Maintenance**

The overall implication of the results reported above is that volatilization of ammonia from animal manure is a large flow of N associated with agriculture. The data in Table 13 and the reported investigation of ammonia losses are important in management of N from manure because the manager must consider ammonia volatilization losses through the same sequence from defecation to land application. Inadequate management of the manure during handling before land application as well as after can result in substantial losses of N by ammonia volatilization. It is apparent that utilization of N from manure is inefficient not solely because of large scale confinement of animals, but because the opportunity for ammonia loss is inherent in the processes of manure production. For the latter reason, if maximum loss is desired to reduce pollution potential then promotion of volatile losses of ammonia would be relatively easy to achieve.

Clearly, the desirability of volatile losses of ammonia depends on the management objective. At present, land application is part of almost every manure management scheme. In situations where there are large concentrations of animals and little land is available for spreading of the manure, ammonia volatilization is desirable because the potential is lessened for nitrate accumulation in groundwater in the area. Maximum efficiency of N utilization from manure in crop production, however, is also a desirable management objective. Therefore, where manure is spread on cropland and the ratio of land area to animal numbers is favorable, minimization of volatile ammonia losses is the preferred manure management objective. Assessment of either of these two manure management situations is aided by quantitative knowledge of the processes and dynamics of ammonia volatilization from manure.

**Implications of Ammonia Volatilization to Methane Production
from Animal Manure**

In addition to management of manure for disposal with minimal environmental impact and for utilization as a fertilizer, a third management situation arises in connection with anaerobic digestion to produce methane. The management objective in methane production from manure is to provide energy to partially offset energy demand on farms. This objective, however, is not mutually exclusive from the two discussed above. During anaerobic digestion microbial action releases methane and leaves a residue containing ammoniacal N in greater concentration than the original manure. The residue in the form of a slurry must be removed from the digester and disposed or utilized. The anaerobic digestion residue containing a significant concentration of ammoniacal N is subject to volatile losses of N from the point of exposure to air. When the residue is spread on the soil surface, ammonia volatilization will likely occur in a manner exactly analogous to that reported above for raw dairy manure. The physical condition of the slurry may cause some deviation from the observed results with raw manure. There are possibilities of some soil contact by infiltration of the slurry or crusting of the slurry surface which may be a physical impedance for diffusion of ammonia into the air. These deviations may change the rate of ammonia loss but not the fact that volatilization will occur. A rate effect would be an advantage if volatile losses were slowed sufficiently to allow incorporation of the residue into the soil.

There are two management techniques proposed for use in disposing of the slurry residue from methane production from manure that are particularly negative in the context of soil fertility management. One proposed method of getting rid of the residue is dilution and distribution through an irrigation system on the land. Disregarding for the moment any potential water management problems it seems apparent that such a system would promote large volatile loss of ammonia during the actual irrigation process. The degree of ammonia loss with irrigation of this slurry residue has not been investigated. Because of the high solubility of ammonia in water, infiltration of a diluted slurry may be more efficient in contacting soil than a surface application of undiluted material. Losses of ammonia from the spray droplets during transit from the irrigation nozzle to soil must be balanced against possible ammonia retention from infiltration. Should the infiltration segment of the process be dominant in determining ammonia flows then water management to maximize infiltration is desirable. There is, however, potential for losses of N intended as fertilizer using spray irrigation as a method of distribution.

A second proposed technique in connection with methane production is harvest of crop residues to provide an additional source of carbon to augment carbon supplied from the manure. Crop residue management is an important aspect of maintaining the fertility status of soils. Returning crop residues to the soil partially offsets the removal of N in the portion of the crop harvested for traditional economic incentives. The organic forms of N in the crop residue provide reasonable protection against loss as microbial action assimilates the fresh residue into the organic N pool of the soil. Harvest of this crop residue for methane generation, however, transforms N into the volatile ammonia form. Even the best managed return of the digestion residue to the fields where the crop residue was obtained is not sufficient to prevent a potentially significant loss of N as volatilized ammonia from the soil-plant system. Long term operation of a methane generator augmented by harvested crop residues may well cause a general decrease in the N fertility status of the soil. The energy cost of external inputs of N fertilizer to make up this deficit could exceed the additional energy from methane obtained from harvested crop residues.*

SUMMARY AND CONCLUSIONS

Manure is a descriptive term that derives its meaning from use of a material to fertilize soil. Since animal waste was one of the most commonly used materials for this purpose, manure and animal waste have become almost synonymous terms. The original meaning of manure, once equated with fertilizer, has become increasingly associated with animal waste. The use of manufactured fertilizer over the last few decades has largely replaced manure in its role as a fertilizer. Problems of large quantities of accumulated manure and increasing cost of manufacturing fertilizer has renewed interest in using nutrients from manure as fertilizer. Manure is also being investigated regarding the feasibility of producing energy via methane generation with a realization that fertilizer values may be retained simultaneously.

Evaluation of replacing inorganic fertilizer with manure nutrients requires an awareness of the processes involved in fertilizer, crop and animal production and the relationship of these processes. There are several processes associated with each of these production activities which are all interlocked by the flow of nutrients essential to plants and animals. Fertilizer production at present is characterized by exploitation of nonrenewable energy and raw fertilizer material resources and is an external input to soil-crop systems. Crop and animal production are closely integrated through the flow of nutrients in feed and return of nutrients to the soil-plant system from manure.

Examination of the manure sources in the United States reveals that cattle as an aggregate are the largest producers of manure. A similar overview of land resources shows that about two-thirds of the total land surface in the United States is used either for growing crops or grazing of animals. Feed and food grains occupy over 50% of the cropland and receive applications of over 50% of the fertilizer consumed in the United States. From this a picture emerges of where the manure nutrients are derived and what crops require the majority of nutrient input. Corn, soybeans, wheat, and cotton as an aggregate receive nearly two-thirds of the nutrients

---

* Soil injection would minimize the N losses.

consumed from fertilizer. The conclusion to be drawn here is that closer integration of cattle production and production of these four crops would increase the probability of feasibly replacing fertilizer with manure.

Regional distribution of cropland and consumption of fertilizer nutrients reveals large variations in plant nutrient requirements from region to region. Fertilizer has an advantage over manure in that fertilizer can be formulated and tailored to specific nutrient requirements whereas, nutrients in manure must be taken all together. Use of manure to correct a single nutrient deficiency may cause excesses and imbalances in nutrients not deficient.

A major distributional characteristic of manure production results from the deposition of manure by grazing animals. An estimated 57% of the total manure production is by grazing animals. This means that only the manure produced in confinement (43%) is available for direct substitution for fertilizer. There are large regional differences in quantities of manure produced in confinement with a likelihood that intraregional distribution may be even more variable.

Nutrient contents in manure are extremely variable because of many factors. This is a major disadvantage relative to fertilizers which are precisely formulated. Concentrations of nutrients in manure, in addition to being variable, are generally much lower than in fertilizer which increases transportation costs for manure nutrients. Water and organic matter are the major components of manure that are non-nutritive. These components not only decrease nutrient concentration but they also make the handling characteristics of manure less desirable than fertilizer.

Nitrogen is the major plant nutrient contained in manure, whereas phosphorus is relatively deficient. Potassium in manure ranks second to N content, but K is not required by plants in as large a quantity as N. Neither is K required so universally in a geographical sense as N. Considering only manure produced in confinement, manure has nutrient contents that could potentially replace about 42% of the N, 29% of the P, and 57% of the K consumed as fertilizer. Regional distributions of manure nutrients deviate widely from these national figures with intraregional deviations expected to be even greater. One significant fact is that in the North Central region, which contains almost 60% of the cropland, about 49, 31 and 59% of the N, P and K from fertilizer could be replaced by manure N, P and K.

Nitrogen is a key nutrient element in manure because N exhibits the largest mineral-nutrient flow in soil-plant and animal systems. Nitrogen also exhibits the greatest potential for movement in soil-plant systems by water, volatilization and other avenues of mass flow. These processes have implications in potential pollution of water and efficiency of utilization of N in soil-plant systems as well as animal and human nutrition in general.

Volatilization of ammonia from animal manure is a large flow of N that decreases the value of manure as a replacement of inorganic fertilizers. The flow of N as volatilized ammonia is significant in the total context of food production. Experimental measurements of ammonia loss from dairy manure spread and left on the soil surface show that under favorable conditions 70-100% of the ammoniacal N in the manure volatilizes. This method of land application is commonly practiced. Furthermore, there is evidence that ammonia volatilization begins shortly after defecation and may account for an estimated 50% of the total N content in manure. Applying this 50% figure only to N from manure produced in confinement results in a flow of N equivalent to 12 kg N/ha of cropland or about 22% of the total fertilizer N consumed (55 kg N/hectare). Relative to human nutrition, ammonia volatilization from manure produced in confinement amounts to 7.6 kg N/person/year which exceeds the dietary protein consumption from both plant and animal sources combined (5.8 kg N/person/year). If all manure N were brought under these assumptions the quantities of N flowing by ammonia volatilization from manure would be more than doubled. Clearly ammonia volatilization from animal manure has implications for pollution potential from manure, efficiency of food production, methane generation from manure and associated management systems. This inherent leakage of N from the agricultural N cycle may well be the final determinant of substitutability of manure for fertilizer. The N inefficiency in the present crop and animal production systems may dictate changes which decrease the flow of N for food production.

REFERENCES

1. Hargett, N. L. 1974. 1974 Fertilizer summary data. National Fertilizer Development Center, Tennessee Valley Authority. Muscle Shoals, Alabama. 132 p.
2. U.S. Department of Agriculture. 1973. Agricultural Statistics 1973. U.S. Govt. Printing Office, Washington, D. C. 617 p.
3. Bruns, E. G. and J. W. Crowley. 1973. Solid manure handling for livestock housing, feeding and yard facilities in Wisconsin. University of Wisconsin. Extension Bulletin No. A2418.
4. Azevedo, J. and P. R. Stout. 1974. Farm animal manures: an overview of their role in the agricultural environment. California Agricultural Experiment Station Manual 44. 109 p.
5. McVickar, M. H., G. L. Bridger and L. B. Nelson. Editors. 1963. Fertilizer technology and usage. Soil Science Society of America; Madison, Wisconsin. 464 p.

6. U.S. Department of Agriculture. 1973. Livestock and meat statistics. Economic Research Service, Statistical Reporting Service and Agricultural Marketing Service Statistical Bulletin No. 522.
7. Morrison, F. B. 1958. 9th ed. Feeds and feeding, abridged. The Morrison Publishing Company; Claremont, Ontario, Canada. 696 p.
8. Salter, R. M. and Schollenberger. 1939. Farm manure. Ohio Agricultural Experiment Station; Wooster, Ohio. Bulletin 605.
9. Hutchinson, G. L. and F. G. Viets, Jr. 1969. Nitrogen enrichment of surface water by absorption of ammonia volatilized from cattle feedlots. Science 166:514-515.
10. Elliot, L. F., G. E. Schuman and F. G. Viets, Jr. 1971. Volatilization of nitrogen-containing compounds from beef cattle areas. Soil Sci. Soc. Amer. Proc. 35:752-755.
11. Luebs, R. E., K. R. Davis and A. E. Laag. 1973. Enrichment of the atmosphere with nitrogen compounds volatilized from a large dairy area. J. Environ. Quality 2:
12. Luebs, R. E., K. R. Davis and A. E. Laag. 1974. Diurnal fluctuation and movement of atmospheric ammonia and related gases from dairies. J. Environ. Quality 3:265-269.
13. Lauer, D. A., D. R. Bouldin and S. D. Klausner. 1975. Ammonia volatilization from dairy manure spread on the soil surface. J. Environ. Quality. (submitted for publication).
14. Fisher, L. J. 1974. Influence of feeding systems, digestibility of ration and proportion of concentrate consumed on the quantity and quality of excreta voided by lactating cows. pp. 283-290. *In* Processing and Management of Agricultural Waste. Proceeds of the 1974 Cornell Agricultural Waste Management Conference (Rochester, NY).

30.

Energy Use and Economics in the Manufacture of Fertilizers

John L. Sherff*

Under contract to the Federal Energy Administration (FEA), Arthur D. Little, Inc., studied the economic impact of shortages on the U.S. fertilizer industry. This chapter presents some of the findings of that study which pertain to the importance of energy to the manufacture of fertilizers.

ENERGY CONTENT OF FERTILIZERS

In order to determine the energy contents and costs for the manufacture of fertilizers and for the fertilizer industry, as well as to determine fuel switching capabilities, we estimated the quantities of energy, both fuels and electric power, required to manufacture each of the fertilizer raw materials and end products. Since the energy requirements are functions of plant design and the skills of the operators, there is variability, sometimes considerable, from plant to plant. We tried to select typical or average conditions. Fortunately, since there has been so much new plant construction in the past ten years, the typical plant to produce nitrogen or phosphatic fertilizer tends also to be a fairly modern one, built as a unit rather than pieced together over a number of years, and fairly easily generalized as to energy use. Notable exceptions are mining operations and the old superphosphate and mixing plants. Energy requirements for these individual processes are represented

---

* Arthur D. Little, Inc., Cambridge, Massachusetts.

in Table 1. These factors were then used to determine the energy content of individual fertilizer materials, total energy use by the fertilizer industry, and costs of energy to the industry.

Table 1. Energy use in fertilizer production.

Product or Operation	Electric (kWh)	Natural Gas or Fuel Oil (MM Btu)	Steam (M Pounds)*
Ammonia	45.5	39.5	—
Urea Solution	153	1.75	—
Urea Prilling	27	1.75	—
Nitric Acid	45	—	(0.7)
Ammonium Nitrate Solution	6	—	—
Ammonium Nitrate Prilling	35	2.3	—
Sulfur Mining	20.1	6.81	—
Sulfuric Acid	12	—	(—1.9)
Phosphate Rock Mining and Beneficiation	50	—	—
Phosphate Rock Drying	—	2.2	—
Phosphate Rock Grinding	13	—	—
Phosphoric Acid (P_2O_5 Basis)	120	—	4.44**
Normal Superphosphate	20.8	—	—
Triple Superphosphate (Granular)	40	1.3	—
Diammonium Phosphate	20	1.5	—
Potash Extraction (Flotation Method)	90	2.26	—
Wet Mixing	35	0.5	

* Parentheses indicate credit.
** Usually provided by the sulfuric acid plant.

Far and away the largest energy requirement for the fertilizer industry is for the manufacture of ammonia. Since synthetic ammonia is the source of almost all nitrogen fertilizers in the country, it is the nitrogen component of fertilizers which requires the greatest amount of energy. Depending on the product, nitrogen fertilizers require 49-62 million Btu per ton of nitrogen in their manufacture. Phosphate fertilizers, on the other hand, require only 12-19 million Btu per ton of P_2O_5, including the energy required to produce the phosphate rock and sulfur raw materials. Potash fertilizers require only about 6 million Btu per ton of K_2O. In deriving these figures, we utilized a factor for electric power of 10,400 Btu per kilowatt hour, which reflects the approximate input of fuel to a generating plant to produce one kilowatt hour at the point of use. The energy contents of fertilizer materials are presented in Table 2.

Fuels and electric power constitute roughly 12% of the cost of fertilizer manufacture, as shown in Table 3. We were asked to determine what proportion of the price of fertilizers was represented by energy. However, for the past year and a half, the prices of fertilizers have been rising so rapidly that no one in the industry can agree on what they are or what they were at any given time. Thus,

Table 2. Energy content of individual fertilizers (million Btu per ton of nutrient).

	N	P_2O_5	K_2O
Anhydrous Ammonia	49		
Urea			
Solution	58		
Prilled	62		
Ammonium Nitrate			
Solution	54		
Prilled	62		
Nitrogen Solution	55		
Normal Superphosphate		15	
Triple Superphosphate		19	
Diammonium Phosphate	49	12	
Potassium Chloride			6
Mixed Fertilizer	52	16	6

we worked with a normalized price by adding up all the costs and allowing the manufacturers a 30% return on fixed assets and working capital to cover interest, income taxes, and profits. In order to estimate fuel prices, we reviewed the natural gas prices being paid by most ammonia plants. Discussions with about 10 major companies established fuel oil and electric power costs and trends on a geographic basis. The manufacture of nitrogen fertilizers in 1973 required the expenditure of $172 million for fuels, largely natural gas, and $12 million for electric power. The manufacture of phosphate fertilizers required the expenditure of $63 million for fuels, primarily for the mining of phosphate rock and sulfur, and $25 mil-

Table 3. Cost components of fuels in basic fertilizer manufacture (1973).

	Nitrogen Fertilizers		Phosphate Fertilizers		Total	
	$ Million	%	$ Million	%	$ Million	%
Raw Materials						
Phosphate Rock	—	—	492	36	492	22
Sulfur	—	—	172	13	172	8
Other	26	3	15	1	41	2
Fuels						
To make Raw Materials	—	—	55	4	55	2
Other	172	19	8	1	180	8
Electric Power						
To make Raw Materials	—	—	14	1	14	1
Other	12	1	11	1	23	1
Other Manufacturing Costs	199	22	268	20	467	20
Plant Costs	409	45	1,035	75	1,444	63
General Selling & Administration	81	9	50	4	131	6
Margin (30% of Assets)	419	46	286	21	705	31
Selling Price (Theoretical)	909	100	1,371	100	2,280	100

lion for electric power. In total, then, these two major divisions of the fertilizer industry spent $235 million for fuels and $37 million for electric power in 1973. Additional areas of the industry, such as potash mining and mixing, required additional fuels, but on average about 10% of the theoretical sales dollar was spent on fuels and another 2% on electric power.

The fertilizer industry has a heavy reliance on natural gas. The total use of fossil fuels was 574 trillion Btu in 1973. All but 16 trillion was as natural gas, as can be seen from Table 4. Fuel oil has been used only to a limited extent along the East Coast, where imported residual fuel oil was available. The Florida phosphate industry, traditionally a user of natural gas, has now almost completely converted to fuel oil because of the shortage of natural gas in Florida and the relative ease with which the phosphate industry could make the shift.

Table 4. Fuel and electric power use by the fertilizer industry (1973).

	Fuel (10^{12} Btu)			Electric Power 10^6 kWh
	Natural Gas	Fuel Oil	Coal	
Ammonia — Feedstock	298	—		584
— Fuel	193	1		—
Urea	8	—		460
Ammonium Nitrate	9	—		390
Phosphoric Acid	—	—		647
Triple Superphosphate	1	1		110
Diammonium Phosphate	4	1		286
Normal Superphosphate	—	—		65
Mixtures	5	—		350
Subtotal	516	4		2,893
Phosphate Rock	3	12		1,643
Sulfur	29	—		76
Potash	10	—		379
Subtotal	42	13		2,098
Total	558	16		4,991
National Energy Use for				1,800,000
All Purposes	23,400	12,100		
Proportion Fertilizers	2.4%	0.1%		0.3%

Considering the size of the fertilizer industry, it is a small energy user in the national context. It consumes 0.3% of the nation's electric power, 0.1% of the nation's distillate and residual oil, and 2.4% of the nation's natural gas.

EFFECTS OF ENERGY SHORTAGES ON
FERTILIZER AVAILABILITY AND COST

There are shortages of all the major fertilizer raw materials—phosphate rock, potash, sulfur and ammonia or natural gas—in the United States as well as world-wide. Except for natural gas, these shortages are due not to shortages of energy but to the cyclical nature of the industry.

Several years of overcapacity, depressed prices and low profits discouraged investment in new capacity. Supply and consumption came into balance much more rapidly than those in the industry expected, and it was not possible to expand quickly enough to meet the demand. Such shortages in phosphate rock and sulfur may be alleviated in 2 to 3 years, and possibly those industries could go through another overcapacity cycle. Political problems in Canada may retard potash expansion to meet market demands. If these problems are ironed out, however, there are ample reserves of potash which can be quickly developed.

The ammonia shortage was caused by the same cyclical factors. The shortage of natural gas did not cause it. The shortage of natural gas may prolong the ammonia shortage, however.

That fuel on which the fertilizer industry relies the most is in shortest supply. Natural gas production is in decline, and nothing will happen to turn this around very quickly. So far, we have been saved from catastrophic shortages of natural gas by two consecutive warm winters. Nonetheless, natural gas production was off 6% in 1974 and may well decline a similar amount in 1975.

So far, the nitrogen industry has had an unusually successful record in holding onto its gas supplies. According to The Fertilizer Institute, 337,000 tons of ammonia production will be lost this year because of a shortage of natural gas. This represents only about 2% of total production capability. The Federal Power Commission has granted a high priority to process uses of gas and this has benefited the ammonia industry. Furthermore, hard work on the part of the producers to assure themselves of gas supplies has contributed to this factor of success. Nonetheless, the supply of natural gas is expected to continue to decline, and the ability of existing ammonia producers to maintain their supplies will diminish. This will force shifts to using fuel oil for reformers which will require significant investment on the part of the industry as well as to require the use of much more expensive fuels.

Already the phosphate industry of Florida has made the conversion to fuel oils. Since fuel oil prices have quadrupled in the last year and a half and since fuel oil had previously been more expensive than natural gas, this has resulted in a very large increase in their fuel bill. Fuel, however, was not an important part of their cost of manufacture and so has not increased their costs significantly.

While fuels and electric power are extremely important to the industry, the cost of these fuels has not been of outstanding importance. The major escalations in fuel costs have been for fuel oil and in some instances electric power. These are major inputs to the phosphate fertilizer industry, where the cost of fuel is not particularly important. Much lesser escalations have occurred in the price of natural gas, although some individual users have had their prices more than trebled. Thus, the nitrogen industry, for which fuel costs are a very critical component, has on average suffered lower escalations in prices than has the phosphate industry. Nonetheless, the costs of certain ammonia producers have increased significantly, and in a shortage situation the price can be set on the basis of the highest cost marginal supply.

Future escalations in the price of gas to ammonia plants on average will be very great. Intrastate gas will sell at market clearing levels. Unless I miss my guess, interstate gas to industrial customers, even though regulated at the federal and state levels, will likewise sell at market clearing levels. This means $2-3 million Btu—much higher than at present.

The price for natural gas at the wellhead has risen very fast. Gas prices which a few years ago were between 15¢ and 25¢ per MCF are now at between $1.50 and $2.00 per MCF, an increase of eight-fold. Prices for natural gas purchased from interstate pipelines or their dstributor customers have been insulated from these rapid escalations by the large amount of gas purchased on long-term contract some years ago, as well as by federal regulation which prevents interstate pipelines from paying going market prices for new gas. These same regulations which maintain prices at a low level also are preventing interstate pipelines from competing for new gas supplies, and, therefore the customer is faced with the predicament of being told that his natural gas is cheap, but unavailable. As gas supplies to interstate pipelines continue to decline, existing ammonia plants being served by interstate pipelines will lose their gas supplies unless they are allocated it. Under existing regulations, new ammonia plants cannot be built based on interstate gas.

Intrastate gas is still available to new and existing customers, but at a price. The price for new intrastate gas is very high—between $1.00 and $2.00 per MCF. Nearly all new ammonia plants which have recently been announced will be based on intrastate gas. These plants will pay a minimum of 60¢ per MCF and perhaps as high as $2.00 for their gas. There is still high variability in the intrastate pricing, and new ammonia plants are being sited so as to take ad-

vantage of specific situations where the gas has not yet reached the
$1.50-$2.00 level. Some exitsing plants using intrastate gas are at a
greater disadvantage than the new ones. Those which have contracts expiring are at the mercy of their suppliers and may be
forced to pay prices up to $2.00 per MCF for gas. Some of those
which have gas under contracts which do not expire for some time
are finding that their suppliers are not able to honor those contracts.

ALTERNATIVES

For those subject to curtailment of natural gas, the first reaction
is to develop arguments why they should have high priority use of
that gas. Proposals have been made to Congress to allocate natural
gas to fertilizer plants. Except for the manufacture of ammonia,
this is not necessary. Other fertilizer plants can shift to fuel oil, although not instantaneously. The fuel is used to raise steam or in
furnaces for drying. Historically, the industry has selected cheaper
fuel, which, except in a few instances, was natural gas. Conversion
to the use of fuel oil requires little more than changes in burners
and piping and the installation of storage tanks. In some instances,
soot removal will be necessary. There are, of course, exceptions.
Severe problems have been encountered in plants using bag filters
on their dryers. Collection of hygroscopic dusts is a problem under
any conditions, and operators have been unable to adjust to the use
of fuel oil for such dryers.

In most instances, though, the major incentive for staying with
gas is fuel economics. The actual cost of operations or of converting
plants to the use of other fuels is minimal. Thus, if the fuels in
question were at equal cost, more fertilizer plants would quickly
shift to liquid fuels from natural gas. The investment, when amortized over its useful life, is small, as are increased operating costs.
The total cost per ton of product, exclusive of the incremental fuel
cost due to differences in the prices of fuels, for most products is
only 10-50¢ per ton as follows:

Boilers	10-50¢ per ton
Dryer Furnaces	12-17¢ per ton
Ammonia Reformer Fuel	25¢ per ton

Why haven't fertilizer manufacturers shifted to alternate fuels
when it is so inexpensive to do so? By and large they haven't been
forced to do so, and the differential in the cost of fuels remains
very large in most cases in spite of increasing cost of natural gas.
It is evident, however, that as soon as they are convinced that
natural gas will be unavailable, they will shift very quickly to the
use of fuel oil. In most cases, the cost of fuel is not a significant part
of total manufacturing cost and therefore, they can afford the
increased cost. The actions of the phosphate industry in Florida in-

dicated that the industry will not hesitate to put in the investment in order to assure that production is maintained and that its existing investment is not jeopardized by a shortage of fuels.

The manufacture of ammonia is a different story, however. You will note that in Table 4 we have separated ammonia's use of natural gas into categories of feed stock and fuel. Natural gas is a feedstock providing hydrogen for the synthesis of ammonia. As such, it is very difficult to convert the use of natural gas to fuel oil or even to lighter fractions such as LPGs or naphtha without, as a minimum, reducing the capacity of the plant and, as a maximum, requiring the scrapping of the plant altogether. As a fuel, natural gas is used to fire high temperature reformers. Because of technical requirements, it is much more difficult to convert reformers from natural gas to fuel oil than to convert boilers or furnaces, but it can and is being done.

Thus, while the fertilizer industry has a very heavy reliance on natural gas, over half of this natural gas use is not convertible to other fuels, and only a very small proportion of the total is easily convertible to other fuels.

Neither the FPC nor state regulatory agencies have developed a mechanism for helping to secure supplies of natural gas to new ammonia plants. In the face of declining natural gas supplies, gas for new customers can only be provided by taking gas supplies away from existing customers. This could happen by the price mechanism if regulatory bodies would allow free market pricing for natural gas. Since they do not, legislation is needed to reduce or prohibit the use of natural gas by other users who can use alternative fuels. At the present price of natural gas, ammonia can be produced more cheaply from natural gas than from other feedstocks. However, somebody must use the more expensive fuels. Why should ammonia producers have a privileged position?

No new ammonia plants using liquid or solid hydrocarbon feed-stocks are being built in the United States. Given the high prices for petroleum, companies considering non-natural gas plants will more likely consider technology to produce ammonia from coal. The technology is not nearly so well developed as for petroleum-based plants, but the potential price difference between petroleum and coal may provide the impetus to develop better coal-based technology.

Two factors are discouraging the investment in ammonia facilities based on non-natural gas feedstocks: the pending apparent overcapacity in ammonia, and the threat of foreign competition.

New ammonia capacity is going in at a significant rate. If all capacity recently announced goes ahead, U.S. ammonia capacity will increase by 33% between 1974 and 1978 from 17.6 million tons to 23.3 million tons. Additionally, we expect an additional 2.2 million tons in Alberta by 1978, mostly slated for the U.S. Plans for even greater amounts of ammonia in Alberta have been announced, but the Province has decided to limit the quantity of natural gas used for ammonia manufacture. One million tons per year of new capacity aimed at the U.S. market is also being built in Trinidad and Mexico. There is no guarantee that all of the U.S. capacity will be built. Nor is there a guarantee that existing plants will be able to continue to operate at high levels. But the threat of this pending ammonia surplus and the knowledge that companies are still able to find natural gas in a period of general shortage makes investors wary of investing in coal or oil-based plants, which will have costs greater than those of natural gas plants.

Another perceived threat is the large amounts of gas available in the Middle East and Venezuela. If this gas is used to provide the basis for a massive investment in ammonia plants, it could provide a worldwide surplus. Once such plants are built, the output would be sold at almost whatever price will be necessary. Once again, a high cost U.S. producer could be put out of business under such circumstances.

Probably not all of the capacity will be built in the U.S. because not all plants which have been announced will be able to get gas supplies. The ammonia capacity in the Middle East and elsewhere will not be as great as it potentially could be. Also, the existing U.S. industry will have difficulty maintaining high operating rates. Nonetheless, the threat of these occurrences is making investors timid.

If the U.S. ammonia industry expands to meet domestic needs, the U.S. could save the importation of roughly 21 million tons of ammonia equivalent over a five-year period. At $150 per ton, this is equivalent to over $3 billion in foreign exchange savings. This much ammonia will require an additional 832 billion cubic feet of natural gas in a five-year period. In a period of declining supplies of natural gas, the amount required for these incremental ammonia supplies will have to be taken from other users, such as electric utilities or large industrial boilers, who will have to replace this fuel with fuel oil or coal. This implies the increase by such users of approximately 148 million barrels of oil, or 35 million tons of coal. If these users or the ammonia producers can use coal, it implies a net savings in foreign exchange of about $3 billion over a five-year period. If they cannot use coal, but must use oil, fuel oil or crude oil imports must be increased. At $12 per barrel, U.S. oil imports would have to increase by $1.8 billion over a five-year period. This still would be preferable to the importation of $3 billion worth of ammonia or derivatives.

31.

Waste Management Systems in Relation to Land Disposal/Utilization

K. Robinson*

In Britain there appears to be a reluctance by farmers to use biological processes for the stabilization of animal wastes because they have yet to be convinced that such processes will work satisfactorily and that the capital and operating costs can be justified. In the past decade extensive programs of study to examine the ability of biological stabilization processes to reduce the pollution potential of animal wastes (1) have shown that while they do bring about considerable reductions in the BOD and COD the residual levels of these and other components such as ammonia and suspended solids are still too great to permit the liquid fraction of the waste to be discharged directly to a watercourse. The only satisfactory alternative at the present time is to return the stabilized waste to the land. The steeply rising cost of manufactured fertilizer and the need to conserve dwindling natural reserves have given additional impetus to the recycling of wastes.

The problem facing the farmer therefore is that he must maintain or improve the availability of food supplies, utilize as much of the waste as possible and observe the regulations for environmental protection.

* Bacteriology Division, School of Agriculture, Aberdeen, Scotland.

The question must be asked, "How can these objectives be achieved most effectively?"

The purpose of this chapter is to use the results obtained during waste stabilization studies at Aberdeen to compare a variety of waste management-stabilization systems, each giving different degrees of stabilization, with a system of no stabilization process, in relation to the recycling of the waste through land.

METHODS AND RESULTS

In the period since 1968 stabilized wastes have been produced using a number of aerobic and anaerobic processes either alone or in combination, sited in a field-scale complex (2-5) (Figure 1).

The types of waste available from this complex of land utilization and used in this comparison are:

1. Swine waste collected and stored in slurry channels. This waste

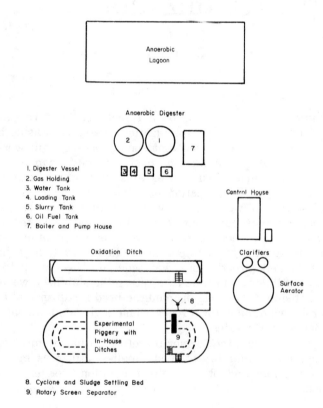

1. Digester Vessel
2. Gas Holding
3. Water Tank
4. Loading Tank
5. Slurry Tank
6. Oil Fuel Tank
7. Boiler and Pump House

8. Cyclone and Sludge Settling Bed
9. Rotary Screen Separator

Figure 1. Stabilization systems in the experimental field complex.

has been used after adequate mixing and with or without adjustment of the suspended solids as the input to some of the stabilization processes.

2. Anaerobic lagoon supernatant.

3. Mixed liquor produced by aerobic stabilization of lagoon supernatant in an oxidation ditch (retention = 10 days).

4. Mixed liquor produced by aerobic stabilization of slurry in an oxidation ditch (retention = 17 days).

5. Mixed liquor produced by aerobic stabilization of slurry in an in-house oxidation ditch (retention = 68.5 days).

6. Mixed liquor produced by second-stage denitrification of the mixed liquor from 4.

7. Anaerobic digester effluent.

The composition of these wastes is shown in Table 1.

Aerobic stabilization has the effect of reducing the concentration of ammonium-nitrogen and increasing the concentration of oxidized nitrogen, particularly with long retention times, compared with raw waste. Loss of total nitrogen occurred with all processes except anaerobic digestion and was most marked with second-stage denitrification. Potassium is unaffected by stabilization and its concentration has been used to assess dilution by rainfall or evaporation. Phosphorus is present mainly in the suspended solids, these settle out in the lagoon which accounts for the apparent reduction in phosphorus.

Land Utilization

In order to simplify comparison of the wastes shown in Table 1 in terms of their fertilizer value it will be assumed that they were produced by stabilization systems associated with a 1000-head swine fattening unit, that the nutrient requirements for the crops are 6000, 3000 and 3000 units of N, P_2O_5 and K_2O respectively; that pollution control legislation would be infringed by application of nutrients in excess of these levels; and that waste could not be applied during a 6 month winter period. In this way factors such as housing costs, feed, labor and waste storage are common and can be ignored.

The nutrient values of the wastes as available N, P_2O_5 and K_2O are shown in Table 2. It can be seen that except for the raw waste and digester effluent the wastes are unable to satisfy the crop requirement for nitrogen and this imbalance would therefore need to be supplied from an alternative source. Phosphate values exceed requirements except for lagoon or lagoon-oxidation ditch processes. There appears to be a trend for phosphate to increase in ditch systems, possibly because of evaporation during long periods of operation. A reduction in phosphate might have been expected because

Table 1. The composition of wastes from a range of waste management systems.

SYSTEM	COD total	COD dissolved	Kj-N	NH_4-N Mean, g/l	NO_2-N	NO_3-N	P	K	Suspended Solids	Cu mg/l
Raw Waste	57.20	16.56	3.49	2.25	—	—	0.5	1.5	33.35	26.46
Lagoon Supernatant	5.39	2.89	1.56	1.29	—	—	0.21	1.02	—	5.42
Oxidation ditch — stabilized lagoon supernatant	13.07	1.54	1.13	0.03	0.08	0.17	0.25	1.09	6.38	11.82
Oxidation ditch mixed liquor, 17 days retention	33.50	3.87	2.29	0.64	—	—	0.72*	0.95*	25.98	3.07*
In-house oxidation ditch mixed liquor, 68.5 days retention	35.11	1.74	2.01	0.04	0.06	0.35	0.97	1.74	23.8	42.16
Anaerobic digester effluent*	28.91	5.1	3.41	2.39	—	—	—	—	49.13	—

* Limited analysis.

of the phosphorus present in suspended solids removed from the ditches by mechanical screen and cyclone. Potash is also available in excess assuming that application rates are based on supplying sufficient nitrogen.

Economics

The cost of providing stabilization facilities and/or storage for a 1000-head swine fattening unit has been based on costs prevailing in Northeast Scotland during 1974/75 (6, 7), and the operating costs for power are extrapolated from the power consumed by equivalent facilities operating in the field complex. The distribution of annual costs for each of these facilities is shown in Table 3. In the case of capital and equipment costs the values shown take into account depreciation over 10 years and interest at 18% on half the capital.

The lagoon and the combined oxidaton ditch provide both storage and stabilization and therefore the cost of storage for 6 months for 4.5 liters of waste per pig day can be credited to these facilities. The power costs for ditches used for the stabilization of slurry with retention times of 17 and 68 days are high and would probably be smaller for the size of ditches used in this comparison because in the smaller experimental systems the motors used with the aerators have not been operated with maximum efficiency. The aerator for mixing and aeration of the contents of the ditch used for second-stage denitrification was operated for only two hours/day and more work is needed to show whether such a system could be operated with fewer rotors and therefore lower operating costs. Capital costs are highest for the anaerobic systems, that is the lagoon and digester, the latter having the advantage that the total nitrogen content is approximately the same as in the raw waste. The aerobic systems in contrast have relatively high power costs and suffer from the disadvantage that they reduce considerably the nitrogen content of the input waste. A further disadvantage resulting from the loss of nitrogen is that additional nitrogen will need to be purchased to make up the deficit between the nitrogen available from the stabilization system and that required by the crop.

The total cost of providing stabilization facilities and where necessary additional nitrogen can be seen from Table 4. The costs represent losses varying from £2-£5 per pig place and only in the case of raw waste recycling is there a credit balance, this being the value of the surplus nitrogen.

DISCUSSION

As a result of aerobic or anaerobic stabilization the amount of available nitrogen in stabilized waste is different from that in the

Table 2. N.P.K. values of waste from a 1000-head swine fattening unit with treatment facilities.

SYSTEM	Volume of Waste m³/yr	Available Nitrogen units/yr	P_2O_5 units/yr	K_2O units/yr	*Nitrogen Balance
Raw waste	1642.5	7391.5	3777.9	5978.8	1391.5
**Lagoon	2190	5670	2102.4	5387.6	−350
**Lagoon-Ditch	2190	1499	2496.6	5781.7	−4501
Ditch (17 day retention)	1642.5	2102.6	5420	3777.9	−3897.4
Ditch (68 day retention)	1642.5	1346.9	7292.9	6931.5	−4653
Denitrification	1417	178.6	5186.3	5158	−5821.4
Anaerobic digester	1642.5	7851.5	—	—	1851.5

* Surplus or deficit of nitrogen from requirement.
** Animal production calculated at 6.0 l/pig day to allow for rainfall.

Table 3. Distribution of annual stabilization costs (£).

SYSTEM	Storage	Lagoon	Oxidation Ditch	Denitrification	Digester	Rotors pumps etc.	Power
Raw waste	468	—	—	—	—	—	—
Lagoon	−468	2280	—	—	—	250	—
Lagoon-Ditch	1468	2280	521	—	—	278.50	365
Ditch (17 days retention)	—	—	332	—	—	142.50	3650
Ditch (68 days retention)	—	—	1339	—	—	142.50	3650
Denitrification	—	—	—	1158	—	142.50	304
Anaerobic Digester	—	—	—	—	1900	—	725

Table 4. Annual cost of waste management systems.

SYSTEM	*Value of available N £	Stabilization cost £	Balance £
Raw Waste	111	0	111
Lagoon	−28	2062	−2090
Lagoon-Ditch	−360	2976	−3336
Ditch (17 day retention)	−312	4125	−4437
Ditch (68 day retention)	−372	5135	−5507
Denitrification	−466	1604	−2070
Anaerobic digestion	148	2645	−2497

* Value of surplus or deficit nitrogen at 8p/unit (*i.e.*, difference from 6000 units required).

original unstabilized waste and this difference must be taken into account when the stabilized waste is recycled as a plant nutrient. In unstabilized waste the available nitrogen is present as ammonium-nitrogen and the same is true for waste which has been stabilized anaerobically in a lagoon or digester. Anaerobic stabilization although it does not change the form of available nitrogen does lead to nitrogen loss and this loss is apparently greater in a lagoon than a digester but it is probable that a lagoon, because of its longer retention time, may lead to a greater production of available nitrogen by decomposition of insoluble nitrogen compounds in the settled solids. During aerobic stabilization some of the ammonium nitrogen is oxidized to nitrate and/or nitrite and the relative concentrations of each of these vary with retention time. In addition to the changes which occur in the status of available nitrogen loss of nitrogen occurs and this loss is greater than that occurring with anaerobic stabilization. It is possible that nitrogen loss occurring during aerobic stabilization does so because inefficient operation allows denitrification to take place; such losses could be avoided by elimination of the inefficiencies (8). Not unexpectedly nitrogen loss is considerable if aerobic stabilization is followed by a second-stage denitrification process. In addition to nitrogen loss both aerobic and anaerobic stabilization reduce the COD of a waste, have no effect on potassium or copper and only the lagoon, because of solids settling, reduces the amount of phosphorus available for recycling.

In order to assess the role of stabilization processes in the conservation of energy and natural resources, pollution control and the productivity and profitability of an animal production unit the capital and operating costs for a range of stabilization systems, associated with a 1000-head swine fattening unit have been compared. In making the comparison it has been assumed that land pollution and its associated effects such as ground and surface water contamination would not occur if the rate of application of recycled waste was restricted to that required to satisfy the crop requirement for available nitrogen and that the results obtained using experimental field-scale stabilization systems could also be obtained using equivalent scaled-up systems.

The results show that the animal unit can only be self-sufficient for nitrogen by recycling unstabilized waste, anaerobic digester effluent or lagoon supernatant and that for the first two of these it would be necessary, within the constraints imposed, to transport nitrogen from the unit. The disadvantages of over-production of unstabilized waste are that a regular customer must be available, that transport of a partially stabilized odorous waste may not be acceptable if it occurs too frequently and there is a possibility of transferring disease from one farm to another. The lack of sufficient nitrogen which results from the use of aerobic systems eliminates these disadvantages but introduces others such as the need to buy-in nitrogen and the operating costs associated with aeration. The total annual cost of providing stabilization range from £2-£5 per pig place depending on the system used and this must be seen as the cost of nitrogen elimination and odor control. The cost is high and must be considered within the profitability of each individual animal unit. The difference between these costs and those for unstabilized waste, in the absence of the disadvantages noted earlier, suggest that transport of waste within a limited area may be the most viable form of waste management.

The comparison highlights other areas which should be given further consideration in the selection of a stabilization system and in the formulation of policy. For example only a stabilization system associated with a lagoon is capable of reducing the phosphorus concentration to less than that required for plant nutrition. This deficiency could be overcome either by removing some of the settled sludge from the lagoon or by purchase of phosphorus. Potassium levels are in all cases higher than those required and steps would need to be taken to neutralize the effects of excess potassium. The accumulation of copper in the soil is undesirable and is high with the recycling of wastes from animals fed on copper-supplemented diets. Like phosphorus, copper is contained primarily in the suspended solids and therefore lagoons, by removing suspended solids, have a further advantage over other systems of stabilization.

The comparison which has been made is dependent upon the constraints which were imposed and it is interesting to consider whether the same results would be obtained if the constraints were changed. For example let it be assumed that the unit *must* be self-sufficient for nitrogen and that productivity *must* be increased. Table 5 shows the stocking densities which would be required to supply either 6000 units N, 3000 units P_2O_5, or 3000 units K_2O. Limiting the rate of application to the crop requirements for either phosphate or potash would decrease and not increase productivity. Although a nitrogen-based application rate does permit increased stocking density it is doubtful whether the very high densities calculated for a stabilization system comprising an aerobic stage with secondary denitrification are practical, particularly when it is considered that the application of sufficient nitrogen would lead to massive over-application of both phosphate and potash.

Table 5. Stocking density limitations based on nitrogen, potash or phosphate application.

	N 6000 units	P_2O_5 3000 units	K_2O 3000 units
Raw waste	812	321	203
Lagoon	1062	577	225
Lagoon/Oxidation Ditch	4002	486	210
Oxidation Ditch	2854	224	321
Oxidation Ditch (68.5 day retention)	4455	166	175
Denitrification	28980	202	203
Anaerobic Digester	764	—	—

ACKNOWLEDGMENTS

This paper has drawn upon the results obtained during several years of investigation and involving the participation of many persons. Their contribution to the work is acknowledged. I would like to acknowledge particularly the contribution of Mr. A. M. Robertson of the Scottish Farm Buildings Investigation Unit who has had special responsibility for the design operation and maintenance of the systems in the field complex.

REFERENCES

1. Robinson, K. Aerobic processes for the stabilization of animal wastes. CRC Press, Critical Rev. in Environ. Control 4: 193 (1974).
2. Robinson, K., J. R. Saxon, and S. H. Baxter. Microbiological aspects of aerobically treated swine waste. Proc. Symp. on Livestock Wastes, Am. Soc. Agric. Eng., PROC-271: 225 (1971).
3. Report. The treatment of piggery wastes. North of Scotland College of Agric. Aberdeen (1974).
4. Robinson, K. Waste Treatment with a protein bonus. Proc. Conf. on Animal Waste Management, Cornell Univ., Ithaca, N.Y. 415 (1974).

5. Robertson, A. M., G. A. Burnett, P. N. Hobson, S. Bousefield, and R. Summers. Bioengineering aspects of anaerobic digestion of piggery wastes. Proc. Symp. on Livestock Wastes, Am. Soc. Agric. Eng. ISLW-75 (In press).
6. Wight, H. J. Farm Building cost guide. Scottish Fm. Buildings Investigation Unit, Bucksburn, Aberdeen, Scotland (1975).
7. Wight, H. J. Initial costs of setting up a slurry system. Fm. Building Progress. Scottish Farm Buildings Investigation Unit, Bucksburn, Aberdeen, Scotland 39: 13 (1975) .
8. Robinson, K., and D. Fenlon. Biologically-controlled loading of aerobic stabilization plants. Proc. Symp. on Livestock Wastes, Am. Soc. Agric. Eng. ISLW-75 (In press).

32.

Utilization of Plant Biomass as an Energy Feedstock

J. A. Alich, Jr. and R. E. Inman*

The Project Independence task force, in their final Project Independence blueprint report, projected that under the accelerated implementation plan bioconversion might provide as much as 8% of our energy requirements in the year 2000. They further speculate that under this scenario bioconversion will be the single most important solar energy concept. Under bioconversion, four sources of biomass are included: urban solid waste, agricultural residues, and energy crops—both terrestrial and marine.

SRI has evaluated the potential of terrestrial energy crops for electric power generation and substitute natural gas production. This chapter is derived from a project funded by the National Science Foundation in the Research Applied to National Needs Program entitled, "Effective Utilization of Solar Energy to Produce Clean Fuel." This multidisciplinary study was a combined effort of the Institute's Energy Technology Department, Department of Food and Plant Science, and Department of Food and Agricultural Industries. The study was designed to make a preliminary investigation of:

- Type or types of vegetation best suited for a solar conversion facility.
- Type and availability of lands required for a solar conversion facility.

---

* SRI, Menlo Park, California.

453

- Logistics and economics of growing the desired crops, including all aspects of farming—from land preparation through crop growing and harvesting.
- The energy budget for a biomass plantation.
- Firing the crops directly for electric power generation and converting them to clean fuel gas (methane or low-Btu gas) either at the farm site or at selected markets.
- The cost of power generated at the farm site and transported to consuming areas, and the cost of power generated at the marketplace.
- Overall project feasibility—both technical and economic—including a sensitivity analysis of the many contributing cost factors and the outlook for improving the economics in the future.
- Research needed in key technical and economic areas.

A technical and economic evaluation was made of the concept of producing and collecting plant biomass as an energy feedstock and the conversion of this feedstock into usable forms of energy. The study was performed in two phases; the first phase concerned the production of biomass, and the second phase concerned biomass conversion. Major findings of the study, as well as later work, are summarized below.

PRODUCTION OF BIOMASS

Several potential sources of plant biomass are available and conceivably useful as feedstock supplies. A formal plantation entailing the cultivation of selected high-yield plant species for the sole purpose of biomass production would be the most practical source, especially on the basis of yield per unit area-time. Terrestrial plantations may be considered initially more practical than water plantations, since the state-of-the-art is much more advanced in the former. Costs of installing artificial ponds and harvesting water plants such as algae, for instance, are currently much higher than terrestrial plantation costs. However, in view of the high biomass yields that have already been demonstrated with single-celled algae and those believed to be possible for species such as water hyacinth, the concept of landlocked hydro-plantations deserves continued interest and research. As technology develops in this area, especially in regard to harvesting methods, this knowledge may also be extended toward the development of controlled marine plantations. The concept of marine plantations is especially intriguing, in that it is essentially free of the restrictions imposed on the terrestrial plantation by the expense and limited availability of land and water. The giant kelp deserve special attention in this area. Problems in containing a marine plantation so that yields may be optimized by fertilizer applications and in harvesting the marine biomass appear to be the major points of difficulty.

Vegetation

Other possible biomass sources are natural land vegetation produced under natural conditions and agricultural wastes and residues. The major disadvantage of natural vegetation is its relatively low yield compared with that possible on a formal plantation. The significance of lower yields is that a greater land area would be needed to provide a given quantity of feedstock. Moreover, this land area would have to be situated in regions where natural rainfall and temperatures were conducive to optimum yields within the range of yields possible. Such regions in general are more densely populated and more heavily farmed than are regions where conditions are less favorable for natural growth, with the result that areas supporting high annual increments of natural productivity would be heavily interspersed with land devoted to agriculture and other uses. Hence, in addition to lower yields per acre, the acres would be widely dispersed, tending to compound the costs of biomass harvesting and transporting the harvested biomass to a central conversion facility. Natural vegetation is also less amenable to manipulation and, hence, less sensitive to advances in technology that under plantation situations could improve yields and decrease costs. Man's most potent tool in biomass production is his mechanical and agronomic genius. In the taking of wild harvests, however, there would be little opportunity to exploit this genius. Finally the environmental implications of harvesting wild vegetation, especially as concerns possible deleterious effects on wildlife, soil erosion, and the integrity of natural watersheds, would warrant close inspection.

Agricultural Wastes and Residues

Agricultural wastes and residues constitute an impressive tonnage of biomass when viewed on a national scale. Largely, these include residues from field and row crops left in the field following harvest, manures from confined livestock and poultry operations, and forestry residues deposited either at the mill or on the forest floor. It has been estimated that 650 million dry tons of such residues are generated in the United States each year, representing enough feedstock to fire 130 electric power plants of 1000 MW capacity or to satisfy roughly 30% of the nation's current natural gas demands. Hence, considerable potential for energy conversion would seem to reside in this source of biomass. The use of agricultural residues is encouraged by several apparent advantages: residue production is not restricted by geography since it occurs anywhere crops are grown, costs of residue production are defrayed by the sales value of the food or fiber crop harvested, sales of residues could represent a source of added income to farmers, residue use in many cases would eliminate costly and environmentally sensitive disposal problems, and no new land or water development would be required.

On the other hand, disadvantages are just as readily apparent. Cash uses have already been found for a great quantity of these residues produced. For instance, crop residues are used for livestock feed, field mulch, and to improve soil tilth. Some forest residues are already used as fuel and soil amendments and for manufacture of various wood products. Manures are used as fertilizer where the economics of transport permit. Hence, the net quantity of residue available is but an as yet undetermined fraction of the total produced. Moreover, residue production, especially crop residues, is heavily dependent on season, resulting in a discontinuity of supply. Quantities produced in localized areas from year to year depend also on market conditions and normal crop rotation practices, resulting to some degree in unreliable supplies. Compounding the problems of discontinuity and unreliability of supply are the poor drying conditions that normally exist during the season when most crop residues become available and the high water content of animal manures. High moisture residues cannot be efficiently burned; neither can they be stored for any appreciable time, nor dried artificially because of the large energy requirement. Wet residues would be amenable to gasification only via anaerobic digestion. Finally, because of their diffuse distribution, costs of collection and transport in many cases may exceed the economic value of the residues as an energy feedstock, especially if the value is determined by the price of other available feedstock, such as coal. In such a case, their value as an energy feedstock would depend on their potential for use in on-site, "backyard" miniconverters.

Choice of Species

The most likely species of plants for use on a biomass plantation at this time are selected crop plants, some of which have already been essentially developed for biomass production. Among these are sugarcane, sorghum, kenaf, sunflower, forage grasses, and certain tree species. Other suitable species, either wild or cultivated, undoubtedly await only an effort to search them out. The choice of species should be made primarily on the basis of potential biomass yield. Other criteria would include: adaptability to cultivation under climatic conditions prevalent in the United States, water and nutritional requirements, ease of cultivation, Btu content, conversion suitability, and perhaps the fit of the species with other biomass species and their farming schedules under plantation conditions (see Table 1).

Yields of up to 30 tons per acre-year are anticipated using species currently being considered. Such yields conceivably could be increased through selective farming practices, strain selection or plant breeding, or the optimum matching of species and plantation sites. Mechanical devices may also be developed to make the most of production potentials by individual species. These might include planting a shade-loving understory crop beneath the primary biomass crop, both crops to be harvested simultaneously, thus making optimum use of available land area and adding a few tons per acre to the harvest. Other such devices might include high-density plantings to afford early canopy closure and alleviate weed control problems, and the development of root-harvesting techniques for annual crops, which could be expected to increase the harvest by 5 to 10%.

The practice of multiple cropping of annual species has already been incorporated into the plantation concept as developed in this study. Short season crops with determinate growth such as sunflower would be especially suited to this practice, since the biomass crop *per se* would develop in a much shorter time than required to mature the traditional seed crop. It is anticipated, for instance, that a near optimum sunflower biomass crop would be harvested 3 to 4 months after planting, contrasted to the 5 to 6 months required for the sunflower seed crop. Appropriate planting and harvesting schedules could be established in field testing programs. Species such as kenaf could also be multi-cropped, but might require different scheduling than sunflower. Perennial crops such as sugarcane, sorghum, and forage grasses could be harvested several times during the growing seasons, especially if this practice would induce higher annual yields than a single harvest performed at the end of the season. This practice would also afford a more continuous supply of biomass through the year. Additional costs of multiple harvest would have to be offset by yield or logistical gains.

Energy Consumption

Inspection of the energy consumed by use category indicates the operations that are the most energy sensitive and, hence, areas in which significant energy savings might be realized by changes in practice or technology. The most energy sensitive category is that of farm chemical manufacture, which accounts for over 68% of the energy used on a plantation with a local supply of irrigation water. The high energy consumption in this area is due largely to the production of anhydrous ammonia fertilizer in the amounts needed to obtain the high biomass yields visualized. Although no

Table 1. Above-ground, dry biomass yields of selected plant species or complexes.

Species	Location	Yield (tons/ acre-year)
Annuals		
Sunflower x Jerusalem artichoke	Russia	13.5
Sunflower hybrids (seeds only)	California	1.5
Exotic forage sorghum	Puerto Rico	30.6
Forage sorghum (irrigated)	New Mexico	7-10
Forage sorghum (irrigated)	Kansas	12
Sweet sorghum	Mississippi	7.5-9
Exotic corn (137-day season)	North Carolina	7.5
Silage corn	Georgia	6-7
Hybrid corn	Mississippi	6
Kenaf	Florida	20
Kenaf	Georgia	8
Perennials		
Water hyacinth	Florida	16
Sugarcane	Mississippi	20
Sugarcane (state average)	Florida	17.5
Sugarcane (best case)	Texas (south)	50
Sugarcane (10-year average)	Hawaii	26
Sugarcane (5-year average)	Louisiana	12.5
Sugarcane (5-year average)	Puerto Rico	15.3
Sugarcane (6-year average)	Philippines	12.1
Sugarcane (experimental)	California	32
Sugarcane (experimental)	California	30.5
Sudangrass	California	15-16
Alfalfa (surface irrigated)	New Mexico	6.5
Alfalfa	New Mexico	8
Bamboo	South East Asia	5
Bamboo (4-year stand)	Alabama	7
Abies sacharinensis (dominant species) and other species	Japan	6
Cinnamomum camphora (dominant species) and other species	Japan	6.8
Fagus sylvatica	Switzerland	4.3
Larix decidua	Switzerland	2.2
Picea abies (dominant species) and other species	Japan	5.5
Picea omorika (dominant species) and other species		6.4
Picea densiflora (dominant species) and other species	Japan	6.1
Castanopsis japonica (dominant species) and other species	Japan	0.3
Betula maximowicziniana (dominant species) and other species	Japan	3
Populus davidiana (dominant species) and other species	Japan	5.5
Hybrid poplar (short-rotation)		
Seedling crop (1 year old)	Pennsylvania	4
Stubble crop (1 year old)	Pennsylvania	8
Stubble crop (2 years old)	Pennsylvania	8
Stubble crop (3 years old)	Pennsylvania	8.7
American sycamore (short rotation)		
Seedlings (2 years old)	Georgia	2.2
Seedlings (2 years old)	Georgia	4.1
Coppice crop (2 years old)	Georgia	3.7
Black cottonwood (2 years old)	Washington	4.5
Red alder (1-14 years old)	Washington	10
Eastern cottonwood (8 years old)		3
Eucalyptus sp.	California	13.4
Eucalyptus sp.	California	24.1
Eucalyptus sp.	Spain	8.9
Eucalyptus sp.	India	17.4
Eucalyptus sp.	Ethiopia	21.4
Eucalyptus sp.	Kenya	8.7
Eucalyptus sp.	South Africa	12.5
Eucalyptus sp.	Portugal	17.9
Miscellaneous		
Algae (fresh-water pond culture)	California	8-39
Tropical rainforest complex (average)		18.3
Subtropical deciduous forest complex (average)		10.9
Puckerbrush complexes (average)	North Carolina	2.2
Puckerbrush complexes (average)	Maine	4.4
World's oceans (primary productivity)		6

data base is available on energy consumed in the production or collection of nitrogenous fertilizers from sources other than ammonia, the use of mined nitrates (Chilean nitrate, guano) in place of anhydrous ammonia might result in energy savings. This opportunity warrants further investigation.

Electrical power used in irrigation, the next most energy-sensitive category, could be decreased only by reducing the amount of water needed. Since the water requirement is a function of both climate and crop species, this requirement would be reduced either by locating the plantation in a region where natural rainfall can satisfy a larger proportion of the water requirement or by selecting crops that need less water to produce optimum yields. Locating the plantation so as to better accommodate water needs, however, may unduly compromise other facets of biomass production such as length of growing season and sundrying capability. The selection of biomass species on the basis of their low water yields would also result in most cases in slower growing crops and lower yields. The questions of potential yield versus water needs of specific crops and of potential yield versus climate and location also warrant further inspection and research (see Table 2).

The last significant category of energy consumption is that of fuel consumed by field equipment during the performance of field tasks. Within this category, 60% of the fuel consumed is used in the transportation of harvested biomass. Conceivably, energy savings in transportation might be realized by substituting various forms of rail transport or automated transport systems for the trucking system used in the model, or by compacting the biomass before transport. Hence, research on the development of biomass transport systems would be warranted should the plantation concept be implemented.

Biomass Production Cost

On the basis of the cost assessments made in 1973, which included an analysis of both general farming costs information and projected costs for a hypothetical plantation, the cost of producing and harvesting 30 tons of dry biomass per acre-year was estimated to be between $9.50 and $10.00 per ton. The cost of fertilizer alone represents 23% of total cost. Most of this cost is for anhydrous ammonia, which is currently selling at record prices with no sign of easing off. A possible method of decreasing the price to the plantation of this increasingly costly commodity might be to couple an ammonia production operation to the biomass plantation, using a portion of the gas produced from the biomass as a raw material for ammonia production. The obvious areas of tradeoffs would warrant close examination. Development of new labor- and energy-saving prac-

Table 2. Energy consumed in biomass production on a hypothetical plantation.
(Btu per Acre-Year)

Consumption Category	Rate	1000 Btu/ Acre-Year
Farming Operations		
Field tasks per crop		
Herbicide applications	0.305 gal diesel/acre	
Plant and fertilize	0.757 gal diesel/acre	
Fertilize (side-dressing)	0.305 gal diesel/acre	
Cut and chop	1.627 gal diesel/acre	
Fresh haul	2.540 gal diesel/acre	
Turn and dry	0.028 gal diesel/acre	
Dry haul	2.178 gal diesel/acre	
Pesticide application	0.017 gal aviation fuel/acre	
Total field tasks per crop	7.757 gal/acre	
Total field tasks, three crops at 138,690 Btu/gal	23,271 gal/acre	3,228
Irrigation	77 kWh/acre-foot	
At 5 acre-feet	385 kWh/acre-year	
At 3,413 Btu/kWh	1,314,000 Btu/acre-year	
At 0.333 thermal efficiency		3,942
Miscellaneous electricity use	5 kWh/acre-year	
At 3,413 Btu/kWh	17,065 Btu/acre-year	
At 0.333 thermal efficiency		51
Total farming operations		(7,221)
Farm chemicals manufacture		
Nitrogen (1000 lb NH$_3$/acre)	15,602 Btu/lb	15,602
Phosphorus (50 lb P$_2$O$_5$/acre)	6,019 Btu/lb	301
Potassium (100 lb K$_2$O/acre)	4,158 Btu/lb	416
Herbicide (6 lb/acre)	43,560 Btu/lb	261
Insecticide (3 lb/acre)	43,560 Btu/lb	131
Fungicide (2 lb/acre)	43,560 Btu/lb	87
Total farm chemicals manufacture		(16,798)
Farm machinery manufacture (10,000 acres)		
6 tractors at 4.6 tons, 6-yr life	4.6 tons steel/year	
3 planters at 2.0 tons, 2-yr life	3.0 tons steel/year	
3 fertilizers at 2.0 tons, 2-yr life	3.0 tons steel/year	
3 herbicide rigs at 1.0 ton, 5-yr life	0.6 tons steel/year	
6 combines at 8.0 tons, 6-yr life	8.0 tons steel/year	
32 fresh haul units at 5.0 tons, 10-yr life	16.0 tons steel/year	
10 dry haul units at 6.0 tons, 10-yr life	6.0 tons steel/year	
1 turner at 5.0 tons, 10-yr life	0.5 tons steel/year	
8 irrigation pumps at 1.0 tons, 20-yr life	0.4 tons steel/year	
20.6 mi feeder lines at 13.2 tons, 20-yr life	20.2 tons steel/year	
Sprinkler system at 34.6 tons, 20-yr life	1.7 tons steel/year	
Total steel (10,000 acres)	64.0 tons/year	
At 18,865 kWh/ton	1,207,360 kWh/year	
At 3,413 Btu/kWh	4.12 × 10^9 Btu/year	
Total farm machinery manufacture, per acre		(412)
Seed production (0.3% of above total)		(73)
Total energy consumption		(24,504)
Total energy yield (30 tons @ 15 × 10^6 Btu/ton)		450,000
Energy balance		425,496
Energy input; energy output		1:18

tices, especially in the areas of biomass transport, could also decrease costs. Yields higher than 30 tons per acre-year might also be provided through species selection and new research on biomass crops, which would also result in decreased costs. An updated estimate would show costs of $13.00 to $16.00 per ton.

Potential Plantation Locations

It is believed that the primary potential region for the location of biomass plantations is the southwestern United States, principally the states of Arizona and California below 35 degrees latitude and between 113 and 117 degrees longitude. A region of secondary potential is south coastal Texas below 29 degrees latitude and between 97 and 100 degrees longitude, although public lands there are scarce. Areas in eastern Arizona and New Mexico might also be considered, but relatively more severe winter (November to February) temperatures would compromise the potential for year-round production in those areas.

Excluding water availability, southwest Arizona and southeast California have all the inherent attributes most capable of maximizing biomass production in the United States. These attributes include:

- Plentiful acreage of publicly owned Class 1, Class 2, and Class 3 land.
- Highest intensity of solar radiation and greatest number of mean annual hours of sunshine (*i.e.,* minimum cloud cover) for any area in the United States.
- Longest growing season in the United States.
- Very mild winters with high minimum monthly temperatures for November through February.
- Most rapid increase in solar radiation intensity in the spring, which is essential in getting the first crop each year off to a good start.
- Rapidly expanding economic base.

The secondary potential region for the solar farm, that of south coastal Texas, has the following deficiencies:

- Lack of public lands.
- Relatively low springtime solar radiation.
- Low annual solar radiation compared with that of the Southwest.
- Quite low mean annual sunshine, meaning a significant number of days with cloud cover.
- Potential for catastrophic weather events.

The region, however, does have some positive attributes:

- Excellent winter temperature regime.

- Good potential water supply with the importation of 12 to 13 million acre-feet per year from the Mississippi River into the Texas water system (proposed). The coastal division would bring a significant quantity of water into the south coastal area.
- Fertile, deep soils.
- A significant quantity of range land (privately owned), which could be upgraded to a higher land classification.

Undoubtedly, other potential sites in the United States may be available for relatively small solar farms (depending on the requirements of the conversion facility); however, in this general overview, the attempt has been to demonstrate the needs of the biomass plantation concept in terms of land and climatic requirements by idealizing the site selection process.

An outstanding drawback to biomass production in the Southwest is lack of water. As previously discussed, the Southwest is already a water-deficient area, with current annual consumption exceeding the rate of replenishment. Consequently, biomass production on a large scale in this region is not feasible without the development of a large supplemental supply of water. The most effective potential supply would probably be interbasin water transfer from areas of water surplus. Any such development would be a long time in coming and would require extensive assessment far beyond the scope of this study. While acknowledging that water availability is limiting to plant growth, it is also realized that it is not necessarily limiting to the concept of plant biomass production.

BIOMASS CONVERSION

Biomass is a suitable fuel for conversion to SNG or for electric power production. It has less ash than coal or municipal refuse. Biomass is low in sulfur content and does not contain the harmful corrosive compounds found in many municipal refuses. Biomass has a lower Btu content per pound than coal but has a higher content than that of municipal refuses.

As late as 1880, wood was the largest source of energy in the United States. However, in 1970 wood produced only about 1% of the U.S. energy requirements. Currently, a major portion of this wood is used by wood product processors, while bagasse is used by sugarcane processors. Although the concept of raising steam from biomass is not entirely new, the scale and method of biomass production and utilization proposed here are new.

Firing the Crops Directly

As a base case, a large (1000-MW), modern, high-temperature and high-pressure station with reheat is analyzed. It is recognized that these are not the parameters of current systems that utilize biomass. As a result, development work is required.

A 1000-MW station operating at an annual load factor of 80% would require a land area of ⁻245 square miles (heat rate = 10,000 Btu per kWh based on higher heating value) to provide the biomass required.

Site water requirements would be about 700 million gallons per day (~2150 acre-feet per day). It is anticipated that irrigation water would first be used for cooling and then for irrigation. Recycle of ash to the biomass plantation shows promise; however, the method and suitability of recycling ash require further consideration. Although the concept of increasing localized carbon dioxide concentration by recycling flue gas shows merit on a theoretical basis, it does not appear practical at this time.

Figure 1 shows the effect of fuel cost on electrical power genera-

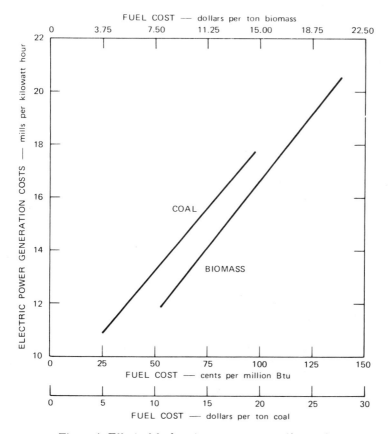

Figure 1. Effect of fuel cost on power generation cost

tion costs. Comparing biomass with a mine-mouth coal plant (coal at 35 cents per million Btu) results in an advantage of three to four mills per kWh for coal (biomass at 100 cents per million Btu). The importance of fuel costs in the comparison is obvious.

On the basis of energy input to energy output for electric power production, biomass compares well with coal as well as with other fossil fuels. The efficiency of electric power production from biomass and coal is summarized in percentages in Table 3.

Table 3. Efficiency of electric power production from biomass and coal.

	Deep Mined Coal[a]	Surface Mined Coal[a]	Biomass
Extraction/production			
Resource recovery	57.0%	80.0%	100.0%[b]
Energy input	0.8	0.8	5.6
Net available	56.2%	79.2%	94.4%
Processing and transport	[c]	[c]	[d]
Conversion	38.0	38.0	34.1
Transmission	91.2	91.2	91.2
Net delivered	19.5%	27.5%	29.4%

a. Source of coal data, "Energy and the Environment—Electric Power," Council on Environmental Quality, p. 40 (August 1973).
b. Excludes photosynthetic efficiency.
c. Mine-mouth plant.
d. Included in energy input: extraction and production.

Low-Btu Gas/Combined Cycle

An alternative method of power generation from coal is through preparation of low-Btu (100 to 150 Btu per scf) gases from coal and utilization of the gases in a combined gas turbine-steam cycle to produce electricity. In the gasification process, sulfur in the fuel is converted almost completely to H_2S, which is removable before combustion using existing technology. This concept has been developed because of its satisfactory sulfur removal, as well as its promise of attaining high efficiencies in the future. Since SO_2 removal is not a problem with biomass, this concept is evaluated because it shows promise as an efficient and environmentally acceptable method to use coal as well as a method to utilize biomass that is potentially more efficient than direct firing.

Based on current technology (1800°F-gas turbine inlet temperature), an overall heat rate for biomass on the order of 11,400 Btu per kWh would be expected. This would result in a 1000-MW station operating at an annual load factor of 80%, requiring a land area of ~280 square miles to supply the needed biomass. The cooling water/irrigation system synergisms and the potential for ash and

CO_2 recycle are similar to those discussed for direct firing. In addition, a potential by-product of the gasification portion of the system is ammonia.

As was the case with direct firing, coal has an economic advantage. On a similar basis, biomass would show a greater energy efficiency. While the low-Btu gas/combined cycle systems based on current technology do not now show an economic advantage over direct firing, future systems including advanced gasification and gas turbine technology should be competitive with alternative power generating technologies.

Substitute Natural Gas Production

Biomass can be converted to SNG via biological or chemical processes. In either case, the SNG efficiency (SNG efficiency = Btu value of SNG output/Btu value of biomass and utility inputs) based on currently available technology would be about 60%. The selection of a conversion process will depend on biomass characteristics and farming techniques. The current agricultural logistics indicate terrestrial farming and hence a product that can be field-dried. For this case, the most suitable process at the current time is pyrolysis because of both biomass characteristics and the current degree of development of the pyrolysis process. Although hydrogasification and biological digestion have not reached the same level of development for handling this type of material as pyrolysis, the prospects for future development and the fact that it could be some time before this concept is implemented might lead to selection of one of these processes. The adoption of aquaculture as the means of biomass production and developments in the area of anaerobic digestion could result in this technology playing the major role in SNG from biomass production. In addition to the technologies already mentioned, concepts such as catalytic methanation and advanced fluidized bed technology are being considered and, depending on the time frame of the introduction of the biomass concept, could also be of importance.

Based on the SNG efficiency of 60%, a 76-million scf per day plant with an operating factor of 90% would occupy a land area of ~140 square miles to produce the biomass required. Site water requirements would be about 400 million gallons per day or ~1220 acre-feet per day, with almost all of it being required for irrigation. Current SNG from coal plants have water requirements of only a few million gallons per day. The potential for ash and CO_2 recycle is similar to previous cases. As in the low-Btu gas/combined cycle case, the plant could be designed to produce ammonia as a by-product.

Based on early 1975 economics, SNG produced from coal (mine mouth at 35¢ per million Btu) would cost from $2.00 to $2.50 per million Btu, while from biomass (biomass at 100¢ per million Btu) costs from $3.00 to $3.50 per million Btu are projected.

Transportation

An important consideration in development of a biomass plantation and conversion facility is locating the complex close to potential demand centers. Flexibility in this regard is limited by the suitability of demand areas as sites for biomass production. The southwestern United States, chosen as the best area for the development of this concept, is already rich in fuel resources, negating to a great extent transportation savings. Also, it is a water-deficient area where there is great competition for available water.

While the cost of electric power production from biomass is reasonable by current standards, the economics of transportation favor its use in areas that are proximate to the biomass plantation. This requirement together with the desirability of locating the biomass plantation in the Southwest, limits to a great extent the development of this concept for electric power production. The transportation economics of SNG production, unlike the electric power generation case, permit a broad spectrum of user areas to be economically serviced from a southwestern location.

33.

Protein and Energy Conservation of Poultry and Fractionated Animal Waste

Gerald M. Ward* and David Seckler**

Present day emphasis upon energy conservation has directed attention to the massive amount of energy represented by wastes produced by U.S. livestock. Many schemes have appeared recently for utilizing this waste either as fuel, feed or fertilizer. The latter is the oldest and most acceptable, but the use of the protein in manure as a nitrogen fertilizer is of the same order of efficiency as the practice of placing a fish under each hill of corn as practiced by North American Indians. If the nutrients in animal waste can be recovered in an acceptable form this should be the most economic use of manure and such is the result of an analysis of the data available (1).

In this period of world-wide protein shortage, there is a serious problem of finding protein supplies to support animal production. A largely unused and almost completely neglected source of high quality protein is available in sufficient quantities to supply much of the supplemental protein requirements for livestock production. The source of this protein is animal waste. It has been common to consider that nutrient constituents in manure are indigestible. This

---

* Department of Animal Sciences, Colorado State University, Fort Collins, Colorado.
** Department of Economics, Colorado State University, Fort Collins, Colorado.

is a serious fallacy particularly in the case of protein because about 5 g of nitrogen or 30 g of endogenous protein is excreted in the feces of ruminants for each kg of feed dry matter consumed (2). As illustrated in Figure 1, about 50% of the protein in ruminant feces consists of microbial protein and another 25% of the protein consists of tissue cells from the digestive tract. Some controversy exists concerning the digestibility of microbial cells but more recent studies indicate a high digestibility (3, 4).

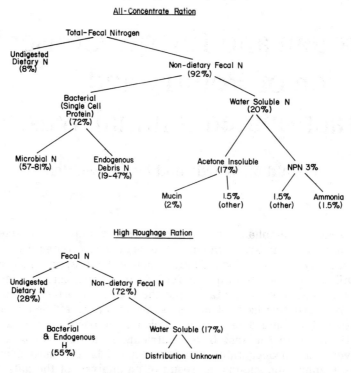

Figure 1. Nitrogen distribution in the feces of sheep (4, 10).

When fed similar rations, protein distribution in cattle manure would be similar to that of Figure 1. The high quality of protein in cattle manure and its relatively low percentage of non-protein-nitrogen (NPN) makes it a very desirable source of protein for non-ruminant animals and a source of protein for which there is no competition from man. Although cattle manure contains protein of high quality and availability, the protein is mixed with large and variable amounts of ash and fiber largely indigestible by nonruminants.

To be of value for poultry and swine feeding, it is necessary to fractionate cattle manure to increase the concentration of protein and decrease the content of ash and fiber. Recently many suggestions, proposals, and patented processes have appeared to convert animal waste to useful products. The process with which we have worked and for which nutritional information has been presented recently is the "Cereco" process which accomplishes a separation by producing three fractions from manure; one high in fiber, while one is high in protein, and one high in ash.

For the Cereco process, feedlots are partly covered with concrete so that the waste can be gathered in all types of weather before biological decomposition begins. Typically, the manure is gathered every 4-10 days, depending on climatic conditions which affect decomposition. Manure is ground, slurried with water and other ingredients and fermented in large concrete pits. Fermentation encourages the growth of microorganisms which convert nitrogen and carbon into true protein (single cell protein). Fermentation reduces pH which aids subsequent processing and pasteurization. After fermentation, the slurry is transported into the plant for processing into the three Cereco fractions.

NUTRITIONAL STUDIES

Digestibility and metabolic studies of the C-I and C-II fractions were conducted with Hereford steers and the results for C-II are summarized in Table 1. The results indicated C-I to be equal in feed value to corn silage and the C-II supplement had about 75% the value of soybean meal. Three lots of 100 steers each were fed either C-I, C-II or a control ration and feed consumption and growth were essentially the same for all three rations. The details of these experiments have been published elsewhere (5).

Nutrition trials with broilers indicated that C-II was well utilized as 5% of the diet by broilers 0-4 weeks of age and 10% for 5-8 week old birds. Adult White Leghorn layers had egg production and quality

Table 1. Protein digestibility and nitrogen balance of steers supplemented with C-II or soybean meal.

	Protein Supplement	
	C-II	Soybean Meal
D. M. Digest. (%)	65.0	72.9
O. M. Digest. (%)	68.6	74.5
Protein Digest. (%)	50.0	60.9
N Balance (g/day)	21.2	39.0
N Balance—Coefficient of Variation (%)	48.1	37.3
% N Retained	16.2	25.3
% Digest N Retained	31.8	41.5

equal to a standard layer ration when either 15 or 30% C-II was sub-stituted in the ration (6). Rainbow trout were reared for a 150-day period with 14% of a commercial ration replaced by C-II and no sig-nificant differences were observed in growth rate (7). No health prob-lems associated with feeding C-II have been observed in birds or fish. Eggs, broilers or fish produced have been indistinguishable in flavor from products produced on control rations.

The feed-conserving potential of the Cereco system is indicated by the data in Table 2, which is based upon the assumption that the solid waste produced by feedlot cattle and dairy cattle could all be recovered and processed although this is obviously an overestimate especially for dairy cattle. However, the results provide an idea of the magnitude of current wasted resources. If all the manure could be processed, it represents the equivalent of 7.1 million tons of corn and 6.1 million tons of soybean meal, or about 5% of the total U.S. corn crop and about 25% of the soybean meal production. The acre-age required totals 14 million acres (although more than one-half the return from the soybeans is usually from oil). Capital annual pro-duction costs to produce this amount of corn and soybeans on Iowa farmland is estimated to be about 16 billion dollars.

The ability to produce a high protein (20-30%) low fiber product

Table 2. Corn and soybean equivalents of the manure production of U.S. feedlot steers and dairy cattle.

	Feedlots	Dairy	Total
No. of Cattle/yr (M)	12.5	10.0	22.5
Manure (M tons DM/yr)	16.0	17.0	33.0
Potential Cereco-I (M tons DM/yr)	6.25	8.8	15.0
Potential Cereco-II (M tons DM/yr)	4.69	5.20	9.9
Corn equivalent of Cereco-I (M tons)	2.9	4.2	7.1
Corn equivalent (M acres)	1.1	1.6	2.7
Soybean meal equivalent of Cereco-II (M tons)	2.9	3.2	6.1
Soybean meal equivalent (M acres)	5.4	5.9	11.3

Calculations for the two Cereco products and their crop equivalents as shown in Table 2 are:
1. C-I (silage) a fermented corn silage type product equivalent to 47% of No. 2 shelled corn.
   C-II (protein) a dry pelleted feed with the similar amino acid quality as soy-bean meal calculated at 62% of the value of soybean meal.
2. Manure statistics from a survey by Richard Graber, "Agricultural Animals and the Environment," published by Feedlot Waste Management Regional Extension Project, Oklahoma State University, Stillwater, Oklahoma, March, 1973.
3. The crop production and cost statistics were taken from "Selected U.S. Crop Budgets: Yields, Inputs, and Variable Costs," Volume II, North Central Re-gion, U.S. Department of Agriculture, Economic Research service, April, 1971. North Central and East Central Iowa Area E. Yields: corn, 91.4 bushels or 2.6 tons per acre; soybeans, 27.8 or .83 tons per acre.

from cattle manure makes possible an integrated recycling of nutrients between species to form a symbiotic relation between cows and hens. Poultry manure contains a high crude protein content but two-thirds of this crude protein consists of NPN mostly in the form of uric acid. However, NPN can be used by ruminants and uric acid, the principal NPN component of poultry manure, has been shown to be equal or superior to the common NPN feed source, urea (8). Cattle manure needs to be fractionated to be of value to poultry and pigs but poultry manure can either be dried and fed directly to cattle or processed.

The enormous potential for protein and energy recovery in a cycle involving milk cows and laying hens is illustrated in Figure 2. The number of animals and their production as indicated in Figure 2 are intended only as illustrations of the potential; cows, of course, do not produce milk all year nor do hens lay one egg per day for a year.

The potential under U.S. conditions is indicated by the fact that in 1974, 23 million cattle were marketed from U.S. feedlots. With a turnover of two times per year, there are about 12 million head of cattle in feedlots at any one time. At 200 g of fecal protein per head per day, the yearly production of C-II protein would be about 1 million tons per year. Assuming a ratio of 30 laying hens to each beef animal in the above protein cycle, these cattle could provide supplemental protein to the equivalent of 120 million hens or about 40% of the total laying flock of the United States. These hens could, in turn, provide most of the supplemental protein requirements for feedlot cattle. These hens would produce roughly one-fifth of crude protein equivalent of the one million tons of urea used in ruminant feeds last year.

To demonstrate the potential significance for less developed countries, the above dairy-hen cycle is applied to the case of India. India has the largest cattle population in the world—some 230 million head. Of this number, 16 million are producing dairy cows (9), which produce 21.4 million tons of milk, resulting in an average per capita consumption of 38.7 kg. If these 16 million dairy cattle were linked to laying hens at the above ratio of 30 hens per cow, 510 million hens would be sustained in their supplemental protein requirements. The results of these calculations indicate production of 300 grams of milk and almost 1 egg per capita per day without the use of other supplemental protein.

Livestock wastes represent a tremendously under utilized feed resource. An integrated effort to better utilize the nutrients in animal wastes will not only reduce pollution and replace valuable crops used for animal production, but it also has the potential of forming the basis of a viable system for providing high quality animal products for the world's poorer people.

The implications for energy conservation inherent in the recovery of nutrients from animal waste are apparent from the examples above. The reduction in acreage that would be required for crop production also implies a major reduction in fuel energy requirements. Utilization of manure as a feed source would mean some reduction in nitrogen and phosphorus for fertilizer. However, any nutrient recovery system would probably provide more efficient recovery of the nitrogen in manure than is usually the case when used as fertilizer.

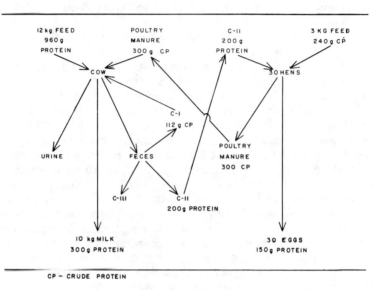

CP – CRUDE PROTEIN

Figure 2. Daily protein cycle between Cereco processed dairy waste and poultry waste.

Assumptions Used for the Calculations Presented in Figure 2.
(All data on a daily basis)

—One cow requires 12 kg of feed dry matter (DM) containing 1300 g of crude protein and produced 10 kg of milk containing 300 g of protein and produced (at 70% DM digestibility) 3.6 kg of feces. Estimated distribution in Cereco products is 40% C-I, 30% C-II, and 30% C-III or 1.44 kg of C-I at 8% CP = 112 g CP and 1.08 kg of C-II at 25% CP = 270 g CP or 203 g of true protein (75%).

—One hen requires 100 g of feed containing 15 g of protein and produces 50 g of eggs containing 5 g of protein and produces waste containing 10 g of crude protein. Waste contains about 30% true protein and 70% NPN.

—Hens and cows fed grain and roughage feeds averaging 8% protein. One hen fed 100 g of 8% crude protein feed = 8 g or a protein deficit of 6 g per day. The 203 g of true protein produced by one cow would support about 30 hens.

—Thirty hens produce waste containing 300 g crude protein containing 120 g true protein; and 280 g CP as NPN.

—One cow fed 12 kg of feed @ 8% protein = 960 g—a deficit of 340 g CP which is more than satisfied by the 300 g of CP from the waste of hens plus 112 g of CP from C-I.

Finally, of course, any system for recovery of the feed nutrients from manure is dependent upon a change in the regulations that presently ban such practices. A change in regulations is needed in the interests of improved efficiency of animal production and for promoting control of environmental pollution.

SUMMARY

The Cereco process developed by Ceres Ecology Co. processes manure into three fractions; one high in fiber for feeding to ruminant animals, a second high in protein (20-30%) for feeding to either ruminants or nonruminants and a third high in ash intended as a soil amendment. Nutritional studies have been made of fraction 1 as a feed for cattle and sheep. Fraction 2 has been evaluated as a protein supplement for cattle, sheep, broilers, layers and rainbow trout. Fraction 1 has been shown to be equal in feed value to average corn silage and fraction 2 approaches the protein value of equivalent amounts of soybean meal.

The ability to fractionate a high protein fraction makes possible a fully integrated cycle utilizing the crude protein of poultry manure to provide supplemental protein for cattle and the high quality microbial protein of cattle manure to support poultry. Assuming a ratio of 30 laying hens per head of feedlot cattle in the U.S., these cattle could provide the supplemental protein for 120 million hens; about 40% of the layers in the U.S. These hens in turn could provide the supplemental crude protein requirements of 4 million feedlot cattle.

The replacement value of corn by C-I and soybean by C-II if all the waste from feedlot and dairy cattle in the U.S. were processed is estimated to equal 14 million acres of cropland. Nutrient recovery from animal waste also allows significant savings of fuel energy by replacing other feed sources.

REFERENCES

1. Harper, J. H., and D. W. Seckler. Engineering and economic overview of alternative livestock waste utilization techniques. Presented at International Symposium on Livestock Wastes, Urbana, Illinois, April 21-24, 1975.
2. Agricultural Research Council. The Nutrient Requirements of Farm Livestock No. 2 Ruminants. London (1965).
3. Hoogenraad, J., F. J. R. Hird, R. G. White, and R. A. Leng. Utilization of ^{14}C labelled bacillus subtilis and escherichia coli by sheep. Br. J. Nutr. 24:129 (1970).
4. Mason, V. C., and R. Palmer. Studies on the digestibility and utilization of the nitrogen and irradiated rumen bacteria by rats. J. Agric. Sci. 75:567 (1971).
5. Ward, G. M., D. E. Johnson, and R. D. Boyd. Nutritional value for steers of Cereco silage and Cereco protein produced from feedlot manure. Feedstuffs, Vol. 47, pg. 19 (1975).
6. Kienholz, E. W., J. M. Navallo, G. M. Ward, and M. C. Pritz. Nutritional value of Cereco for poultry. Feedstuffs, Vol. 47, pg 21 (1975).

7. Post, G., and G. M. Ward. The use of Cereco II in a rainbow trout ration. Feedstuffs, Vol. 47, pg. 24 (1975).
8. Oltjen, R. R., R. A. Dinius, M. I. Poos, and E. E. Williams. Na urate 25% urate and uric acid as NPN sources for beef cattle. J. Nutrition 35:272 (1972).
9. FAO Agricultural Commodity Projections 1970-1980. Volume 1, p. 114.
10. Hecker, J. F. The fate of soluble mucin in the gastro-intestinal tract of sheep. J. Agric. Sci. Camb. 80:63 (1970).
11. Mason, V. C. Some observations on the distribution and origin of nitrogen in sheep feces. J. Agric. Sci. Camb. 73:99 (1969).
12. National Academy of Sciences. National Research Council. Nutrient requirements of beef cattle. (1970).
13. Smith, L. W. Nutritive Evaluations of animal manures in processing agricultural municipal wastes. Ed. G. E. Inglett. Avi. Publ. Co., Westport, Conn. (1973).

34.

Protein Production Rates by Algae Using Swine Manure as a Substrate

L. Boersma,* E. W. R. Barlow,** J. R. Miner,*
H. K. Phinney,* and J. E. Oldfield*

Water power, wood, coal, petroleum products, and natural gas have provided the energy needed to fuel the industrial revolution. With it profound social and ethical changes have occurred. Now the fuels which were used to support these changes are running low. The accumulation rate of natural energy resources is so slow that they must be considered finite when viewed in relation to the rate of present consumption. More importantly, energy use is increasing at an accelerated rate. As world-wide communication has increased so has the desire of many people to share the standard of living based on the availability of inexpensive energy resources. The latent demand for these finite resources can be ascertained from Table 1.

It is a great challenge to preserve and enhance the present standard of living of the western societies and allow the people of less developed civilizations to achieve similar standards. New sources of energy must be found and energy producing processes must be used with greater efficiency. Present practices reject more than half of the energy produced from fuels as waste heat. Electric power plants

---

* Oregon State University, Corvallis, Oregon. Departments of Soil Science, Agricultural Engineering, Botany and Plant Pathology and Animal Science, respectively.
** MacQuarie University, North Ryde, N. S. W. 2113, Australia. School of Biological Sciences.

Table 1. Comparison of energy consumption rates in different countries (1).

Country	kg per Capita (Coal Equivalent)	Energy Use as Multiple of Nigerian Use
United States	11,244	191
United Kingdom	5,507	93
West Germany	5,223	89
USSR	4,535	77
France	3,928	67
Japan	3,267	55
Italy	2,682	45
Mexico	1,270	22
China	561	10
Brazil	500	8
Philippines	298	5
India	186	3
Indonesia	123	2
Pakistan and Bangladesh	96	2
Nigeria	59	1
World Average	1,927	33

reject two units of waste heat for every unit of electricity produced. Much of the machinery operated by electricity is less than 50% efficient. There does not appear to be much that can be done to improve the efficiency of steam plants or internal combustion engines.

Energy loss via waste heat production could be reduced if energy-consuming processes were arranged in an orderly cascade of decreasing input-output temperatures. It may well be that in the future we cannot afford to design power plants for maximum efficiency of generating electricity only. Power plants may have to be designed for maximum energy utilization with a final recovery of the energy input much greater than the present 35%.

Of equal importance is the provision of an adequate diet to all inhabitants of the earth at a reasonable cost. The agricultural industry of the United States has been successful in producing an abundance of food at low cost. The efficiency of food production has increased by the use of better equipment, manufactured fertilizers, high yielding plant varieties, disease control, and pest control. These developments have made modern agriculture strongly dependent on the availability of low cost energy (2-7). The energy is used for manufacturing farming equipment, fertilizers, operating the farming equipment, pumping irrigation water, drying crops, and manufacturing agricultural chemicals. The cost of food production in all parts of the world will increase as this resource becomes scarce and more expensive.

Alternatives to present day agricultural production methods must be considered before energy costs have increased dramatically.

Needed change can only be brought about by new concepts and radical departures from the existing methods—not by gradual improvements in established techniques.

This report describes a system for recycling nutrients contained in animal manure by growing algae using the digested waste as a substrate. Nutrient recovery by algae was chosen because 1) algae present a protein and vitamin rich feed of potentially high value for animals; 2) under favorable conditions algae can be cultivated all year long; 3) algae, being photosynthetic organisms, present minimum requirements for supplemental oxygenation, although aeration will still be required; and 4) algae are autotrophic organisms which can utilize inorganic carbon sources (carbon dioxide, bicarbonate) and thus do not require high organic carbon levels, such as needed by yeasts or fungi.

ADVANTAGES OF SINGLE CELL PROTEIN

Microorganisms such as algae, yeasts, or bacteria are a high quality, protein and vitamin rich food, often referred to as "single cell protein." This avoids unpleasant connotations relative to food which might be associated with such terms as "bacterial" or "microbial." In addition to their high protein content the advantages of these organisms are the possibility of growing them during the entire year, and the potential for technological management in a continuous flow industrial production process (8-12). The protein production rate of algae is more than 25 times that of soybeans (Table 2). Caution should be exercised when comparing yields of soil grown crops with those of water grown crops. Algae growing in heated basins can be cultured throughout the year, while soybeans can only be grown economically during the summer months. This accounts for about a fourfold difference in dry matter production. The algal cell is harvested without leaving parts behind, while the soybean plant roots, stems, and leaves are left in the field. This accounts for about a twofold difference in dry matter production. The protein content of algae is about three times higher than that of soybeans. These three factors combine for the better than 25-fold higher rates

Table 2. Yields per hectare of several crops. The indicated values were obtained from indicated literature reports (13).

Crop	Nitrogen Content	Protein Content	Dry Matter Yield	Protein Yield
	%	%	kg/ha/year	
Soybeans (14)	2.60	16.25	13,708	2,225
Corn (14)	1.20	7.50	29,680	2,227
Sugarcane (14)	0.29	1.78	124,970	3,224
Algae (15)	8.10	51.20	112,000	51,344

of protein production per unit surface area by algae. Algae can obtain their nutrients from digested domestic sewage or animal manure.

PRINCIPLES

System Design

A system for reuse of animal waste was developed on the premise that energy and food supply problems can be solved in part through the development of integrated food production systems in which resources are recycled. Our project was designed to demonstrate the feasibility of using waste heat from thermal power plants to sustain a food producing complex in which nutrients are recycled. Microorganisms will be used to convert swine manure into a high-protein animal feed and a methane-rich fuel gas. Waste heat such as that available from steam electric plants will be used as a low cost source of energy for maintaining stable, elevated, temperatures in anaerobic digesters and units producing single cell protein.

The proposed waste management system (Figure 1) consists of a livestock confinement building where the manure is quickly removed from the animal quarters and routed to a nutrient recovery unit maintained at elevated temperatures (7). Flushing is done with sufficient frequency to prevent anaerobic decomposition and associated odors by the discharge from a syphon activated storage tank. The manure slurry flows to a pump from which it is lifted with an airlift pump to a solid-liquid separator. The solid fraction falls into an anaerobic digester while the liquid flows into the basins for the culture of algae. The algal suspensions from the basins are passed through a centrifuge for harvesting. Clarified liquid is stored in a reservoir from which it is pumped for reuse in the flushing process.

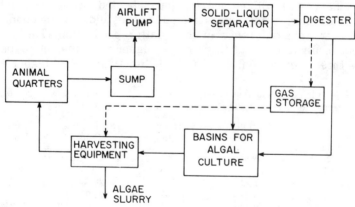

Figure 1. Schematic diagram of the proposed animal waste management system.

The hydraulic manure transport system was chosen because it requires little labor and allows removal of manure from the animal quarters at frequent intervals. The solid-liquid separator minimizes the size of the anaerobic digester and provides an input to the ponds for algal culture. The anaerobic digester decomposes the organic solids. The energy needed for maintenance of the temperature required for efficient digestion (35°C) is to be provided by cooling water from the power plant condenser. In certain situations the temperature of the cooling water may not be high enough to maintain a 35°C basic temperature. Gases from the digester may be accumulated and used for additional heating.

Benefits of Temperature Control

The temperature of the water in which the algal cells are growing is one of the most important growth controlling variables. Low temperatures reduce the efficiency with which algae can use available light energy. The growth rate of algal cells is controlled by both a light-dependent process and a light-independent process. The rate at which compounds can be produced by the light-dependent process is controlled by the rate at which these products can be converted to insoluble cell constituents (*e.g.*, starch) by the light-independent process. When this conversion rate is low, light utilization is limited. The rate of the light-independent process is controlled by the culture temperature. The rate of cell doubling which is a measure of growth rate is two per day at 25°C but nine per day at 35°C (16).

Maintaining a constant high basin temperature has additional important consequences. The desired species in the culture are those which have a rapid growth rate at the higher temperatures. These species do not compete well with undesirable organisms in a medium with daily temperature fluctuations. The dominance of desired algal species is greatly aided by providing a constant temperature.

The algal system of protein production has not been used heretofore because of problems encountered in the development of a reliable, large scale, continuous flow, production process. Biological stability, meaning that the desired organism continues to predominate in the culture and grows at a constant rate, is greatly improved by temperature control obtained by heating the basins with waste heat.

Composition of Swine Manure

The swine manure used in experiments to be discussed later was obtained from the Oregon State University swine center, where hogs are kept on partially slotted floors. The manure drops into pits from which it is periodically flushed to a nearby anaerobic lagoon. Samples were obtained from the drain leading to the anaerobic lagoon (Table 3).

The mineral composition of these samples is representative of that which can be expected in the effluent from the solid-liquid separator. A literature report (17) of the mineral composition of swine waste is included in Table 3.

Table 3. The mineral composition of swine waste in mg/liter compared to that of several nutrient media used in the mass culture of algae.

| Element | Swine Waste | | Mass Culture Media | | |
	OSU Swine Center	Literature (17)	A (18)	B (19)	C (20)
		mg/liter			
Nitrogen					
(NH$_3$ + NH$_4$)	282	330	173	139	139
Phosphorus (PO$_4$)	90	84	285	62	171
Magnesium	45	48	346	25	70
Potassium	189	228	842	543	600
Sulfur (SO$_4$)	—	81	325	33	64
Calcium	96	342	30	9	1
Sodium	88	—	—	67	—
Chloride	—	—	60	119	—
Iron	—	168	10	—	1
Zinc	0.11	—	20	0.05	0.005
Manganese	0.28	—	4	0.05	0.05
Copper	0.01	—	4	0.02	0.002
Boron	0.14	—	20	0.35	0.05
Cobalt	—	—	1	0.02	0.001
Molybdenum	—	—	4	0.06	0.001

The mineral composition of several substrates which have been used for the mass culture of algae are shown in Table 3 for comparison. The concentrated *Chlorella* culture medium A used by Meyers (18), which can support cell concentrations of 30 g/liter or greater, illustrates the adaptability of *Chlorella* to high concentrations of mineral salts. *Scenedesmus* also tolerates a wide range of mineral salts concentrations (13). The salts concentrations of the swine waste are therefore unlikely to affect the growth of *Chlorella* or *Scenedesmus*.

The concentration of nitrogen in the liquified swine manure can be maintained at the levels required to sustain the growth of algae at dry matter concentrations of 1.0 to 3.0 g/liter by adjusting the flow rate in the proposed recycling system. In most outdoor algal cultures the cell density ranges from 0 to 2 g/liter (18) and more dilute nutrient media such as those of Hemerick (19) or Krauss and Thomas (20) may be used.

The other macronutrients—phosphorus, sulfur, magnesium, and

potassium—are unlikely to become limiting in the proposed design, since only part of the nitrogen is removed on each pass through the algal basins. Although the potassium concentration of the swine waste is much lower than used in the nutrient media, it appears to be sufficient because the potassium concentrations of the nutrient media are far in excess of algal requirements (Table 4). The concentration of the micronutrients—zinc, manganese, copper, and boron—in the swine waste also appears adequate to support algal growth.

Table 4. The mineral composition of swine waste compared with values of the mineral composition of *Scendesmus* and *Chlorella* from the indicated literature sources.

| | Composition Based on Dry Weight* | | | | | Relative Composition | |
Element	A	B	C	D	E	Algae**	Waste***
			%			% of N	
Nitrogen	8.14	7.90	6.10	8.00	6.80	100	100
Phosphorus	2.22	1.72	1.23	1.10	1.42	21	26
Potassium	0.92	1.60	0.74	1.50	0.41	16	67
Magnesium	1.60	0.57	0.89	0.50	1.54	11	12
Sulfur	—	1.12	0.91	1.10	0.34	9	23
Calcium	1.93	—	0.76	—	0.06	12	53

  * A mixture of *Scenedesmus* and *Chorella* (21)
    B *Chlorella* (22)
    C *Chlorella* (23)
    D *Chlorella vulgaris* (24)
    E *Scenedesmus obliquus* (20)
 ** Mean of A-E.
*** Mean of values presented in Table 3.

A more rational method of determining the nutrient requirements of mass cultures is the replenishment method of Krauss and Thomas (20). In this method, nutrients are added to the media at the same rate in which they are removed in the harvested algae. This approach has been used to compare the relative mineral composition of the swine waste to that of *Scenedesmus* and *Chlorella* in Table 4. It is apparent that potassium, calcium, and, to a lesser extent, sulfur will not be fully utilized and therefore will eventually accumulate. The moderate accumulation of these elements in the recycled water is unlikely to affect the growth of *Scenedesmus* and *Chlorella,* because of their tolerance to high salt trations (Table 3).

SPECIFIC EXPERIMENTS

Swine Manure Compared with other Substrates

Much concern has been expressed about the limited light penetration in the swine manure suspension. It was suggested that algal

growth in this medium might be improved by clarification. Two clarification techniques were evaluated: 1) ferric chloride flocculation,
and 2) filtration through activated carbon. Figure 2 shows that dilution gradually improved light transmission. Dramatic improvement
in light transmission was obtained by the $FeCl_3$ treatment and the
activated carbon treatment. Results of the last treatment are not
shown. Both treatments also reduced the phosphorus content of the
waste (Table 5). The ferric chloride treatment removed essentially
all the phosphorus.

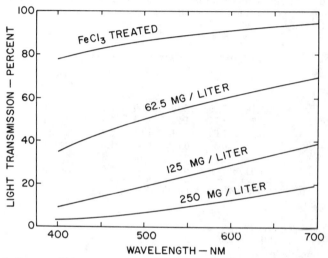

Figure 2. Percent light transmission as a function of wavelength through 1 cm
cells of swine manure obtained from the swine center and diluted to
contain (i) 250 mg N/liter (ii) 125 mg N/liter, (iii) 62.5 mg N/liter and
a sample of the waste containing 250 mg N/liter clarified with $FeCl_3$.

Table 5. Nutrient content (mg/liter) of samples of swine waste treated to improve light transmission.

Variable Measured	Treatment			
	Not Treated	Carbon Filter	$FeCl_3$ 0.25%	$FeCl_3$ 0.33%
pH	7.7	8.5	7.1	6.3
Conductivity (mmhos/cm)	6.3	5.2	7.1	6.0
NH_4 nitrogen	549.8	537.3	569.4	692.8
P	30.0	19.6	0.2	trace
K	360.6	321.4	360.6	360.6
Ca	26.0	38.0	116.0	136.0
Mg	42.6	11.6	54.7	54.7
Na	142.6	165.6	151.8	149.5
B	0.20	0.01	0.18	0.17
Cu	0.04	0.07	0.02	0.03

The manure suspension was evaluated as a substrate for algal growth in culture flask experiments. Sixty milliliters (ml) of substrate were placed in flat bottom, 125 ml, flasks and inoculated. The flasks were placed on a reciprocating shaker in a growth chamber at 30°C and 220 μE/m^2/sec from fluorescent and incandescent lamps. All organisms tested grew poorly on the FeCl$_3$ clarified swine waste (Table 6). The lower pH and phosphate content may have restricted growth. Untreated swine manure used without dilution was the best medium in these and similar experiments. This superiority was attributed to the higher organic matter content of the medium, since levels of inorganic nutrients differed little from the other media. The organic matter may promote algal growth in two ways. Small organic molecules such as sugars or amino acids could be directly utilized by the algae growing heterotrophically. The organic matter also acts as a substrate for growth of bacteria whose respiratory activity provides CO$_2$ for algal photosynthesis and growth.

Table 6. Concentration of algal material (*Chlorella vulgaris* 211/8K) in mg/liter present in culture flasks after three and seven days.

Culture Medium	pH	Age of culture—Days	
		3 Days	7 Days
		mg/liter	
Untreated swine manure	7.6	738	610
Inorganic substrate I	6.5	87	218
Inorganic substrate II	6.5	63	173
FeCl₃ treated swine manure	—	98	100

Screening of Algal Species

The species selected for initial screening were:
Chorella pyrenoidosa Chick I.U. #1230 (high temp TX-71105)
Chlorella "vulgaris" Beij. Strain 211/8K Cambridge
Scenedesmus quadricauda (Turp) Breb. I.U. #614
Scenedesmus obliquus (Turp) Kruger. Gaffron's ScD3WT
Selenastrum capricornutum Printz. (E.P.A. strain)
Anabaena flos-aquae (Lyng) Brebisson. (E.P.A. strain)
Brotryococcus braunii Kutz. I.U. #572
Ankistrodesmus falcatus (Corda) Ralfs. I.U. #188
Spirulina major Kutz. I.U. #552
Stichococcus Bacillaris Nageli. I.U. #314
Euglena Viridis Ehrenberg. I.U. #85

These organisms were tested for their ability to grow on the swine manure obtained from the conduit draining the O.S.U. swine barns. The manure suspension was filtered through two layers of cheese cloth before use to remove large solid particles. Sixty ml of substrate were placed in flat bottom, 125-ml culture flasks which were inoculated with the algal species. The flasks were placed on a reciprocating shaker in a growth chamber set at 30°C and a light intensity of 220 $\mu E/m^2/sec$. After seven days $NaHCO_3$ was added to each flask to provide CO_2 for further growth. At this time the medium was adjusted to pH 7 using dilute HCl. The flasks were then returned to the shaker for three days. Measurements of growth rate and final cell density were obtained by using two sampling times. A three day sample was obtained for an estimate of the growth rate as the algae are in the exponential growth phase during this time (Table 7). The seven day sample was obtained for an estimate of the final cell density attainable, as the algae had reached a plateau at that time. The addition of HCO_3 after seven days was designed to give an estimate of algal growth when the CO_2 limitation within the flask was removed. It also provided a useful comparison of each culture medium's potential to supply CO_2 and fixed carbon to the algae.

Table 7. Cell densities in mg/liter of four algal species growing in manure suspension at 30°C after three, seven, and ten days. $NaHCO_3$ was added and the medium buffered to pH=7 on the seventh day, initial cell density was about 50 mg/liter.

Organism	Age of culture—Days		
	3	7	10
		mg/liter	
Chlorella vulgaris 211/8K	956	945	1330
Scenedesmus obliquus ScD3WT	520	700	1270
Scenedesmus quadricauda	—*	340	800
Selenastrum capricornia	—*	455	950

* Cell density not increased over initial value.

Chlorella vulgaris 211/8K was the most promising species. It was used in all subsequent experiments. This species grew nearly twice as fast during the log phase of growth as the *Scenedesmus obliquus,* while *Scenedesmus quadricauda* and *Selenastrum capricornia* grew extremely slowly. The *Chlorella vulgaris* 211/8K is a high temperature strain with a temperature optimum at 39°C. Even though growing at a slower rate, *Scenedesmus obliquus* ScD3WT also appeared to be well adapted to the swine manure substrate and warrants further evaluation.

In experiments where an alga other than *Chlorella* was inoculated into the swine waste, but did not grow, often a culture of *Chlorella* or a *Chlorella*-like organism resulted. It appears that this organism is in the swine waste when it comes from the building. *Spirulina* and *Botryococcus* did not grow well enough in preliminary tests to be included in the experiments reported here.

Dilution of the Manure

Experiments were conducted to determine the growth rate of algae at several dilutions of the manure suspension obtained from the drain at the OSU swine center. The composition of this material was described above (Table 3). It was diluted with distilled water to obtain ammonium nitrogen concentrations of 250, 188, 125, and 63 mg/liter.

The growth rate of *Chlorella vulgaris* 211/8K at these dilutions was measured in cultures contained in mini-ponds maintained in a growth chamber. The ponds were rectangular fiberglass dishes, 12 cm deep, with a surface area of 0.1 m². The cultures were aerated and gently stirred by bubbling devices consisting of small diameter plexiglass tubes with air outlet holes drilled at 1 cm centers with a #80 drill. Each basin was provided with one tube which was secured to the bottom and ran along the center line.

Experimental conditions were: 1) culture depth: 10 cm; 2) culture temperature: 37±1°C; 3) continuous incandescent and fluorescent light at 381 μE/m²/sec; 4) aeration rate approximately 1 liter per minute from a compressed air supply.

Waste was obtained at the swine barns, filtered through cheesecloth, analyzed for ammonium nitrogen, and adjusted to a concentration of 250±10 mg/liter NH$_4$ nitrogen with distilled water before use. Algae were harvested by flocculation with Al$_2$(SO$_4$)$_3$ and subsequent filtration through a Whatman #1 filter. The algae were dried at 80°C for 1 day. Three liters or 30% of the total basin volume were withdrawn daily during an 8-day period. Withdrawal was done once a day and the volume was immediately replaced with fresh, diluted, waste.

Algal growth rate increased linearly with increases in the concentration of waste (Table 8). The growth response to concentration was probably not due to differences in nitrogen concentrations because the algae from the low waste concentrations had higher crude protein contents (Table 8). The algal growth response to increased

Table 8. Effect of swine waste concentration on the growth rate of *Chlorella vulgaris* 211/8K at 37 ± 1°C. Lights were on continuously at 381 μE/ m²/sec.

Nitrogen Concentration	Rate of Dry Weight Increase	Nitrogen Content of Dry Wt.	Crude Protein Content	Total Amino Acid Content
mg/liter	g/m²/24 hr	%	%	%
250	45.4	8.00	50.0	37.8
188	35.8	8.10	50.5	39.1
125	22.3	8.70	54.4	39.0
63	8.90	12.6	55.3	40.8

waste concentration may be due to a limiting nutrient other than nitrogen or to the higher organic matter content of the more concentrated wastes.

Amino Acid Spectrum

The amino acid spectrum was compared with that from other protein sources (Table 9). The content of essential amino acids of *Chlorella vulgaris* 211/8K grown on the swine manure compares favorably with *Scenedesmus, Spirulina,* soy, and milk protein. The *Chlorella* protein appears to be equivalent to milk protein, except for a deficiency in methionine.

Table 9. Essential amino acid content of *Chlorella vulgaris* 211/8K grown on swine waste at 37 ± 1°C, compared with *Scenedesmus, Spirulina,* milk, and soy protein spectra obtained from the literature.

Amino Acids	Chlorella vulgaris 211/8K*	Scenedesmus (25)	Spirulina maxima (26)	Soybeans (27)	Milk (27)
	grams per 100 grams protein				
Valine	6.2	7.2	6.5	5.2	7.0
Leucine	9.3	9.3	8.5	8.4	9.9
Isoleucine	4.8	4.4	6.0	5.3	6.4
Threonine	5.3	5.2	4.6	4.4	5.4
Methionine	1.8	1.4	1.4	1.7	2.5
Phenylalanine	6.6	4.6	5.0	5.8	4.8
Lysine	8.0	5.7	1.4	5.6	7.7
Tryptophan	N.A.	1.4	4.6	1.3	1.4

* Mean value for samples grown in waste dilutions containing 63, 125, 188 and 250 mg N/liter. There were no significant differences in the amino acids spectra of the *Chlorella* grown in these four waste concentrations.

Seventy-five percent of the total amino acids comprised the crude protein. This is to be expected in microorganisms where much nitrogen is also present as nucleic acids and other non-protein compounds.

Feeding Trials

Only preliminary feeding trials with the algal material grown on swine manure have been conducted. Two algal samples obtained by different harvest techniques were compared with a limited number of other protein sources (Table 10). One algal sample was harvested by centrifugation while the second sample was harvested by precipitation with $(Al)_2(SO_4)_3$ and filtration of the flocculate.

Rats were fed over a 2-week period with the protein sources shown in Table 10 included to provide 12% of the dietary protein requirements. Of these, casein is a high quality protein, while cottonseed meal is of relatively poor quality.

Table 10. Protein efficiency ration (PER) and average daily weight gain of rats fed with the indicated protein sources, included to provide 12 percent of the dietary protein requirement.

Protein Source	Average Gain	Average Feed Intake	PER*
	g/day	g/day	
Casein	3.70	13.37	2.30
Fungus (from Y. W. Han)	1.96	13.17	1.44
Torula yeast	1.76	12.29	1.17
Brewers yeast	1.96	12.94	1.40
Cottonseed meal	2.23	16.26	1.13
Algae (centrifuged)**	2.29	13.29	1.44
Algae (alum pptd)**	1.80	16.20	0.91

* PER (protein efficiency ratio) is of weight gained per g protein consumed.
** OSU swine waste.

In considering both the PER and the growth rate, the algal material harvested by centrifugation appears to be a good protein source. It compares favorably with cottonseed meal which is used extensively as a livestock feed.

The alum precipitated algae did not give good results. These results strongly suggest that harvesting should be accomplished by centrifugation, air flotation, or some other method which does not add toxic material.

Retention Time

The culture volume harvested each day from a fixed pond volume affects algal yield per unit area. Harvest volume determines the retention time of the waste and strongly influences the density of the culture within the pond. The cost per pound of algae harvested by centrifugation depends on the density of the culture. Consequently there are economic trade-offs to be considered in the selection of harvest rate.

Experiments were conducted in which different culture volumes were withdrawn each day. Harvest rate is best expressed as a mean retention time of the waste in the pond. At a culture depth of 10 cm the basins contained 10 liter. Thus a withdrawal rate of 1 liter/day corresponds to a mean retention time of 10 days.

Product yield decreased with increasing mean retention time over the range of retention times tested (Table 11). This decrease results from the fact that a smaller volume is harvested as the mean retention time increases while the concentration does not increase proportionally. The yield must decrease as the mean retention time approaches one. At that point the entire basin volume would be removed once each day leaving few algal cells to inoculate the new substrate.

Table 11. Dry matter and protein yield of *Chlorella vulgaris* 211/8K growing on swine waste adjusted to 250 ppm ammonium nitrogen at 37 ± 1°C with a 12-hour photoperiod, using a light intensity of 381 μE/m²/sec.

Daily Harvest Volume	Retention Time	Crude Protein	Dry Matter Yield	Protein Yield	Cell Density
1/pond	days	%	g/m²/12 hr		g/liter
1.5	6.66	45.2	17.9	8.09	1.006
2.0	5.00	45.6	18.2	8.29	0.984
2.5	4.00	45.5	20.9	9.51	0.952
3.0	3.33	46.7	21.5	10.05	0.874
3.5	3.00	46.1	22.4	10.35	0.796
4.0	2.50	46.4	25.1	10.92	0.771

These experiments only considered retention time as a variable. Product yield and cell density will be different at different light intensities, day length, culture temperature, and frequency of harvesting during the day. These variables are being further evaluated in current experiments.

Nitrogen Balance

An ammonium nitrogen balance indicated that much of the nitrogen was lost, probably by volatilization as ammonia (Table 12). Mean retention time had a significant effect on the ammonium nitrogen remaining in solution at harvest time. Part of the nitrogen remaining in the pond outflow would be recovered in a recycling system.

The amount of nitrogen recovered by the algae decreased from 34% for the retention time of 6.66 days to 17.5% for the retention time of 2.50 days. At the long retention times the system functions more like a batch system.

Table 12. Daily balance of ammonium-nitrogen of the ponds with *Chlorella vulgaris* 211/8K growing on swine waste at 37 ± 1°C. Initial nitrogen concentration was adjusted to 250 mg/liter.

Daily Harvest Volume	Retention Time	by to pond	Remaining in Solution		Recovered by Algae		Volatilized	
1/day	days	mg	mg	%	mg	%	mg	%
1.5	6.66	375	28	7.5	129	34.4	218	58.1
2.0	5.00	500	60	12.0	133	26.6	307	61.4
2.5	4.00	625	110	17.6	152	24.3	363	58.1
3.0	3.33	750	149	19.9	161	21.5	440	58.7
3.5	3.00	875	207	23.7	166	19.0	502	57.4
4.0	2.50	1000	285	28.5	175	17.5	540	54.0

The high rate of ammonia volatilization is a potential shortcoming of the proposed system (Table 12). There are several reasons for these high volatilization rates. Among these are: 1) high temperature of the medium; 2) high pH of the substrate; and 3) high aeration rate used in these trials. The extent to which volatilization rates can be reduced remains to be investigated. The control of pH should help to ameliorate this problem.

Culture Depth

When using inorganic culture media, the optimum culture depth is the one in which all the cells are exposed to some light, but little light passes through the culture. The principle is to avoid an unlit "dead" space at the bottom, where fixed carbon is lost by respiration thereby decreasing the net dry matter production rate. This could be particularly important at elevated temperatures where respiration proceeds at faster rates. On the other hand, it has been suggested that when a material such as animal waste is used as the substrate, an unlit layer at the bottom of the pond may be needed to provide an environment where bacterial breakdown of dissolved organic material can proceed.

Because of diurnal and day to day fluctuations in light intensity under field conditions it will not be possible to precisely regulate the interaction of culture depth and light intensity. However information with respect to culture depth is needed for managerial decisions with respect to interactions of retention times and culture depth.

Production rates for culture depths of 6, 9, 10, 12 and 15 cm were obtained using a constant retention time of 3.0 days (Table 13).

Table 13. Growth rates of *Chlorella vulgaris* 211/8K at 37 ± 1°C for several culture depths and a 3.0-day retention time. A photoperiod of 12 hours was used with light intensity at 387 $\mu E/m^2/sec$.

Pond Depth	Daily Harvest Volume	Retention Time	Growth Rate	Cell Density
cm	l/day	days	$g/m^2/12$ hr	g/liter
6	2.0	3.00	12.8	1.61
9	3.0	3.00	15.5	1.36
10	3.3	3.00	22.4	1.36
12	4.0	3.00	23.2	1.37
15	5.0	3.00	21.9	1.09

For a retention time of 3.0 days the optimum culture depth is approximately 12 cm. It was established in earlier experiments that production rate increased with shorter retention times. Thus a relationship between growth rate, retention time, and pond depth exists.

The experiments necessary to establish this relationship have not been completed.

CO_2 Addition

Actively growing algal cultures consume large amounts of dissolved carbon dioxide. Chlorella cultures when bubbled with air containing 5 to 10% carbon dioxide rather than the ambient level of 0.03% showed increased growth. The addition of CO_2 may also serve to lower the pH of the culture. The optimum pH for the growth of the *Chlorella vulgaris* 211/8K is in the range of 6.5 to 7.0. The pH of the *Chlorella* cultures growing in untreated swine waste ranged from 8.2 to 8.9. Lowering the pH would have the additional benefit of decreasing the volatilization of the ammonia gas.

An experiment was conducted in which air containing 5% CO_2 was discharged through the bubbling device used in all experiments. Substrates used were: 1) swine waste adjusted to a nitrogen content of 250 mg/liter; 2) swine waste diluted with distilled water to a nitrogen concentration of 125 mg/liter; and 3) swine waste adjusted to 125 mg N/liter with urea added to obtain a nitrogen concentration of 250 mg/liter (Table 14).

Table 14. Effect of CO_2 injection on pH, cell density, and growth rate of *Chlorella vulgaris* 211/8K at 37 ± 1°C. Light continuous at 381 $\mu E/m^2/sec$.

Treatment	Nitrogen Content	pH	Cell Density	Growth Rate	Increase Due to CO_2
	mg/l		g/l	g/m²/24 hr	%
CO_2 added					
standard waste	250	7.6	1.458	50.8	20
50 percent dilution	125	7.3	1.373	49.0	42
50 percent dilution, urea added	250	7.7	1.457	51.0	48
No CO_2 added					
standard waste	250	8.4	1.260	42.4	
50 percent dilution	125	8.9	1.039	34.5	
50 percent dilution, urea added	250	8.6	1.036	34.3	

Injection of the CO_2-enriched air reduced the pH of the cultures significantly, although the desired level of pH= 7.0 was not achieved. CO_2 injection was most effective in reducing the pH of the diluted waste. This was probably due to the higher organic matter content of the undiluted waste providing a greater buffer capacity.

Waste concentration or nitrogen content had no effect on production rates when the cultures were aerated with air containing 5% CO_2. Without CO_2 added, waste concentration had a large effect on production rate.

The response to CO_2 addition was much greater in the diluted waste, regardless of nitrogen level (Table 14). The much smaller increase in yield resulting from injection of CO_2 in the standard waste may indicate that the concentrated waste enhances algal growth by providing soluble carbon. This observation was further substantiated in a second experiment where the cultures were exposed to a 12-hour photoperiod rather than continuous light (Table 15). No increased yield resulted from CO_2 injection, indicating that CO_2 accumulated during the dark period in the control treatment in sufficient quantity to satisfy CO_2 demands of the culture during the 12-hour light period. During the dark period CO_2 was not injected and the pH of the experimental culture rose to that of the control.

Table 15. Effect of CO_2 enrichment on *Chlorella vulgaris* 211/8K culture at 37 $\pm$ 1°C. CO_2-enriched air was used during the light period only. A 12-hour photoperiod with light intensity at 30 $\mu E/m^2/sec$ was used.

Treatment	pH 8 am	8 pm	Cell Density g/l	Growth Rate g/m²/24 hr
Standard waste	8.4	8.0	1.218	36.6
CO_2	8.2	7.2	1.238	37.1

CONCLUSIONS

The most encouraging aspect of the studies to date has been the stability of the *Chlorella vulgaris* 211/8K cultures growing in fresh swine manure diluted to an ammonium nitrogen content of 250 mg/liter. Further dilutions reduced the growth rate. Manure solutions with higher nitrogen concentrations were not tested. Clarification of the manure by filtration through activated carbon or ferric chloride flocculation improved the light transmission dramatically. However, the algal growth rates were substantially lower in the clarified solutions. The superiority of the untreated swine manure was attributed to the higher organic matter content. It was suggested that the organic matter provides CO_2 for algal photosynthesis by acting as a substrate for growth of bacteria. This observation was further substantiated by experiments in which CO_2 was added to the cultures.

The average growth rate for a 12-hour photoperiod at 37°C and a retention time of 3.3 days was about 22 g/m²/day. Higher growth rates were obtained with shorter retention times. The constantly high yield of cell dry matter and protein has been encouraging.

Development of optimum management techniques requires additional experimentation to establish interactions between temperature, light intensity, retention time, and culture depth. Yields of 25 to 30 g/m²/day for a 12-hour photoperiod appear easily attainable. Mean culture densities of 1.0 to 2.0 g/liter may be achieved in the field. These high cell densities would reduce the cost of harvesting per unit product.

Neither the nitrogen nor phosphorus content of the waste appears to be limiting algal growth. There was no response to nitrogen levels in any experiment. The algal pond system reduced the total and ammonium nitrogen content of the waste by about 90%. Nitrate did not accumulate in the ponds and in all cases was <1 mg/liter. The algae accounted for 20-30% of the nitrogen taken from the waste media. About 50% of the nitrogen was volatilized from the system, probably as ammonia gas.

The content of essential amino acids in grows per 100 grams of protein compares favorably with milk and soy protein. Short term feeding trials with rats indicated that the algal material harvested by centrifugation is an excellent protein source, even when fed without pretreatment for improvement of digestibility.

* * *

This investigation was supported in part by matching grant funds, Office of Water Resources and Technology, USDI.

REFERENCES

1. United Nations. Statistical Yearbook (1972).
2. Pimentel, D., L. E. Hurd, A. C. Bellotti, I. N. Forster, I. N. Oka, O. D. Sholes, and R. J. Whitman. Food production and the energy crises. Science 182:443 (1974).
3. Heichel, G. H. Energy needs and food yields. Technology Review 76:2 (1974).
4. Hirst, E. Food related energy requirements. Science 184:134 (1974).
5. Anonymous. Energy down on the farm. Rural Research (Australia) 85:4 (1974).
6. Steinhart, J. S. and C. E. Steinhart. Energy use in the U.S. food system. Science 185:307 (1974).
7. Boersma, L., E. W. R. Barlow, J. R. Miner, and H. K. Phinney. Animal Waste Conversion Systems Based on Thermal Discharges. Special Report 410. Oregon Agricultural Experiment Station. Corvallis 54 pp. (1974).
8. Tamiya, H. Mass culture of algae. Ann. Rev. Plant Physiol. 8:309 (1957).
9. Oswald, W. J. and C. G. Golueke. Large-scale production of algae, pp. 271-301. In: R. I. Mateles and S. R. Tannenbaum (eds.), Single Cell Protein. Mass. Inst. Techn. Press, Cambridge, Mass. (1968).
10. Oswald, W. J. and C. G. Golueke. Harvesting and processing of waste-grown microalgae, pp. 371-389. In: D. F. Jackson (ed.), Algae, Man, and Environment. Syracuse University Press, Syracuse, N. Y. (1968).
11. Soeder, C. J. and W. Pabst. Gesichtspunkte für die Verwendung von Mikroalgen in der Ernährung von Mensch and Tier. Ber. Dtsch. Bot. Ges. 83:607 (1970).
12. Stengel, E. Anlagen und Verfahen der Technischen Algenmassen Produktion. Ber. Dtsch. Bot. Ges. 83:589 (1970).
13. Bickford, E. D. and S. Dunn. Lighting for Plant Growth. The Kent State University Press, Kent, Ohio. (1972).
14. Wilcox, O. W. Footnote to freedom from want. Journ. Agr. Food Chem. 7:813 (1959).
15. Mattoni, R. H. J., E. C. Keller, Jr., and H. N. Myrick. Industrial photosynthesis, a means to a beginning. Bio. Science 15:403 (1965).

16. Sorokin, C. and R. W. Krauss. The effects of light intensity on the growth rates of green algae. Plant Physiol. 33:109 (1958).
17. Benne, E. J., C. R. Hogland, E. D. Longnecker, and R. L. Cook. Animal manures—what are they worth today? Agric. Expt. Sta. Bull. No. 231, Michigan State University, East Lansing (1961).
18. Meyers, J. Algal culture, pp. 649-668. In: R. E. Kirk and D. F. Othmer (eds.) Encylopedia of Chemical Technology, First Supp. Interscience Publ. New York (1971).
19. Hemerick, G. Mass culture. In: Handbook of Phycological Methods. (In press).
20. Krauss, R. W. and W. H. Thomas. The growth and inorganic nutrition of *Scenedesmus obliquus* in mass culture. Plant Physiol., 29:205 (1954).
21. Hintz, H. F., H. Heitmann, W. C. Wier, D. T. Torell, and J. H. Meyer. Nutritive value of algae grown on sewage. Jour. Animal Sci., 25:675 (1966).
22. Gromov, B. V. Main trends in experimental work with algae cultures in the U.S.S.R., pp. 249-278. In: D. F. Jackson (ed.), Algae, Men and the Environment. Syracuse University Press, Syracuse, N. Y. (1968).
24. Groghegan, M. J. Experiments with *Chlorella*, pp. 182-189. In: J. S. Burlew (ed.), Algal Culture: From Laboratory to Pilot Plant. Publ. No. 600, Carnegie Inst., Washington, D. C. (1953).
25. Bock, H. D. and J. Wünsche. Möglichkeiten zur Verbesserung der Proteinqualität von Grünalgenmehl. Stizunsberg. Dtsch. Akad. Landw. Wiss., 16:113 (1967).
26. Clement, G., C. Giddey, and R. Menzi. Amino acid composition and nutritive value of the algae *Spirulina maxima*. Jour. Sci. Food Agric. 18:497 (1967).
27. Souci, S. W., W. Fachmann, and H. Kraut. Die Zusammensetzung der Lebensmittel. I-III Wiss. Verlags - Ges., Stuttgart (1962-1969).

35.

Conservation of Energy and Mineral Resources in Wastes through Pyrolysis

Donald M. Nelson* and Raymond C. Loehr*

The use of mechanization and transportation in agriculture has allowed the centralization and concentration of animal production units. As a result, large quantities of waste material are produced at large-scale feedlots and poultry farms.

Traditionally, agriculture waste was not considered a problem and was disposed of on the land as part of crop production. Today, land disposal continues to be a viable method to dispose of the wastes and utilize the associated nutrients. However, other waste management approaches must be critically evaluated to identify those approaches which have minimum energy requirements, possibilities of energy recovery, and/or possibilities of by-product recovery. When the producer cannot utilize all of the waste or if it cannot be sold or processed into a marketable product, he must consider paying to dispose of his waste material, and then find the most economical means of disposal.

By carefully integrating waste management methods into the production scheme, odor control, nutrient removal, volume reduction,

* Engineer, Westinghouse Electric Corporation, Environmental Systems Department, Pittsburgh, Pennsylvania.
** Director, Environmental Studies Program, and Professor, Cornell University, Ithaca, New York.

weight reduction and increased ease of handling may be achieved. Liquid waste treatment systems which depend on biological activity fall into two groups; aerobic, such as oxidation ditches; and anaerobic, such as anaerobic lagoons and anaerobic digesters. Dry waste treatment may be biological, as in composting, or it may involve drying the waste to reduce its weight and biological activity to a minimum. Incineration and pyrolysis represent chemical and physical means of treating waste.

Pyrolysis has potential as the waste disposal method because, if operated properly, the pyrolysis of waste may:

1. involve minimal pollution
2. offer a great deal to the conservationist

In order to evaluate pyrolysis as a mode of waste treatment a comprehensive study was undertaken. The first objective was to develop simple and relatively inexpensive equipment capable of pyrolitically treating waste material on a pilot plant scale. The second objective was to examine the products of pyrolytic waste treatment with special attention given to chemical content, properties hazardous to health, water pollution potential of the char, and the potential effect on land from the application of the chars.

It would be difficult to discuss the subject of pyrolysis without defining certain terms. The terms used in the paper are identified in the Appendix.

LITERATURE REVIEW

A careful review of the literature was conducted before and during the study. Theoretical concepts and design concepts were gathered from the industrial engineering, chemical engineering and chemistry textbooks of the early 1900s. Such information is not available in more modern texts; for example, the 1963 *Encyclopedia of Chemical Technology* does not mention destructive distillation nor pyrolysis (1). The early texts were concerned with coal carbonization and the by-product industries. Some work has been done in Europe and Asia on the coking of brown coal and peat and some texts on the subject were reviewed (2-4). The best source of information in the area of design of pyrolytic equipment and processes was the U.S. Patent Library. The best sources of information on the state of the art and recent research were the publications of the U.S. Bureau of Mines.

Pyrolysis as a method of waste management is relatively new. The advocates of pyrolytic handling have proposed it as a conservative measure and as a money-making proposition (5-9). In the last few years, some interest has developed in the idea of considering destructive distillation as a waste treatment method with a negative profit being the cost of treatment (10, 11).

The earliest efforts did stimulate interest and research on the idea of treating waste by destructive distillation. It was found that if plastics were subjected to low-temperature destructive distillation, hazardous materials might be produced (12). Because municipal solid wastes and some industrial solid wastes include plastics, research has been conducted in the high temperature range.

On the subject of pyrolytic handling of animal manures only three reports were found. None of these referred to any other pyrolytic work with animal manures (6, 13, 14).

The literature on peat and coal carbonization was reviewed in terms of minerals and fertilizer constituents. The review suggested that:

1. No phosphorus is driven off by pyrolysis and that the phosphorus remains in the char as phosphates (15).

2. Fifty to 65% of the original sulfur remains in the char (15), 25 to 30% of the sulfur goes to hydrogen sulfide, some of which reacts with iron to form a ferrous sulfide lining which protects iron retorts (15, 16).

3. Forty to 60% of the original nitrogen is combined in the char (3, 15). Low-temperature char may contain an even higher percentage (16).

None of the research done on destructive distillation of animal manures reported the presence of the before-mentioned minerals in the off-gases and tars. Nitrogen, carbon, oxygen and hydrogen were the sole constituents of the off-gases and tars reported in the studies of the destructive distillation of animal manures. Consequently, very little ash loss should be expected.

Another consideration is the heat value of the waste. It has been suggested that 11,400 Btu were needed to incinerate the waste properly and that only 7300 Btu were available in the raw waste (17). Calculations were made on the off-gases of the high-temperature destructive distillation of poultry manure. It was reported that the heat values of poultry wastes would produce enough heat to vaporize the moisture, allowing for the total solids in the waste. Destructive distillation at 800°C could be self-sustaining if the egg farm wastes were partially dewatered beforehand (13).

Farms must also dispose of dead animals. The old process of subjecting animal bones to destructive distillation and recovering animal charcoal suggested that destructive distillation could be used to treat the animal carcasses as well (18).

GENERAL

In order to evaluate pyrolysis as a means of treating animal waste, a pilot study was undertaken. The waste materials studies were egg farm wastes which included chicken feces mixed with feathers, woodshavings used as animal bedding and dead birds and rodents. Egg farm waste is a good prospect for pyrolytic treatment because egg farms tend to be located near urban areas where storage and disposal of odorous wastes may be difficult. On the other hand, egg farm wastes tend to have a low caloric value and high ash content when compared to other agricultural wastes.

A black iron, continuous flow, horizontal destructive distillation device was designed, built and operated. The unit handled between 180 to 400 hen/days of waste in an 8-hour work day.

Mice were used in carcinogenic studies to evluate the potential hazard of the machine's products to men and animals. Land application disposal of the char was examined during the course of the investigation. The char produced by the device was evaluated to establish guidelines for the pyrolytic treatment of egg farm waste. The leachates produced from the chars were studied to determine the potential water pollution hazards involved in their handling and storage.

Equipment

A destructive distillation device capable of conveying egg farm waste through a pyrolytic zone was needed as were temperature probes in the reaction zone, a closed char hopper and a waste hopper.

Using available information, a simple device was constructed (Figure 1). The waste was augered through a horizontal black iron tube with eight gas ports (three 3/8-in. and five 1/8-in.), spaced at 4-in. intervals along the top of the tube. High-temperature thermocouples

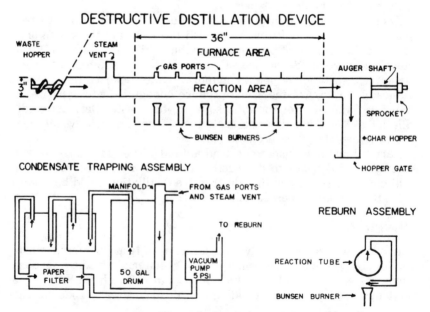

Figure 1. Apparatus.

were strapped to the outside of the tube and inserted through the gas ports to read internal temperatures. The auger was driven by a chain and sprocket from a gear reducer box which in turn was powered by a variable speed motor. The tube was encased in sheet metal which acted as a safeguard and as a heat and flame shield. For convenience and mobility, liquified petroleum gas was the chosen fuel. Seven Bunsen burners with the proper gas jets were placed on a manifold and used to heat the tube directly from the bottom.

The auger extended out the end of the tube and into the waste hopper. It was assumed that the waste covering the auger and filling the end of the tube would effectively seal out atmospheric oxygen. The char was to fall from the auger into a closed hopper, through a hole in the bottom of the tube at the discharge end of the furnace. A hopper wtih a valve at its bottom held the char and shared the atmosphere of the reaction chamber through the opening of the tube. The shaft of the auger emerged from the reaction tube through a hole cut into the pipe cap which acted as the end of the reaction tube. The fittings were not made airtight because a steam blanket was expected to develop inside the tube and the slight positive pressure would keep the atmospheric oxygen out.

The temperature was controlled by varying the number of Bunsen burners used and controlling the heat loss of the device through radiation and convection. The heat lost by the device to its surroundings is a function of its thermal efficiency; the greater the heat loss, the lower the thermal efficiency. Improved thermal efficiency was achieved by means of heat shields. The heat shields were semicircular pieces of sheet metal which fit over the top of the tube inside the furnace. Before a run, the machine was made more or less thermally efficient by adding or removing heat shields. Changing the thermal efficiency was necessary to achieve the higher temperatures, and to ensure even heating at the lower temperatures. Lower efficiencies allowed more burners to be used, which distributed the heat more evenly.

The temperature was read on a meter attached to the chromel alumel thermocouples. Temperature distribution from end to end of the reaction tube was even, but from bottom to top, it was not. The bottom of the metal tube was observed to approach flame temperature, 2200°F, while the top of the tube approximated the temperature of the inside gases.

The steam blanket was controlled by a steam vent located between the furnace and the waste hopper. The off-gases and tars were burned in the flames of the Bunsen burners, or the tar was trapped and only the gases burned in the flames of the Bunsen burners. The tar traps consisted of a 50-gallon steel drum, followed by a series of water-cooled traps and then paper gauze. Though a slight vacuum was maintained in the last traps, a slight positive pressure remained in the tube. The gas ports and steam vents served to connect the tube and the traps, or to feed the reburn by attaching various sets of piping.

Later, it was found that the Standard Oil Company received a patent on a mechanically similar device some 4 months after the device used in this research began operation (19).

Hazardous Residue Study

Though the process is confined, the products may come in contact with the operator. Therefore, the carcinogenic properties of the tars and chars were examined.

A cancer is defined as an abnormal growth or condition of a living cell. A carcinogen is anything that causes a cancer or promotes the growth of a cancer. Four methods of carcinogenesis are considered as probable: viral, genetic, radiation and chemical (20). Of interest to this research were the chemical carcinogens, since the tars produced by the destructive distillation of chicken waste may have carcinogenic properties similar to those of coal tar.

In consultations with Dr. Rickard of the Department of Pathology, New York State Veterinary College, Cornell University, it was decided that there were grounds for concern over the possible similiarities between the tars produced and coal tar, and between the char produced and soot. An experiment was conducted under the guidance of the Pathology Department of the Veterinary College to determine if a potential danger existed. With the help of Dr. Rickard, "Swiss" type mice were chosen for the study. They are hearty, fairly large albino mice, which have been used in laboratory work since the turn of the century. As a group, they respond to many of the same carcinogens as man does, but more readily than man. If not exposed to carcinogens, essentially no cancer occurs among the male mice and the only cancer reported among females is breast cancer at a 1% incidence rate. These caracteristics make evaluation simpler. The mice used were Swiss-Webster ICR mice from the Blue Spruce Farms, Inc., Altamont, New York. They were chosen because they met the requirements and could conveniently be delivered to Cornell. They arrived at 3 weeks of age, in good condition, and the experiment began the same day. The mice were house at the Waste Management Laboratory of the College of Agriculture and Life Sciences, Cornell University.

Four major groups of mice were required by the experiment. The large groups were divided into subgroups to prevent crowding.

1. A negative control group was needed as a base line, or reference group, which was not exposed to the products of the machine. The two subgroups were:

> a. five males and five females,
> b. three males and two females,

These subgroups, *a* and *b*, were treated with toluene which was some-times used as a solvent around the machine and in the tars.

2. A char group was needed to check the effect of exposure to the char produced. The char was used as the bedding and a single-fold towel was provided the mice for nest building. The char was also evaluated as a bedding material.

3.The tar group was exposed to the tar produced by the machine.

4. A positive control was needed to check susceptibility of the mice, and to create a reference point. This group consisted of three males and three females. These mice were exposed to 3,4-benzo(a)-pyrene, a known carcinogen.

Each Thursday, all the mice had the hair on their shoulders and the back of their necks clipped to expose the skin. Even the controls and char mice were clipped to avoid the controversy over the effect of irritation on the development of cancer. After the clip, the mice which were to be exposed to a liquid substance were treated. The treatment consisted of spreading the liquid material on the exposed skin with a camel hair brush. The negative control *a* and *b* subgroups received 0.07 ml of toluene each, in each of 30 applications, for a total of 2 ml each in 210 days. The tar mice received an average of 0.15 grams of tar each, once a week for 30 weeks, for a total of 4.5 grams each in 210 days. Each mouse in the positive control was treated with a total of 16 milligrams of practical grade, 3,4-benzo(a)-pyrene (obtained through the Aldrich Chemical Company), dissolved in 1.7 ml of toluene, in 13 applications.

On the 90th day, treatment of 3,4-benzo(a)pyrene was terminated because all of the positive control mice had developed carcinomas. On the 210th day, the tar and toluene treatment were terminated. On the 217th day, the mice were autopsied at the Veterinary College.

Leachate Studies

The solid products may be stored outside or even disposed of by land application methods. To see if any potential problems might occur, the properties of leachates were examined and a land appli-cation check was made.

Two types of char can be produced: one that is incompletely car-bonized (brown char), and one that is completely carbonized (black char). Carbonization is a dramatic and largely irreversible chemical change occurring when oxygen, hydrogen and nitrogen are driven off, leaving carbon. The material changes from the color brown, characteristic of partially decomposed organic compounds, to black during destructive distillation.

In order to define "proper operation" for the process, an end-point for the process had to be found and defined. The change to a car-bonized material is well defined for the operator by the color change.

Brown and black chars were used in the laboratory while looking for definable differences between chars.

Soaking the brown and black chars in distilled water resulted in dramatically different leachates. In that environment, water can be the vehicle of transport for pollutants. For example, rainwater leaching through a storage pile of the char could run off to surface water or percolate through the soil into the ground water, polluting fresh water. Thus, this discovery supplied a simple means of comparing chars, and demonstrated the potential danger of char to the environment. Therefore, the leachates of brown and black char were compared. Several black chars were used and two brown chars which only showed a trace of brown color were used. Basically, two types of experiments were designed:

1. Soaking equal amounts of char by volume in enough distilled water to cover the sample for 24 to 28 hours; then pouring off the leachate. A second leaching of the same char was also done on one occasion.

2. Weighted samples were placed in filters and distilled water run through at intervals until the leachate from the brown char approached a colorless state.

The leachates were analyzed for clarity, color, pH, COD, TKN, total solids, ash, BOD and ammonia.

THE LAND APPLICATION STUDIES

The char produced may represent a solid waste problem in itself, though its volume and weight would be less than that of the raw waste. If the char is not used as a raw material in industry and if no direct use is found for it, it will eventually have to be disposed of. Landfill of the material is feasible, but this would be a waste of the minerals present. Another alternative is the spreading of this material on the land under the theory that the char would slowly decompose and return the minerals to the soil. The slow release of the chemicals could prevent any dramatic effect on the native plant cover.

By applying char at various loading rates to test plots, the concept that the land could be used as a disposal site was evaluated. By applying ash made from waste materials on parallel plots in quantities approximately equal to the ash content of the char the concept of slow release was examined. An open field was chosen as the site of the experiment. The soil was a disturbed Hudson and Collamer silt loam with a 2 to 6% slope and the field was grown over with the usual assortment of plants found in local pastures. The plots were laid out in the field without disturbing the native vegetation. The char and ash samples were spread over the vegetation and left to settle.

In order to include plants which are sensitive to salts, two tomato plants were obtained. The two plots for the tomato plants were prepared by removing the native vegetation and loosening the soil with a shovel. The tomato plants were planted in the center of the plots one week before the experiment began, and were in good condition. One plot received 1/10-pound of ash per square foot, and the other plot received 2/10-pound of char per square foot.

The eight plots were each 1 foot square; their borders defined by two-by-fours made into boxes. One foot was left between the outsides of the boxes as control areas. The char or ash was applied to the surface, and the natural rainfall was depended on to irrigate the grass plots. The char or ash was worked into the loose soil around the tomato plants but rain was depended on for irrigation (Figure 2). Periodically, the plots were observed and observations recorded.

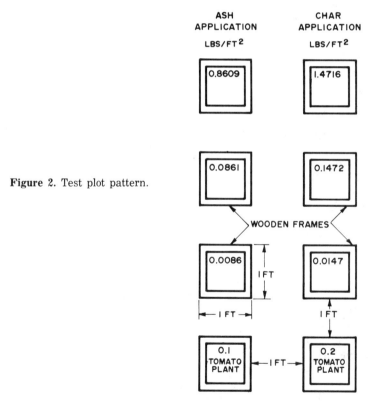

Figure 2. Test plot pattern.

Waste Characteristics

The egg farm waste was a mixture of manure, feathers, spilt water and wasted feed grain (undercage waste). The typical constituents of such waste are noted in Table 1.

Table 1. Undercage waste (21-23).

Constituent	Percent	
Water	54 - 80	wet weight basis
Nitrogen	3.4 - 5.9	dry weight basis
Phosphorus	0.9 - 2.1	dry weight basis
Potassium	0.7 - 2	dry weight basis
Sulfur	0.3 - 1.2	dry weight basis
Calcium	8.0 - 9.3	dry weight basis
Boron	⌣0.02	dry weight basis
Copper	⌣0.006	dry weight basis
Iron	⌣0.2	dry weight basis
Magnesium	⌣1.2	dry weight basis
Manganese	⌣0.04	dry weight basis
Molybdenum	⌣0.02	dry weight basis
Zinc	⌣0.04	dry weight basis

RESULTS OF EXPERIMENTS

The device was operated (run) under various conditions of temperature, furnace configuration (apparatus) and with different waste materials (Table 2). The results were then analyzed and conclusions drawn.

Energy Requirements

To determine the energy requirements for the destructive distillation of these wastes, the furnace was operated without waste material (blank runs) to establish a base line for fuel consumption under each of the three apparatus configurations (runs 9, 10 and 16). The three plots of base line fuel consumption curves were plotted on separate graphs. Additional runs were made with corresponding apparatus and were plotted on these graphs (Figure 3). In runs 6 and 7, the furnace appeared to require less fuel than expected to maintain the temperature of the furnace. Runs 12 and 13 appeared to require more fuel than expected to maintain the temperature of the furnace, while runs 1, 4, 5, 8, 11 and 14 appear to have used the fuel at the expected rate. It should be remembered that the off-gases from the destructive distillation of the wastes were always burned in the furnace and that it was estimated that the off-gases represented enough energy to meet the energy requirements of the destructive distillation of the waste (13). Consequently, within the accuracy of this study, the fuel requirements of the overall process were directly dependent upon the thermal efficiency of the furnace and not on the nature of the waste.

No effort was made to estimate the power required to convey the waste through the reaction unit, therefore conveyance is an additional energy requirement, dependent on the mechanical design of the unit.

Table 2. Summary of machine runs.

Run Number	Operating Temperature, °F	Moisture Content of Raw Waste, %	Apparatus	Amount of Waste, lb	Type Char Produced	Length of Run	LP Gas Used, lb
1	575	55.6	Open to atmosphere	5.5	Brown	3 hr 50 min	3
2	1100	53.4	Open to atmosphere	10.0	Black	4 hr 35 min	12
3	925	58.4	Reburn of all vapors	11.0	Black	3 hr 50 min	7
4	925	59.8	Reburn of all vapors	13.0	Black	3 hr 50 min	9
5	1000	38.0	Condensate trapping	14.5	Black	4 hr 20 min	12.25
6	1200	8.0	Condensate trapping 1 heat shield	14.0	Black	4 hr 7 min	8.25
7	1450	10.0	Condensate trapping 1 heat shield	16.0	Black	6 hr 55 min	17.25
8	900	24.0	Reburn	16.75	Black	5 hr 5 min	9.75
9	a. 925 b. 800		Blank run—no shields			a. 2 hr b. 50 min	a. 4 b. .75
10	a. 1100 b. 1500 c. 1600 d. 1450 e. 1300 f. 850		Blank run—2 heat shields			a. 1 hr 40 min b. 1 hr 40 min c. 20 min d. 2 hr 40 min e. 1 hr f. 1 hr	a. 1.5 b. 3 c. 1.25 d. 3.75 e. 1 f. .5
11	1500	25.0	Reburn of all vapors 2 heat shields	15.75	Black	4 hr 35 min	10.5
12	1000	78.4	Reburn of all vapors 2 heat shields	41.5	Brown	8 hr 2 min	12
13	1000	23.0	Reburn of all vapors 2 heat shields	14.5	Black	4 hr 30 min	5.5
14	1000	27.0	Reburn of all vapors 2 heat shields	4.7	Black	2 hr	2
15	1200		Reburn of all vapors 2 heat shields	Wood shavings & dead birds	Black		
16	a. 1075 b. 1125 c. 925 d. 900 e. 525		Blank run—1 heat shield			a. 40 min b. 40 min c. 1 hr d. 40 min e. 50 min	a. 1.00 b. 1.50 c. 1.25 d. .5 e. .5

It appears that runs 5, 6 and 7 in which the condensate traps were used had an energy advantage over the runs in which the tars were burned. Though the burning of the tars should have contributed energy, the associated steam probably wasted more energy by interfering with the efficiency of the Bunsen burners during the total reburn runs.

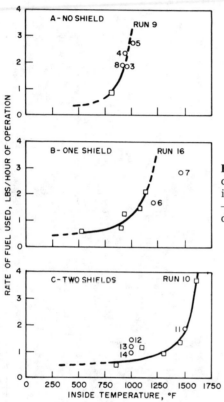

Figure 3. Relative thermal efficiencies of runs made. O — Char producing runs of the associated number; □ — generate the base line and were developed during blank runs.

Carcinogenic Effects

The test for carcinogens was a "negative success." The test was a "success" meaning that the test was valid in that all the necessary data was collected and the controls responded in the desired manner. None of the negative control mice developed any reaction and all the positive control mice developed carcinomas. No mice died during the experiment, with the exception of the positive controls, which were sacrificed when their carcinomas (cancer tissue) threatened their lives. The test was "negative," meaning that the mice subjected to the char and tar did not develop carcinomas. The char mice showed no sign of any adverse reaction. The tar mice developed some lesions, but no carcinomas.

Due to the scale of the experiment, it cannot be inferred that the tars do not have carcinogenic properties. The tests do indicate that carcinogenesis associated with the operation of a destructive distillation device treating undercage hen waste is not a prohibitive hazard.

The lesions developed by the mice suggest that the operator of the device should observe rules of cleanliness because even with complete reburning of the tars and gases, a certain amount of tar is present in and on thte device. The operator is exposed only to the tar build ups during maintenance and repair operations, which further reduces the hazard.

The char caused no ill effects on the mice and lasted longer than the wood chips when used as bedding. The char controlled odors and stayed dry for 3 weeks compared to the wood chips which were damp and odorous after 1 week of use. This suggests a potential for reuse of the char as litter or as an absorbent under cages to make the raw waste easier to handle.

Land Application Studies

In preparation for the land application tests, samples of black char made from undercage hen waste were mixed together and a sample sent to the Soil Testing Laboratory, which reported:

1. 6.1% potassium
2. 7.2% phosphorus
3. pH - 11.6
4. a soluble salt content of 2500 ppm, as determined by the Electrical Conductivity Method

The black chars were also ashed in a 600°C muffle furnace and the ashes submitted to the Spectrographic Laboratory at Cornell.

Resulting data:

Minerals	Content
←———————Percent by weight of ash———————→	
P	8.4 – 10.9
Ca	27.1 – 27.5
K	9.4 – 11.8
Mg	2.45 – 3.00
←———————Milligrams per kilogram of ash———————→	
Na	9,050.0 – 13,350.0
Zn	2,400.0 – 14,000.0
Mn	1,110.0 – 1,260.0
Fe	2,420.0 – 2,620.0
Cu	115.0 – 120.0
B	130.0 – 130.0
Al	890.0 – 1,100.0

The object of the test was only to indicate the magnitude of any hazard created. The most obvious weakness of this portion of the study was its small scale and the resulting lack of replication. Furthermore, no long-term studies of the test plots were possible. Another weak point was that before beginning and after termination, soil and plant studies were not carried out on the test plots.

Soil analyses indicated the soluble salt content of the char sample as 2500 ppm and of the ash as 7500 ppm. It is recommended that the total soluble salts in the soil not exceed 220 ppm for sensitive crops such as tomato plants (24). The tomato plants were observed to respond unfavorably to the application of the char and ash. The reaction of the tomato plant to char was not different than its reaction to ash. These reactions were assumed to be a reaction to the soluble salts.

The highest rate of application of char had no noticeable effect on the native cover, while ash application appeared to have a fertilizer effect. This is probably because the materials within the char were less available than the materials within the ash.

Because of the unusually heavy rainfall and the unusually cool weather during the first part of the experiment, no valid quantitative data can be taken from the test. It appears that the combination of the salt content of the char, the salt content of the soil and the salt sensitivity of the vegetation is the limiting factor controlling the rate of land application of the char as produced by the machine. Therefore, the maximum rate of land application can be determined by calculating the amount of soil needed to dilute the char to a salt concentration acceptable to the desired vegetation.

Leachate Tests

By appearance, the leachates of brown chars were easily distinguished from the leachates of the black chars. Brown chars produced a leachate of a red-orange color, and the black chars produced a leachate which had no measurable color.

Equal amounts by volume of brown char and black char were soaked in enough distilled water to cover the samples. When the samples were drained, the brown char consistently produced a greater volume of leachate. When weighed samples of approximately equal volume were leached for 24 days, the brown char allowed water to pass through at a higher rate and did not retain as much water. The black char was more absorbent and must be exposed to more water than brown char before producing a leachate. Being conservative and estimating black char to retain a quantity of water equal to 25% of its volume, an outdoor storage pile of black char 10 feet deep would not be expected to produce a leachate in most parts of the continental United States during one year of storage.

The standard water pollution tests of BOD, COD, TKN and NH_3 all indicated that the leachate of the brown char had a 5 to 110 times greater pollution potential than the leachate of the black char (Table 3). No quantitative data on BOD, COD, TKN and NH_3 should be taken directly from the long-term leaching tests because of the potential for microbiological activity in the leachate while the leachate was being collected. Though both leachates are potential hazards to the environment, the leachate of the black char is the lesser hazard, and the machine should be operated so as to produce black char.

Table 3. Leaching tests.

| Characteristic | Chars were soaked in distilled water for 24 to 28 hours. | |
	Brown Char Leachate	Black Char Leachate
Clarity	clear	clear
Color: Turner, 590 nm	30%-60%-transmission	97%-100% transmission
pH	7.3-7.4	9.9-10.6
COD	20,000-31,000 mg/l	625-650 mg/l
TKN	2,400-15,000 mg/l	70- 78 mg/l
Total Solids	dry—1.71%-3.4%	2.1%-2.22%
	ash—0.83%-2.45%	1.56%-2.2%
$N-NH_3$	900 mg/l	16.3 mg/l
BOD	22,000 mg/l	160 mg/l

| Characteristic | Chars were leached by a flow of distilled water through the char. | |
	Brown Char Leachate	Black Char Leachate
Clarity	not clear (foam on surface)	clear
Color: Turner, 890 nm	70% transmission	100% transmission
pH	7.8	8.6
COD	650 mg/l	85 mg/l
TKN	50 mg/l	5 mg/l
Total Solids	dry—0.11%	0.17%
	ash—0.06%	0.14%
$N-NH_3$	18 mg/l	none
BOD	260 mg/l	45 mg/l

The leachates of the black chars did not differ radically one from the other, indicating a stable state. Leachates from the two brown chars varied by 50% in terms of COD and by 90% in terms of TKN, suggesting an unpredictable composition in brown chars.

The leachates of black chars had a consistently higher pH than the leachates of brown chars. This characteristic probably would make simple mechanical monitoring of a destructive distillation device a possibility through random sampling of hot char. A less sophisticated method would be to simply have the operator check for traces of brown in the char produced.

The total solids of all the leachates were of the same order of magnitude. The ash content would include the soluble salts referred to in the soil tests, indicating the potential for removing the soluble salt from the char. The BOD tests were positive and without inhibition, indicating a potential for treating both brown and black char leachates in biological treatment units. Therefore, the soluble salts may be removed as a step in preparing the char for use as an industrial raw material or as a soil conditioner, and the resulting leachate treated in a conventional manner.

Economic Estimates

Assuming that the pilot plant model can be scaled up without changing its operating characteristics, design and operating costs can be extrapolated. We will assume a 40,000 hen ranch with 75% egg production and 0.08 pounds of dry waste per bird. The reaction tubes should be two parallel banks of five 10-foot horizontal sections. The raw waste should enter at the top and move through the tubes from top to bottom tube. The major diameter of the auger should be 12 inches, with a minor diameter of 1 inch and a pitch of 12 inches. The shaft speed should be variable between 20 and 26 revolutions per hour. Assuming that the furnace will be efficient in the low temperature range and using the heat capacity of the raw waste as the total energy required, the energy requirements in terms of natural gas as a fuel would be 1110 therms per month. Therefore, the fuel costs would be less than $150.00 per month, or less than two-tenths of a cent per dozen eggs based on 1972 prices.

CONCLUSIONS

1. All the major forms of egg farm waste (hen and pullet undercage waste, dead birds and woodshavings used as litter) can be treated by destructive distillation.
2. The solid residue produced from the waste materials should be black char because it:
   a. represents a stable product.
   b. produces the least hazardous leachate.
3. Results from this study indicated that egg farm wastes should be treated by low temperature pyrolysis (destructive distillation) to:
   a. improve thermal efficiency of the furnace.
   b. allow the use of common materials in the construction of the unit.
   c. reduce the maintenance required on the equipment.
4. By means of a carefully controlled and analyzed carcinogen study using laboratory mice, it was determined that the tar and char produced by the destructive distillation of undercage hen wastes do not represent an undue hazard to men and animals.
5. The chemical fertilizer value of the waste is preserved and concentrated in the char. Even nitrogen, as ammonia, may be sal-

vaged and concentrated as a by-product of the operation by the same methods used in other pyrolytic industries.

6. The black char produced from undercage waste:
   a. represented a 50% to 88% weight reduction, depending upon initial condition of the waste.
   b. represented a 50% to 75% volume reduction, without grinding, depending upon the initial condition of the waste.
   c. contained approximately 3% nitrogen, 7% phosphorus and 6% potassium by weight.

7. The utilization studies were positive and indicated that:
   a. The rate of land application of the char as produced is limited by its soluble salt content, the salinity of the soil used and salt tolerance of vegetation.
   b. The char may be stored in outdoor piles if provision is made to contain the leachate which may result from precipitation.
   c. The char can be prepared for use as a soil conditioner or as industrial raw material by washing it to remove the soluble salts. The wash water may be treated by conventional microbiological methods.
   d. The raw char can be used as an animal litter, and has the desirable properties of being granular and absorbent.

8. The results of this study support the calculations of energy requirements reported in the literature review (13, 17) which suggest that there is almost enough energy in egg farm wastes to support the pyrolysis of the material itself.

9. The ash resulting from the oxidation of chars produced by the destructive distillation of undercage wastes may have value as a mineral fertilizer.

ACKNOWLEDGMENTS

The work reported on in this paper was made possible by the "Management of Animal Wastes," Environmental Protection Agency Training Grant WP-212, administered in the Agricultural Engineering Department, College of Agriculture and Life Sciences, a statutory unit of the State University at Cornell University.

APPENDIX

Pyrolysis: A specific chemical change, either molecularly constructive or destructive, brought about by heat in a closed vessel.

Destructive distillation: The pyrolytic decomposition of a naturally occurring material such as animal waste, peat, coal, wood, bone, etc., which produces char, tar, water and carbonaceous gases.

Char: The solid, carbonaceous material remaining after pyrolysis.

Brown Char: Incompletely carbonized material, still having some of the original compounds present.

Black Char: Completely carbonized material having none of the original compounds present; mineralized material.

Tar: The bituminous material produced during pyrolysis. The two major fractions of tar produced in the destructive distillation of organic matter are:

a. Polar solvent soluble: Tend to be organic acids.
b. Non-polar solvent soluble: Tend to be long-chain carbon compounds.

Gas: Light molecular compounds such as methane, ethane, carbon dioxide.

Carbonization: To reduce to char a carbonaceous material by pyrolysis.

Cracking (also thermal cracking): To rearrange and/or reduce the molecular structure of a liquid or gas by pyrolysis.

Charcoaling: The making of charcoal, wood char, from wood by partial oxidation as of old, or more recently by destructive distillation; and called wood distillation when by-product recovery is included.

Coking: The making of coke, coal char, from coal by destructive distillation, usually with tar and gas collection.

Thermal decomposition: That area of pyrolysis which deals with the reduction of large molecules to smaller molecules.

Thermal composition: (a) That area of pyrolysis that deals with production of more complex molecules; also (b) Part of the field of molecular synthesis.

Thermal rearrangement: The pyrolytic production of a new compound of the same molecular formula, but with a different atomic arrangement than the original; the production of isomers by pyrolysis.

Catalyst: A substance which speeds the approach to equilibrium of a given chemical change with only a slight modification of the free energy of this process.

Oven or retort: A vessel used in pyrolysis.

Low temperature: In pyrolysis, temperatures below 700°C (1300°F). Usually used to infer below 500°C (1000°F).

High temperature: In pyrolysis, temperatures above 700°C (1300°F). Usually used to infer above 900°C (1500°F).

Low pressure (also reduced pressure): When the inside of the pyrolytic vessel is under a partial or nearly complete vacuum.

Atmospheric pressure: In pyrolysis, that range of pressure over which the earth's atmosphere normally fluctuates.

High pressure: In pyrolysis, any pressure above one atmosphere, which is induced by the design of the poeration. Normally reported pressures are between 30 psig and 2000 psig.

Carrier gas: An "inert" gas which fills the pyrolytic vessel and transfers heat to the material; it may also act to exclude oxygen and/ or as a catalyst.

Steam blanket: Water vapor encompassing a material or filling an area to the exclusion of oxygen.

Incineration: Burning; the oxidation of materials, usually with atmospheric oxygen at elevated temperatures.

Zimpro process: Wet oxidation; flameless burning.

Partial oxidation: The burning of some of the material to provide heat and to remove oxygen, so as to pyrolize the remainder.

Carcinogen: Any material which promotes, or causes the development of cancer.

BIBLIOGRAPHY

1. Kirk and Othmer. "Encyclopedia of Chemical Technology." John Wiley & Sons, Inc., New York (1963).
2. Corps of Engineers. Low-temperature carbonization of coal in Japan. Report #74, Natural Resources Section, Supreme Command for the Allied Powers (1947).
3. Christiansson, B. Studies on low-temperature carbonization of peat and peat constituents. Christiansson, Stockholm (1953).
4. Fernald, R. H. Features of producer-gas power-plant development in Europe. Bulletin #4, U.S. Bureau of Mines, Washington, D.C. (1911).
5. Royster, P. H. Process for the destructive distillation of carbonaceous materials. U.S. Patent No. 2,705,697, U.S. Patent Office (1955).
6. Appell, H. R., *et al.* Converting organic waste to oil. A replenishable energy source. Report of Investigation #7560, U.S. Bureau of Mines, Washington, D.C. (1971).
7. Wolfson, D. E., *et al.* Destructive distillation of scrap tires. Report of Investigation #7302, U.S. Bureau of Mines, Washington, D.C. (1969).
8. Sanner, W. S., *et al.* Conversion of municipal and industrial refuse into useful materials by pyrolysis. Report of Investigation #7428, U.S. Bureau of Mines, Washington, D.C. (August 1970).
9. Rodgers, E. S. Apparatus for and method of reducing refuse, garbage and the like to usable constituents. U.S. Patent No. 3,362,887, U.S. Patent Office (1968).
10. Pan American Resources. Pyrolytic decomposition of solid wastes. Public Works, New York, N. Y. (August 1968).
11. Hess, H. V., *et al.* Sewage and municipal refuse liquid phase coking process. U.S. Patent No. 3,652,405, U.S. Patent Office (1972).
12. Oettel, H. Health hazards caused by low-temperature carbonization products of organic material, especially plastics. Abstract #63853s, Chemical Abstracts (1969).
13. White, R. K. and E. P. Taiganides. Pyrolysis of livestock wastes. Paper presented at I.S.L.W., 1971.
14. Midwest Research Institute. The disposal of cattle feedlot wastes by pyrolysis. Missouri (December 1971).
15. Wilson, P. J., Jr., and J. H. Wells. Coal, coke and coal chemicals (first ed.). McGraw-Hill, New York (1950).
16. Hurd, C. D. The pyrolysis of carbon compounds. The Chemical Catalog Co., Inc., New York (1929).
17. Ludington, D. C. Dehydration and incineration of poultry manure. Cornell

University, Ithaca, N.Y.—National Poultry Industry Waste Management Symposium, Lincoln, Nebraska (1963).

18. Michelman, J. Process for obtaining pyrrol, pyrrol derivatives and pyrocoll from animal waste. U.S. Patent No. 1,572,552, U.S. Patent Office (1926).

19. Gutberlet, L. S. Screw-conveying retorting apparatus with hydrogeneration means. U.S. Patent No. 3,658,654, U.S. Patent Office, (1972).

20. Busch, H. Methods in cancer research. Vol. I. Academic Press, New York (1967).

21. Burnett, W. E. Some physical, chemical and biological properties of poultry manure. An unpublished paper prepared at Cornell University (May 15, 1968).

22. Davis, E. G., I. L. Feld, and J. H. Brown. Combustion disposal of manure wastes and utilization of the residue—Bureau of Mines solid waste research program. Technical Progress Report 46, U.S. Department of the Interior, Washington, D.C. (1972).

23. Michigan State University. Animal manures: what are they worth today? Michigan State University, Circular Bulletin #231 (1961).

24. Greweling, T. and M. Peech. Chemical soil tests. Bulletin 960, revised. Cornell University Agricultural Experiment Station, Ithaca, N.Y. (October 1965).

36.

Thermal and Physical Properties of Compost

David R. Mears, Mark E. Singley, Ghulam Ali,
Frank Rupp III*

The problem of properly managing and disposing of farm wastes has received a great deal of attention in recent years. The magnitude of the problem has been well documented, and research has been conducted on a variety of methods for disposing of these wastes. Schulze (1) pointed out the potential advantages of composting relative to other methods of solid waste disposal and noted large-scale plants in operation in Europe and Asia. Wiley and Kochitizky (2) discussed windrow composting which converted raw material into well-disintegrated, ripe, crumbly matter in three to four months in fair weather.

In 1969 the Biological and Agricultural Engineering Department of Cook College, Rutgers University, undertook a research project on the large-scale composting of organic wastes, primarily the wastes from a swine feeding operation which utilized garbage. Large windrows up to 2 m high and 3 m wide were formed of the materials to be composted. These windrows were turned with varying frequencies with a commercial composting machine (Roto-Shredder). In addition to the swine wastes, other materials in varying amounts were added to the windrows to learn which combinations would compost most rapidly. Other waste materials utilized included straw, municipal refuse, sewage sludge and old compost.

* Associate Professor, Professor, former Graduate Assistants, respectively, Biological and Agricultural Engineering Department, Cook College, Rutgers University—The State University of New Jersey, New Brunswick, New Jersey.

The results of this research program have been reported by Singley, Decker and Toth (3). It was found that composting was most rapid when the materials being composted were mixed to provide the proper Carbon/Nitrogen ratio and when there was sufficient bulk in the material to provide enough void space for natural air circulation between turnings by the machine. Addition of 70% straw or 25% municipal refuse by volume to swine waste was found to add sufficient carbon and bulking properties to reduce composting time to about 4 weeks. It was also found that sewage sludge cake composted in the various mixtures composted in much the same way as did the swine waste.

Figure 1. Mechanical turning of compost windrow.

An important aspect of composting is the destruction of pathogens. Savage (4) studied the population changes in enteric bacteria and other microorganisms during composting for the various treatments studied. He found that under the best management conditions temperatures reached 60°C within 3 days and that intestinal bacteria were destroyed within 2 weeks.

Schulze (1) and many others have reported the importance of temperature to the composting process. Based on laboratory studies, various optimal thermophilic composting temperatures have been proposed including 55 to 65°C (5), 60 to 65°C (6) and 75°C (7). In a laboratory composter constructed with a water jacket for cooling, Wiley (8) found the more rapid decomposition occurred in the 54 to 63°C range than in the 66 to 77°C range. In aeration studies he showed that a low rate of 10.4 to 16.6 liters per hour per kg of volatile solids resulted in a slow composting process as compared to an aeration rate of 23.6 to 76.4 liters per hour per kg of volatile solids. At a still higher aeration rate of 85.2 to 202 liters per hour per kg of volatile solids, he found that decomposition was not significant. In this case it is evident that excessive aeration can cause overcooling and dehydration of the composting material. It is important to note that temperature alone is not a good indicator of biological activity especially in a large mass. As total heat generated by biological activity in a compost pile results in a combination of a heat loss to the ventilating air, heat loss by conduction from the pile and temperature rise in the pile, it is clear that one must be aware of the thermal properties of the composting material in any analysis of the energy balance of the system.

In support of the overall project on rapid controlled composting of organic wastes, several studies were undertaken on specific aspects of the composting process and those discussed in this paper are:

1. The determination of the thermal properties of composting materials and the changes in these properties during the composting operation. As the removal of heat is an important factor in decomposition and in large volumes the failure to remove heat may cause the process to become temperature-limited, it is helpful to understand the heat transfer processes within the material. The thermal properties measured were specific heat, thermal diffusivity and thermal conductivity (9). These properties were determined as functions of moisture content.

2. The determination of the particle size distribution of the composting material and the changes in this particle size distribution during the composting operation. As aeration is important both to supply the oxygen required to support aerobic decomposition and to remove surplus heat which cannot be carried away by conduction, a knowledge of the particle size distribution is needed.

3. The determination of the compressibility of the composted material. Because two often-stated objectives of composting are the stabilization and reduction of volume of wastes prior to disposal in a sanitary landfill, the compaction of the final product that can be achieved is of interest. Also, self-compaction in large piles may reduce void spaces enough to create problems in maintaining aeration, thus a knowledge of the compressibility of the material can help in the determination of the maximum feasible size of windrow.

4. The determination of the total volume reduction and changes in bulk density of composting materials. These parameters are important with regard to heat transfer, aeration and the proportion of the volume of waste material that must ultimately be disposed of.

THERMAL PROPERTIES OF COMPOSTING WASTES

For the determination of specific heat, an 850-cc wide-mouth thermos was used as a calorimeter. To account for the effect of heat lost to its body during the experiment, the thermos was calibrated with water to determine the water equivalent of the thermos as a function of depth. A thermometer accurate to $\pm$ 0.1°C was used for temperature measurements. The method of mixtures was used for the determination of specific heat with hot water used as a loser of heat to the material being tested. Mixing of compost and water at the same temperature did not result in any measurable raise of temperature indicating that any heat released in the wetting of the compost is insignificant. This experimental procedure was evaluated by measuring the specific heats of corn, rice and aluminum. The measured results were within 2.8%, 4.9% and 1.4%, respectively, of published values for these materials (10). In all specific heat determinations the material being measured was mixed in the water for 5 minutes before making the final temperature measurement.

The thermal diffusivity of compost was determined by the method of Pflug, Blaisdell and Kopelman (11). The concept of the method is based on the approach of Ball (12) and Ball and Olson (13). The transient temperature rise of a cylinder of cold compost enclosed in a metal can and placed in a constant temperature warm water bath is determined. The cylinders used in this study were made of 0.46-mm-thick aluminum sheet 8.6 cm in diameter and 18 cm long. For each test the compost was sealed between insulated end plugs. Preliminary tests showed that the axial heat flow at the 1/3 points of the cylinder was less than 5% of the radial heat flow, indicating that no serious error would be introduced by assuming the temperature response at the center of the can to be the same as for an infinite cylinder. By plotting the logarithm of the normalized temperature difference between the center of the cylinder of compost and the hot water bath versus time, the termal diffusivity can be calculated if the density of the compost has been previously determined. Using this method the technique was verified by determining the thermal diffusivity of sand and the measured value was found to be within less than 4% of the published value.

The windrow on which the thermal properties studies were focused was composed of swine wastes with an addition of 5% by weight (10% by volume) of straw. The swine waste consisted of feces, uneaten feed, bones, plastics, paper, glass, etc. The windrow was turned four times per day on weekdays and the composting period lasted from March 22 through April 26, 1971, at which time it was judged to have reached maturity relative to the thermophilic phase of composting. Six representative samples were collected at weekly intervals for the thermal properties determinations. Each of the samples was subdivided into at least four subsamples which were air dried to varying moisture contents ranging between 7-12% to 60-65% wet basis.

Due to the size of the containers used for the thermal diffusivity determination it was necessary to exclude large pieces of bone, metal, plastic and glass from the sample. These excluded items totaled between 2 and 10% of the sample weight. At least three thermal diffusivity determinations were made in each moisture content category. The weight and volume of each can of compost were measured for the density determination. Then the samples were sealed in the container for the thermal diffusivity determination. After this was done the contents of each can were divided into subsamples for oven drying for moisture content determination and for the specific heat determination.

For discussion purposes all of the data for the final sample are presented in Table 1. The values for thermal conductivity, K, presented in the last column, are calculated by the relationship:

$$K = .1163 \ a\rho \ C_p$$

where K is the thermal conductivity in watts/m°C, a is the thermal diffusivity in cm²/h, ρ is the density in g/cm³ and C_p is the specific heat in cal/g°C. Statistical analysis of these data and the data from the other five samples showed that in all cases the specific heat and the thermal conductivity varied linearly with moisture content and that in no case was a higher order term significant at the 5% level. The results of the final sample are presented graphically in Figure 2.

A summary of the statistical analysis of the data from all six samples is given in Table 2. Column 2, the intercept, would indicate the value of the property for bone dry compost. It should be noted that

Table 1. Thermal properties of compost (Sample No. 6, April 26, 1971).

Sample No.	Moisture Content (% wb)	Thermal Diffusivity a (cm²/h)	Density ρ (g/cm³)	Sp. Heat C_p (cal/g°C)	Thermal Conductivity K (watts/m °C)
6A₁	47.75	9.75	0.6618	0.5798	0.4351
6A₂	46.25	10.31	0.6618	0.4820	0.3825
6A₃	45.10	11.15	0.6618	0.5257	0.4511
6B₁	38.90	12.17	0.5742	0.4530	0.3682
6B₂	37.60	12.73	0.5742	0.4930	0.4191
6B₃	35.10	14.40	0.5742	0.4425	0.4255
6C₁	10.65	17.93	0.4462	0.3031	0.2820
6C₂	20.42	18.02	0.4462	0.2801	0.2619
6C₃	17.00	18.30	0.4462	0.3476	0.3301
6D₁	9.39	27.87	0.4202	0.2305	0.3139
6D₂	9.35	29.08	0.4202	0.2024	0.2876
6D₃	8.85	22.30	0.4202	0.2474	0.2696

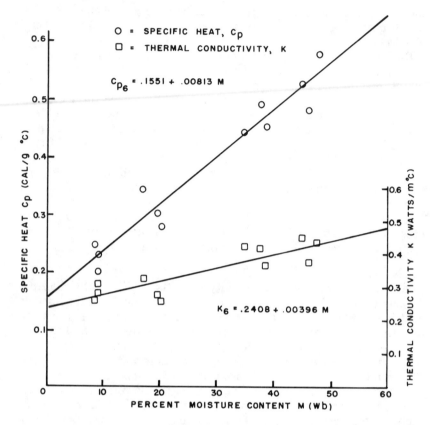

Figure 2. Specific heat, c_p, and thermal conductivity, k, of the compost as a function of moisture content. Sample No. 6 time $=$ 35 days.

this value is based on the extrapolation from determinations made at higher moisture contents. Although it was found that there was no significant heat of wetting associated with any samples tested, an experiment was conducted which showed that there is significant heat of wetting for oven-dried samples. Therefore, if one were to measure the specific heat of very dry material by the method of mixtures with water, this factor would need to be accounted for. The ninth column gives the 95% confidence interval on the regression equations extrapolated to the condition indicating 100% water. The actual specific heat of water, 1 cal/g °C and the actual thermal conductivity 0.654 watts/m °C are seen to lie very close to the calculated extrapolated points and within the 95% confidence limits of the prediction equations.

The usual moisture content range for these composting materials

Table 2. Statistical analysis of the data for specific heat and thermal conductivity of the compost.

	1	2	3	4	5	6	7	8	9
Sample No.	Age of Windrow (days)	Zero Moist. Intercept (I)	Regr. Coef. (S)	Stand. Error on 'S'	100% Moist. Intercept I+100S	Conf. Interval at 2	Conf. Interval at 5	Confidence Limit at 2	Confidence Limit at 5
				Specific Heat (cal/g °C)				**Specific Heat of Water 1.00**	
1	0	.0550	.00940	.00042	.9951	.0403	.0612	.0148,.0954	.9339,1.0563
2	7	.0699	.00921	.00053	.9909	.0520	.0739	.0179,.1219	.0170,1.0648
3	14	.0771	.00922	.00087	.9991	.0779	.1258	−.0080,.1550	.8733,1.1299
4	20	.0834	.00906	.00082	.9894	.0599	.1314	.0235,.1433	.8580,1.1208
5	30	.1289	.00834	.00068	.9629	.0501	.1093	.0788,.1790	.8536,1.0722
6	35	.1551	.00813	.00066	.9831	.0436	.1016	.1122,.1994	.8815,1.0847
				Thermal Conductivity (watts/m °C)				**Thermal Conductivity of Water 0.654**	
1	0	.1017	.00579	.00057	.6807	.0548	.0832	.0469,.1565	.5975,.7639
2	7	.1263	.00550	.00057	.6763	.0560	.0796	.0703,.1823	.5967,.7559
3	14	.1538	.00512	.00078	.6658	.0699	.1126	.0839,.2237	.5532,.7784
4	20	.1559	.00469	.00100	.6249	.0744	.1607	.0815,.2303	.4642,.7856
5	30	.2156	.00445	.00074	.6606	.0550	.1197	.1606,.2706	.5409,.7803
6	35	.2408	.00396	.00074	.6368	.0500	.1163	.1908,.2908	.5205,.7531

is between 20 and 65% wb. The regression relations developed here can be used to determine the thermal properties of the compost at any moisture content in this range. It is to be noted that other mate- rials being composted under other conditions would not be expected to have thermal properties identical to the material tested here. However, since it has been established that in all cases the variations in thermal properties are linear with moisture content, it is possible to predict specific heat or thermal conductivity as a function of mois- ture content by testing a single sample at any convenient moisture content. By using this point and the property for 100% water, the entire relationship is established.

The results presented in Table 2 show that both the specific heat and the thermal conductivity of compost change significantly during the maturation process. At any given moisture content both specific heat and thermal conductivity increase with time. The most likely reason for this increase in these properties is the relative increase in the proportions of inorganic materials such as metals, glass and ash as composting proceeds. The specific heats and thermal conduc- tivities of these components (10) are somewhat higher than the ex- trapolated values for dry compost established here.

These results also show that composting material generally can be regarded as having low thermal conductivity and specific heat. Fur- thermore, the thermal properties of actual composting material may be expected to decrease with maturation as the effect of the reduc- tion in moisture content of the composting material will usually more than make up for the increases in the thermal properties of dry com- post with time. Therefore, a large compost pile which is actively de- composing will be self-insulating and heat removal by conduction will not be rapid. To prevent thermal inhibition of the biological pro- cesses, it is necessary to cool the pile by aeration and/or frequent turning of the pile.

PARTICLE SIZE REDUCTION OF COMPOSTING WASTES

According to Gottas (14), Gray (15), Snell (16), Wiley (17) and a University of California study (18), the initial particle size of the material to be composted is an important factor in the rate of de- composition. Gray (15) indicates that a 3.1-cm average particle size was optimum when using a rotating drum composter with a mixture of plant residue, leaves and paper. In the Rutgers study it was antici- pated that particle size would be continuously reduced during the composting operation due to the combined effects of decomposition and the breakup of particles by the machine turning the piles.

Representative samples of composting material were collected throughout the decomposition process for particle size analysis. Par- ticle size determinations were conducted with a standard set of sieves in accordance with ASAE Standard S 319 (19). Each sample was sub-divided and two samples were analyzed at the moisture content at the time of sampling. The other pairs of samples were air dried to various lower moisture contents before testing to learn the effect of moisture content on particle size.

Variations between pairs of data were usually 5% or less. Some difficulty was encountered in analyzing the particle size of samples that were collected when the moisture content of the material exceeded 45% wb. This material was not particulate in nature but formed large clumps that assumed the shape of the container. If this wet material were air dried, the particle size would depend on the method of drying. When samples that were less than 45% moisture were air dried in thin layers and tested, it was found that there were no significant changes in particle size with changing moisture. Also, repeated sieving of a single sample produced the same particle size analysis after each test, indicating that the sleeving operation itself does not significantly influence the size of a particulate sample.

Analysis of the data showed that particle size follows a logarithmic normal distribution. The geometric mean diameter was determined from the data for each screened sample according to the method of ASAE Standard S 319 (19). The results of the particle size determinations for five of the windrows composted are shown in Figure 3.

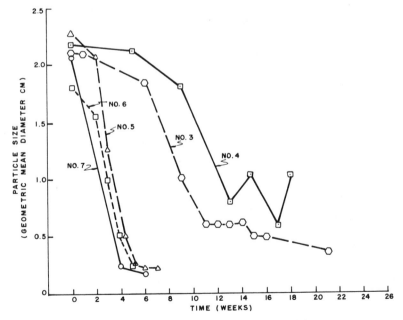

Figure 3. Particle size vs. time, windrows 3-7.

The seven data points shown which have geometric mean diameters greater than 2.0 cm were from samples of moisture content greater than 45% wb. Due to the problems in drying this material for a representative particle size analysis, these points are not considered reliable, but do indicate the general trend.

In comparing these results with other indicators of the rate of composting such as moisture content, temperature and pH, it was found that all indicators followed the same general sequence. Windrow 7, composed of swine waste and straw and turned four times a day, decomposed most rapidly. Windrows 3 and 4, composed only of swine waste and turned one and three times a week, respectively, were the slowest to decompose. Windrows 5 and 6 were turned four times a day, but were composed of swine waste and a 60/40 mixture of swine waste and previously composted material, respectively.

Although particle size is continuously reduced with decomposition, more work would need to be done to establish a reliable relationship between particle size and stage of decomposition. Analysis of the particle size of completely decomposed materials in 5 other windrows composed of various mixtures of materials showed significant differences. For example, the final particle size of composted mixtures of swine waste and municipal refuse were about three times the particle size of mixtures of swine waste alone or mixed with straw. Thus it is clear that for particle size to be a reliable indicator of stage of decomposition, the relationship would have to be established for the particular mixture of materials being composted.

COMPRESSIBILITY OF COMPOSTED WASTES

The compressibility of the composted materials was determined by a method developed by Pollock (20) who studied the relationship of the compressive strength of silage to moisture content. A sample of material is placed in a large steel cylinder with a plunger which is loaded in a universal testing machine. The force applied to the plunger is continuously plotted against the position of the plunger producing an axial force/axial displacement diagram. From this the relationship of applied axial stress and the bulk density of the material can be determined. Pollock (20) found that the relationship between bulk density, ρ, and applied axial stress, σ, could be represented by the equation:

$$\ln \sigma = \ln \sigma_o + C \ln \rho$$

where $\ln \sigma_o$ and C are regression coefficients formed by statistical analysis of the data. It was found that the compressive properties of the composted material could also be adequately described by this expression over the range of densities and applied stresses studied.

The results of the compression studies are given in Table 3. Statistical analysis (Duncan's test) of the data showed that the results for all samples of completed compost from windrows if swine waste and some additive (Nos. 6, 7, 9, 10) were not significantly different from each other. However, the finished compost of windrow 8, which was swine waste with no additive, was significantly less compressible. Also, windrow 12, which was composed of swine waste and straw, but was still in the process of composting, was significantly more compressible than the others.

Table 3. Results of compression studies of compost for selected windrows.

Windrow Number	6	7	8	9	10	12
Composition	Swine waste and old compost	Swine waste and straw	Swine waste only	Swine waste and straw	Swine waste and munic. refuse	Swine waste and straw
Moisture Content % Wet Basis	7.4	8.4	6.1	8.8	9.9	11.4
Compressibility regression coefficient Group Mean, C.	9.24	10.35	13.06*	9.625	8.667	4.725*
Initial Density kg/m³	768	503	571	470	503	313
Density at an applied pressure of 5.5×10^5Pa	1396	882	921	838	913	711
Volume reduction %	45	43	38	44	45	66

* Significantly different from others.

WINDROW VOLUME AND DENSITY OBSERVATIONS

Because reduction of total volume of material is often an important objective of composting, data on changes in total volume and bulk density are important. These items are difficult to measure accurately in the field as the piles are irregular in shape and undisturbed samples for bulk density determinations are difficult to obtain. Total volume in the windrow was determined by measuring the windrow profile at several sections along its length and computing the average cross-sectional area which was multiplied by the length.

Figure 4 shows the volume reduction that was achieved by six of the earliest windrows studied. Note that of these the only one that achieved a satisfactory rate of decomposition was number 7. The bulk density curves for three windrows studied later are shown in Figure 5. All of these windrows were composed of mixtures of materials which composted much more rapidly than pure compost. Note that the bulk density of the material in the windrow is somewhat greater than the initial bulk density of the material used in the compression tests as reported in Table. 3. This is due primarily to the handling of the material as it was loaded into the cylinder in which the compression tests were made.

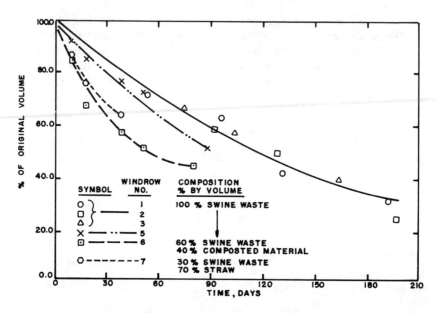

Figure 4. Phases of windrow composting as evidenced by volume reduction.

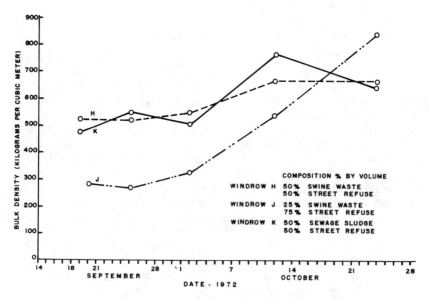

Figure 5. Changes in bulk density with decomposition.

These investigations were supported in part by United States Department of Agriculture Agreements 12-14-100-10078 and D. A. - 11,160. The tests were carried out on the farm of Lester and Bernard Germanio, Belleplain, New Jersey.

Reference to commercial products or trade names is made with the understanding that no discrimination and no endorsement is intended or implied.

Paper of the Journal Series, New Jersey Agricultural Experiment Station, Rutgers University—The State University of New Jersey, Department of Biological and Agricultural Engineering, Cook College, New Brunswick, New Jersy.

REFERENCES

1. Schulze, K. L. Continuous thermophilic composting. Applied Microbiology. 10: p. 108-122 (1962) .

2. Wiley, J. S. and O. W. Kochitizky. Composting developments in the United States. Compost Science. 6:2 (1965).

3. Singley, M. E., M. Decker, and S. J. Toth. Composting municipal refuse and sewage sludge mixtures. ASAE paper No. NA 73-101 (1973).

4. Savage, J. Microbial aspects of .composting hog wastes. Unpublished M.S. thesis, Rutgers University (1972).

5. McBee, R. H. The anaerobic thermophilic cellulolytic bacteria. Bacteriological Review and Proceedings. 14:51-61 (1950).

6. Sui, R. Microbial decomposition of cellulose. Reinhold, New York (1951).

7. Gottas, H. B. Composting. World Health Organization, Geneva (1956).

8. Wiley, J. S. A look at European composting. Public Works. 92:107-110 (1961).

9. Ali, G. Thermal properties of composting material. Unpublished M.S. thesis, Rutgers University (1973).

10. Marks, L. S. Mechanical engineer's handbook. Fifth ed. McGraw-Hill Book Co., Inc., New York (1951).

11. Pflug, I. J., J. L. Blaisdell, and I. J. Kopelman. Developing temperature-time curves for objects that can be approximated by a sphere, infinite plate or infinite cylinder. ASHRAE Trans. 71:238 (1965).

12. Ball, C. P. Thermal process time for canned food. National Research Council Bulletin, No. 37 (1923).

13. Ball, C. P. and F. C. W. Olson. Sterilization in food technology. McGraw-Hill Book Co., Inc., New York (1957).

14. Gottas, H. B. Composting sanitary disposal and reclamation of organic waste. World Health Organization, Palais des Nations, Geneva, p. 205 (1956).

15. Gray, K. R. Accelerated composting. Compost Science. 7:3 (1967).

16. Snell, J. R. Proper grinding—key to efficient composting. Compost Science. 1:4 (1961).

17. Wiley, J. S. Progress report on high rate composting studies. Proceedings of Eleventh Industrial Waste Conference, Purdue University, Lafayette, Indiana (1956).

18. ————. Sanitary engineering research projects reclamation of municipal refuse by composting. Technical Bulletin #9. The University of California (1953).

19. ————. Method of determining and expressing fineness of feed material by sieving. American Society of Agricultural Engineers, Standard S 319, Agricultural Engineers Yearbook (1970).

20. Pollock, K. A. The compressive strength of silage and its relationship to moisture content. Unpublished M.S. thesis, Rutgers University (1969).

Index